Springer Proceedings in Mathematics & Statistics

Volume 491

This book series features volumes composed of selected contributions from workshops and conferences in all areas of current research in mathematics and statistics, including data science, operations research and optimization. In addition to an overall evaluation of the interest, scientific quality, and timeliness of each proposal at the hands of the publisher, individual contributions are all refereed to the high quality standards of leading journals in the field. Thus, this series provides the research community with well-edited, authoritative reports on developments in the most exciting areas of mathematical and statistical research today.

Somnath Bhattacharyya · Hari Shankar Mahato
Editors

Modeling, Analysis and Simulations of Multiscale Transport Phenomena

ICMASMTP 2022, Kharagpur, India, August 25–27

Springer

Editors
Somnath Bhattacharyya
Department of Mathematics
Indian Institute of Technology Kharagpur
Kharagpur, West Bengal, India

Hari Shankar Mahato
Department of Mathematics
Indian Institute of Technology Kharagpur
Kharagpur, West Bengal, India

ISSN 2194-1009 ISSN 2194-1017 (electronic)
Springer Proceedings in Mathematics & Statistics
ISBN 978-981-96-3097-4 ISBN 978-981-96-3098-1 (eBook)
https://doi.org/10.1007/978-981-96-3098-1

Mathematics Subject Classification: 65M22, 65H10, 76D05, 35K57, 65N22, 35A35, 80A32, 82C70

This Springer imprint is published by the registered company Springer Nature Singapore Pte Ltd.
The registered company address is: 152 Beach Road, #21-01/04 Gateway East, Singapore 189721, Singapore

Preface

The International Conference on Modelling, Analysis and Simulations of Multi-scale Transport Phenomena (ICMASMTP 2022) was organized by the Department of Mathematics at the Indian Institute of Technology (IIT) Kharagpur, West Bengal, India, from 22 to 25 August 2022. ICMASMTP 2022 is the second conference on applied mathematics organized recently by the department. The aim of this conference was to provide a platform for academicians, experts and researchers in various disciplines of applied mathematics and its relevant areas exclusively in applications of porous media flow, diffusion–reaction equations, fluid dynamics, multi-scale analysis, electrokinetic transport processes, microfluidics and numerical analysis. Through this conference, we assembled researchers working in applied mathematics, numerical analysis, scientific computation and applications with interfaces in physics, engineering, chemistry, biology and medical sciences. This conference bestowed an avenue to exchange ideas on recent advances in areas related to computational methods, numerical modelling and simulation, as well as their applications in engineering and science. It accommodates presentation on a wide range of topics to facilitate inter-disciplinary exchange of ideas in science, engineering and related disciplines and foster various types of academic collaborations.

The conference provides updates on a wide range of scientific research such as transport in porous networks, evolution equations and their existence theories, homogenization theory, numerical techniques and simulations for transport processes, free boundary problem, multiscale modelling of electrokinetic transport, computational linear algebra, numerical mathematics for nonlinear systems, micro- and nano-fluidics modelling, theoretical and computational fluid dynamics, computational biology along with the machine learning approach. The contributions made by the eminent mathematicians, engineers, scientists and young researchers around the globe are now published in the form of this conference proceedings.

This proceeding published recent advances and shed new light on theoretical development in areas related to computational methods, numerical modelling and simulation, as well as their applications in engineering and science.

This conference was attended by the academicians, scientists, engineers, researchers and most importantly research scholars of various institutions of our country such as IITs, IISc Bangalore, NITs, several central and state universities, and foreign countries as well such as from the USA, the UK, Germany, Austria, Japan, France, Belgium, Sweden and Norway.

We are sincerely thankful to the plenary speakers such as Prof. H. Ohshima from the Tokyo University of Science (Japan), Prof. A. Jüngel from TU Vienna (Austria), Prof. I. S. Pop from Hasselt University (Belgium), Prof. Jérôme F. L. Duval from Université de Lorraine (France), Prof. S. Hardt from TU Darmstadt (Germany), Prof. A. K. Pani from BITS Goa (formerly with the IIT Bombay), Prof. S. Sundar from IIT Madras, Prof. G. D. Veerappa Gowda from the TIFR CAM Bangalore, and Prof. A. K. Nandakumaran from the IISc Bangalore. We would also like to express our sincere gratitude to our invited speakers like Prof. Suman Chakraborty from the IIT Kharagpur, Prof. Gopal Chandra Shit from Jadavpur University, Prof. Lok Pati Tripathi from the IIT Goa, Prof. Manoranjan Mishra from the IIT Ropar, Prof. Adrian Muntean from the Department of Mathematics at the University of Karlstad (Sweden), Prof. Kundan Kumar from the Department of Mathematics at the University of Bergen (Norway), Prof. Satyajit Pramanik from the IIT Gandhinagar, Dr. Michael Eden from ZeTeM at the University of Bremen (Germany), Dr. Deep Ray from the University of Southern California, Dr. Ritesh Dubey from SRM Chennai, Prof. Prashanta Kumar Mandal from Visva-Bharati University, Prof. Akash Anand from the IIT Kanpur, Prof. Ameeya Nayak from the IIT Roorkee, Prof. Shirshendu Chowdhury from the IISER Kolkata, Prof. Arbaaz Khan from the IIT Roorkee, Prof. Swati Mukhopadhyay from the University of Burdwan, Prof. Bhanuman Barman from the NIT Patna and Prof. Sanjib Kumar Pal from Jadavpur University. We would also like to thank all our speakers of contributory talks.

On behalf of the organizing committee, we would like to thank our Chief Patron Prof. V. K. Tewari (Director of the IIT Kharagpur), Chief Guest Prof. Amit Patra (Deputy Director of the IIT Kharagpur) and Guest of Honour Prof. Hiroyuki Ohshima (the Tokyo University of Science, Japan), Prof. Suman Chakraborty (the IIT Kharagpur) and Chair Person Prof. Somnath Bhattacharyya (Head of the Department of Mathematics, the IIT Kharagpur) for their auspicious presence. We would like to thank the members of the scientific committee for their guidance and suggestion. We are also thankful to our faculties for their kind support and heading several sessions as the chairman. We also like to put on record the contribution of our research scholars and postdocs, namely Dr. Manish Chaudhary, Dr. Nibedita Ghosh, Nitu Lakhmara, Arghya Kundu, Dr. Babu Bhaskar for helping us to organize an international event apart from preparing and formatting the manuscript. Finally, we would like to thank the Department of Mathematics at the IIT Kharagpur for facilitating the necessary technical support.

Kharagpur, India

Somnath Bhattacharyya
Hari Shankar Mahato

Contents

Global Martingale Solutions to a Segregation Cross-Diffusion System with Stochastic Forcing

Mrinmay Biswas and Ansgar Jüngel

Abstract The existence of a global martingale solution to a cross-diffusion system with multiplicative Wiener noise in a bounded domain with no-flux boundary conditions is shown. The model describes the dynamics of population densities of different species due to segregation cross-diffusion effects. The diffusion matrix is generally neither symmetric nor positive semidefinite. This difficulty is overcome by exploiting the Rao entropy structure. The existence proof uses a stochastic Galerkin method, uniform estimates from the Rao entropy inequality, and the Skorokhod–Jakubowski theorem. Furthermore, an exponential equilibration result is proved for sufficiently small Lipschitz constants of the noise by using the relative Rao entropy. Numerical tests illustrate the behavior of solutions in one space dimension for two and three population species.

Keywords Population dynamics · Cross-diffusion · Martingale solutions · Tightness of laws · Large-time behavior of solutions

1 Introduction

The segregation of population species can be described by cross-diffusion systems involving quadratic nonlinearities; see, e.g., [3]. This class of models was derived as the mean-field limit of moderately interacting particle systems [9]. In this paper, we analyze these systems taking into account the random influence of the environment.

M. Biswas
Indian Institute of Technology Kanpur, Department of Mathematics and Statistics, Kanpur 208016, India
e-mail: mbiswas@iitk.ac.in

A. Jüngel (✉)
Institute for Analysis and Scientific Computing, Technische Universität Wien, Wiedner Hauptstraße 8–10, 1040 Wien, Austria
e-mail: juengel@tuwien.ac.at

S. Bhattacharyya and H. S. Mahato (eds.), *Modeling, Analysis and Simulations of Multiscale Transport Phenomena*, Springer Proceedings in Mathematics & Statistics 491, https://doi.org/10.1007/978-981-96-3098-1_1

We assume that the dynamics of the population density u_i of the i-th species is modeled by

$$du_i(t) = \operatorname{div}\big(\delta\nabla u_i + u_i\nabla p_i(u)\big)\,dt + \sum_{j=1}^n \sigma_{ij}(u)\,dW_j(t), \quad p_i(u) = \sum_{j=1}^n a_{ij}u_j, \tag{1}$$

in a bounded domain $\mathcal{O} \subset \mathbb{R}^d$ $(d \geq 1)$ with the initial and no-flux boundary conditions

$$u_i(0) = u_i^0 \ \text{ in } \mathcal{O}, \quad \nabla u_i \cdot \nu = 0 \ \text{ on } \partial\mathcal{O},\ t > 0,\ i = 1, \ldots, n, \tag{2}$$

where ν is the exterior unit normal vector to $\partial\mathcal{O}$, $W_1, \ldots, W_n$ are cylindrical Wiener processes in a Hilbert space U (see Sect. 2 for details), and $\delta > 0$, $a_{ij} \geq 0$ for $i, j = 1, \ldots, n$. The density u_i depends on the spatial variable $x \in \mathcal{O}$, the time $t \geq 0$, and the stochastic variable $\omega \in \Omega$. The terms $a_{ij}u_i\nabla u_j$ for $i \neq j$ are called cross-diffusion terms.

The analysis of equations (1) is delicate already in the deterministic case, where $\sigma_{ij}(u) = 0$, since the diffusion matrix is generally neither symmetric nor positive semi-definite. The key idea of the analysis is to exploit the entropy structure associated to (1).

To explain this structure, we need two assumptions. First, we suppose that the diffusion matrix $(u_i a_{ij}) \in \mathbb{R}^{n\times n}$ has only eigenvalues with positive real part (for $u_i > 0$). This means that system (1) is parabolic in the sense of Petrovskii, which is a minimal condition for local solvability [1]. Second, we assume the existence of numbers $\pi_i > 0$ satisfying

$$\pi_i a_{ij} = \pi_j a_{ji} \quad \text{for all } i, j = 1, \ldots, n,$$

which is the detailed-balance condition for the Markov chain associated to (a_{ij}), and $(\pi_1, \ldots, \pi_n)$ is the corresponding reversible stationary measure. Both assumptions imply that the matrix $(\pi_i a_{ij})$ is symmetric positive definite. We introduce the so-called Rao entropy

$$H(u) = \int_{\mathcal{O}} h(u)\,dx, \quad h(u) = \frac{1}{2}\sum_{i,j=1}^n \pi_i a_{ij} u_i u_j, \tag{3}$$

which is used as a diversity measure in population dynamics [18]. Observe that the entropy density $h(u)$ is not the sum of the individual entropies of the species, but it mixes the species. A formal computation, which is made rigorous below (see Lemma 2), shows that

$$\frac{\mathrm{d}}{\mathrm{d}t}\mathbb{E}H(u(t)) + \delta\sum_{i,j=1}^{n}\pi_i a_{ij}\mathbb{E}\int_{\mathcal{O}}\nabla u_i\cdot\nabla u_j\,\mathrm{d}x + \sum_{i=1}^{n}\pi_i\mathbb{E}\int_{\mathcal{O}}u_i|\nabla p_i(u)|^2\,\mathrm{d}x = 0. \tag{4}$$

Since $(\pi_i a_{ij})$ is positive definite and $\delta > 0$, this yields an a priori estimate for ∇u_i in $L^2(\mathcal{O})$. The last integral can be interpreted as the kinetic energy of the system, with $\nabla p_i(u)$ being the partial velocity. Unfortunately, it does not provide any gradient estimate, and this is the reason why we have included the δ-terms. The main results of this paper are the existence of a global martingale solution to (1)–(2) and the exponential decay in expectation of the solution to its spatial average. These results are detailed in Sect. 2.

The analysis of stochastic cross-diffusion systems is rather recent. The existence of martingale solutions to cross-diffusion systems with a positive definite diffusion matrix (including nonlocal diffusion) was shown in [2]. A stochastic population cross-diffusion system with cubic nonlinearities was investigated in [12]. In these works, the quadratic energy structure allows for the use of a stochastic Galerkin method. Combining the theory of quasilinear parabolic equations with evolution semigroup methods, the authors of [14] proved the existence of a unique local pathwise mild solution to the Shigesada–Kawasaki–Teramoto system with stochastic forcing.

For the existence analysis of system (1), we use similar techniques as in [12]. The large-time behavior result is based on the relative entropy method (which becomes here a weighted L^2 norm). Both results are new.

The random influence may also be taken into account on the level of the fluxes, using a conservative noise of the type $\sum_{j=1}^{n}\operatorname{div}\sigma_{ij}(u)\,\mathrm{d}W_j$. This noise can be handled by considering the noise in $H^{-1}(\mathcal{O})$ and estimating in the norm of that space [10]. Writing the noise in Stratonovich form and assuming that $\operatorname{div}(\sigma(u)/u) = 0$, the conservative noise can be reformulated as an Itô noise plus a regularizing Laplacian term; see [17]. The authors of [13] pass to the kinetic formulation of the equation, which yields an equation in which the noise enters as a linear transport. Unfortunately, these techniques cannot be easily applied to equations (1) with arbitrary conservative noise.

2 Notation and Main Results

Let $\mathcal{O}\subset\mathbb{R}^d$ $(d\geq 1)$ be a bounded domain and let $(\Omega,\mathcal{F},\mathbb{F},\mathbb{P})$ be a filtered probability space endowed with a complete right-continuous filtration $\mathbb{F} = (\mathcal{F}_t)_{t\geq 0}$. We write $L^p(\Omega,\mathcal{F};B)$ or simpler $L^p(\Omega;B)$ for the set of all $\mathcal{F}$-measurable B-valued random variables in a Banach space B, such that $\mathbb{E}\|u\|_B^p = \int_\Omega \|u(\omega)\|_B^p\mathbb{P}(\mathrm{d}\omega) < \infty$ $(1\leq p < \infty)$. Let U be a Hilbert space with orthonormal basis $(\eta_k)_{k\in\mathbb{N}}$. The space of Hilbert–Schmidt operators from U to $L^2(\mathcal{O})$ is defined by

$$\mathcal{L}_2(U;L^2(\mathcal{O})) = \left\{F : U\to L^2(\mathcal{O})\text{ linear, continuous} : \sum_{k=1}^{\infty}\|F\eta_k\|_{L^2(\mathcal{O})} < \infty\right\},$$

endowed with the norm $\|F\|_{\mathcal{L}_2(U;L^2(\mathcal{O}))} = (\sum_{k=1}^{\infty} \|F\eta_k\|_{L^2(\mathcal{O})}^2)^{1/2}$. Let β_{jk} for $j = 1, \ldots, n$, $k \in \mathbb{N}$ be independent one-dimensional Brownian motions. There exists a Hilbert space $U_0 \supset U$ and a Hilbert–Schmidt embedding $J : U \to U_0$ such that the series $W_j = \sum_{k=1}^{\infty} \beta_{jk} J(\eta_k)$ converges in $L^\infty(0, T; L^2(\Omega; U_0))$. Moreover, $W_j(\omega) \in C^0([0, T]; U_0)$ for a.e. $\omega \in \Omega$ [15, Prop. 2.5.2].

We impose the following hypotheses:

(H1) Domain: Let $\mathcal{O} \subset \mathbb{R}^d$ $(1 \le d \le 3)$ be a bounded domain with Lipschitz boundary.

(H2) Initial datum: $u_i^0 \in L^2(\Omega, \mathcal{F}_0; L^2(\mathcal{O}))$ satisfies $u_i^0 \ge 0$ a.e. in $\mathcal{O}$, $\mathbb{P}$-a.s. for $i = 1, \ldots, n$.

(H3) Diffusion coefficients: All eigenvalues of $A = (a_{ij}) \in \mathbb{R}^{n\times n}$ have positive real parts and the detailed-balance condition holds, i.e., there exist $\pi_1, \ldots, \pi_n >$ such that

$$\pi_i a_{ij} = \pi_j a_{ji} \quad \text{for all } i, j = 1, \ldots, n.$$

(H4) Noise coefficients: $\sigma_{ij} : L^2(\mathcal{O}) \times \Omega \to \mathcal{L}_2(U; L^2(\mathcal{O}))$ is $\mathcal{B}(L^2(\mathcal{O}) \otimes \mathcal{F}/\mathcal{B}(\mathcal{L}_2(U; L^2(\mathcal{O})))$-measurable and $\mathbb{F}$-adapted, and there exists $C_\sigma > 0$ such that for all $i, j = 1, \ldots, n$ and $u, v \in L^2(\mathcal{O})$,

$$\begin{aligned} \|\sigma_{ij}(u) - \sigma_{ij}(v)\|_{\mathcal{L}_2(U;L^2(\mathcal{O}))} &\le C_\sigma \|u - v\|_{L^2(\mathcal{O})}, \\ \|\sigma_{ij}(u)\|_{\mathcal{L}_2(U;L^2(\mathcal{O}))} &\le C_\sigma \|u_i\|_{L^2(\mathcal{O})}. \end{aligned}$$

Let us comment on these hypotheses. The restriction $d \le 3$ is not essential; our proof works in any space dimension, but we need to choose a larger space to show the Aldous condition, needed for the tightness of the laws of the Galerkin solutions. Let $P = \mathrm{diag}(\pi_1, \ldots, \pi_n)$. Then, by Hypothesis (H3), PA is symmetric. Since the eigenvalues of PA are positive, we conclude that PA is positive definite. The linear growth condition in Hypothesis (H4) is needed to prove the $\mathbb{P}$-a.s. nonnegativity of u_i. It can be weakened to $\|\sigma_{ij}(u)\|_{\mathcal{L}_2(U;L^2(\mathcal{O}))} \le C_\sigma(1 + \|u\|_{L^2(\mathcal{O})})$, but then we cannot prove the nonnegativity property.

Definition 1 (*Martingale solution*) Let $T > 0$. A *global martingale solution* to (1)–(2) is a tuple $(\widetilde{V}, \widetilde{W}, \widetilde{u})$ such that

- $\widetilde{V} = (\widetilde{\Omega}, \widetilde{\mathcal{F}}, \widetilde{\mathbb{P}}, \widetilde{\mathbb{F}})$ is a complete probability space with a filtration $\widetilde{\mathbb{F}} = (\widetilde{\mathcal{F}}_t)_{t\in[0,T]}$;
- $\widetilde{W} = (\widetilde{W}_1, \ldots, \widetilde{W}_n)$ is a cylindrical Wiener process in $U^n := U \times \cdots \times U$;
- $\widetilde{u} = (\widetilde{u}_1, \ldots, \widetilde{u}_n) : [0, T] \times \Omega \to L^2(\mathcal{O})$ is an $\mathbb{F}$-progressively measurable process such that for all $t \in [0, T]$ and $i = 1, \ldots, n$,

$$\widetilde{u}_i \in L^2(\Omega; C^0([0, T]; L^2_w(\mathcal{O}))) \cap L^2(\Omega; L^2(0, T; H^1(\mathcal{O}))),$$

the law of $\widetilde{u}_i(0)$ is the same as for u_i^0, and $\widetilde{u}_i$ satisfies for $t \in [0, T]$ and $\phi \in H^1(\mathcal{O})$,

$$(\widetilde{u}_i(t), \phi)_{L^2(\mathcal{O})} = (\widetilde{u}_i(0), \phi)_{L^2(\mathcal{O})} - \int_0^t \int_{\mathcal{O}} \big(\delta \nabla \widetilde{u}_i(s) + \widetilde{u}_i(s) \nabla p_i(\widetilde{u}(s))\big) \cdot \nabla \phi \, \mathrm{d}x \, \mathrm{d}s$$
$$- \int_0^t \Big(\sum_{j=1}^n \sigma_{ij}(\widetilde{u}(s)) \, \mathrm{d}\widetilde{W}_j(s), \phi \Big)_{L^2(\mathcal{O})} \quad \widetilde{\mathbb{P}}\text{-a.s.}$$

Our first main result is as follows.

Theorem 1 (Existence of a global martingale solution) *Let $T > 0$ and let Hypotheses (H1)–(H4) hold. Then there exists a global martingale solution $(\widetilde{U}, \widetilde{W}, \widetilde{u})$ to* (1)–(2) *satisfying $\widetilde{u}_i(t) \geq 0$ a.e. in $\mathcal{O}$, $\widetilde{\mathbb{P}}$-a.s. for all $t \in [0, T]$, $i = 1, \ldots, n$.*

As mentioned in the introduction, the idea of the proof is the use of the stochastic Galerkin method combined with the entropy method. We project equations (1) on a Galerkin space with finite dimension $N \in \mathbb{N}$. The existence of a pathwise unique strong solution $u^N = (u_1^N, \ldots, u_n^N)$ (up to some stopping time) is shown by means of Banach's fixed-point theorem. Itô's lemma allows us to derive an entropy inequality; see (4). Since we introduced the δ-terms, we obtain estimates for u^N in $H^1(\mathcal{O})$ uniformly in N.

The tightness of the laws of (u^N) in the topological space Z_T, defined in (15), is proved by applying the criterion of [4]. We deduce from Skorokhod–Jakubowski's theorem that there exists a subsequence of (u^N), another probability space, and random variables $(\widetilde{u}^N, \widetilde{W}^N)$ having the same law as (u^N, W^N), and $(\widetilde{u}^N, \widetilde{W}^N)$ converges to $(\widetilde{u}, \widetilde{W})$ in the topology of Z_T. Because of the gradient estimates and compactness, we infer the strong convergence $\widetilde{u}^N \to \widetilde{u}$ in $L^2(\mathcal{O})$ a.s., and we can identify the limits in the nonlinearities. Finally, the a.s. nonnegativity follows from a stochastic Stampacchia truncation argument from [8].

For the second main result, we introduce the relative Rao entropy

$$H(u|\bar{u}) = H(u) - H(\bar{u}) - \frac{\delta H}{\delta u}(\bar{u}) \cdot (u - \bar{u}) = \frac{1}{2} \sum_{i,j=1}^n \pi_i a_{ij} \int_{\mathcal{O}} (u_i - \bar{u}_i)(u_j - \bar{u}_j) \, \mathrm{d}x,$$

where $\bar{u}$ is the solution to

$$\mathrm{d}\bar{u}_i(t) = \sum_{j=1}^n \bar{\sigma}_{ij}(u(t)) \, \mathrm{d}W_j(t), \quad t > 0, \quad \bar{\sigma}_{ij}(u) = \frac{1}{|\mathcal{O}|} \int_{\mathcal{O}} \sigma_{ij}(u) \, \mathrm{d}x,$$

and $|\mathcal{O}|$ is the measure of $\mathcal{O}$. Recalling $P = \mathrm{diag}(\pi_1, \ldots, \pi_n)$ and $A = (a_{ij}) \in \mathbb{R}^{n \times n}$, the relative entropy can be written as the weighted L^2 norm

$$H(u|\bar{u}) = \frac{1}{2} \| (PA)^{1/2} (u - \bar{u}) \|_{L^2(\mathcal{O})}^2,$$

Theorem 2 *(Exponential time decay) Let Hypotheses (H1)–(H4) hold and let u be a martingale solution to* (1)–(2). *Then there exists $c_0 > 0$ such that for all $0 < C_\sigma < c_0$ (see Hypothesis (H4)),*

$$\mathbb{E}H(u(t)|\bar{u}(t)) \leq \mathbb{E}H(u^0|\bar{u}(0))e^{-\eta t}, \quad \eta := c_0^2 - C_\sigma^2,\ t > 0.$$

Usually, the large-time behavior of solutions to stochastic differential equations is analyzed by proving the existence and uniqueness of an invariant measure and studying the ergodicity of the equations. This can be done, for instance, by establishing the strong Feller property and the tightness of the laws of the solutions and by applying the Krylov–Bogolyubov's theorem [11, Chap. 11]. Here, this program is delicate since we are lacking semigroup properties. Therefore, we rely on estimates from the entropy method.

For the proof of Theorem 2, we apply Itô's lemma to the process $e^{-\eta t/2}(PA)^{1/2}(u - \bar{u})$, take the expectation, and use the Poincaré–Wirtinger inequality with constant $C_P > 0$ for the gradient term. This leads to (see Sect. 4)

$$\begin{aligned} &e^{\eta t}\mathbb{E}H(u(t)|\bar{u}(t)) - \mathbb{E}H(u(0)|\bar{u}(0)) - \eta\int_0^t e^{\eta s}\mathbb{E}H(u(s)|\bar{u}(s))\,ds \\ &\quad \leq -\frac{\delta\lambda}{C_P^2}\int_0^t \|u - \bar{u}\|_{L^2(\mathcal{O})}^2\,ds + \frac{1}{2}\mathbb{E}\int_0^t \|(PA)^{1/2}(\sigma(u) - \bar{\sigma}(u))\|_{\mathcal{L}_2(U;L^2(\mathcal{O}))}^2\,ds. \end{aligned} \tag{5}$$

The difficult part of the proof is the estimate of the last integral. Using the properties of $\sigma_{ij}(u(t))$ and $\bar{\sigma}_{ij}(u(t))$, we find that

$$\mathbb{E}\int_0^t \|(PA)^{1/2}(\sigma(u) - \bar{\sigma}(u))\|_{\mathcal{L}_2(U;L^2(\mathcal{O}))}^2\,ds \leq 2C_1C_\sigma^2\|u - \bar{u}\|_{L^2(\mathcal{O})}^2.$$

Thus choosing $C_\sigma > 0$ (see Hypothesis (H4)) and $\eta > 0$ sufficiently small, this integral can be absorbed by the first term on the right-hand side of (5), and we conclude the result.

3 Proof of Theorem 1

The proof of Theorem 1 is split into several steps.

3.1 *Stochastic Galerkin Approximation*

Let $(e_k)_{k\in\mathbb{N}}$ be an orthonormal basis of $L^2(\mathcal{O})$, which is orthogonal in $H^1(\mathcal{O})$. For each $N \in \mathbb{N}$, define the finite-dimensional subspace $H_N := \mathrm{span}(e_1, \ldots, e_N)$ of $L^2(\mathcal{O})$ and the corresponding projection $\Pi_N : L^2(\mathcal{O}) \to H_N$, $\Pi_N(v) = \sum_{i=1}^N (v, e_i)_{L^2(\mathcal{O})}e_i$ for $v \in L^2(\mathcal{O})$. We consider system (1) projected on the subspace $H_N^n := H_n \times \cdots \times H_N$, for $t > 0$,

$$du_i^N(t) = \Pi_N \operatorname{div}\big(\delta\nabla u_i^N + (u_i^N)^+\nabla p_i(u^N)\big)\,dt + \Pi_N\sum_{j=1}^n \sigma_{ij}(u^N)\,dW_j(t), \quad (6)$$

$$u_i^N(0) = \Pi_N(u_i^0), \quad i = 1,\dots,n, \quad (7)$$

where $z^+ = \max\{0, z\}$ denotes the positive part. This truncation is necessary, since the Rao entropy does not provide nonnegative densities. The nonnegativity of the Galerkin limit u_i is proved in Sect. 3.5. Given $T > 0$, we introduce the space $M_T := L^2(\Omega; C^0([0,T]; H_N^n))$ with the norm $\|u\|_{M_T} = (\mathbb{E}\sup_{0<t<T}\|u(t)\|^2_{L^2(\mathcal{O})})^{1/2}$ for $u \in M_T$. For given $R > 0$ and $u \in M_T$, we define the exit time $\tau_R = \inf\{t \in [0,T] : \|u(t)\|_{H^1(\mathcal{O})} > R\}$. Then $\{\omega \in \Omega : \tau_R(\omega) > t\}$ belongs to $\mathcal{F}_t$ for every $t \in [0,T]$, and τ_R is an $\mathbb{F}$-stopping time.

Lemma 1 *(Local existence of u^N) Let $T > 0$, $R > 0$ and let the Hypotheses (H1)–(H4) hold. Then there exists a pathwise unique strong solution $u^N \in M_{T\wedge\tau_R}$ to (6)–(7) such that for any $t \in [0, T\wedge\tau_R]$, $i = 1,\dots,n$, and $\phi = (\phi_1,\dots,\phi_n) \in H_N^n$,*

$$\begin{aligned}(u_i^N(t), \phi_i)_{L^2(\mathcal{O})} &= (u_i(0), \phi_i)_{L^2(\mathcal{O})} \\ &\quad - \int_0^t\int_{\mathcal{O}} \big(\delta\nabla u_i^N(s) + (u_i^N)^+(s)\nabla p_i(u^N(s))\big)\cdot\nabla\phi_i\,dx\,ds \\ &\quad + \Big(\sum_{j=1}^n\int_0^t \sigma_{ij}(u^N(s))\,dW_j(s), \phi_i\Big)_{L^2(\mathcal{O})} \quad \mathbb{P}\text{-a.s.}\end{aligned}$$

Proof We use Banach's fixed-point theorem. Define the fixed-point operator $S : M_T \to M_T$ for $v \in M_T$ and $\psi \in H_N^n$ by

$$\begin{aligned}(S(v)(t), \psi)_{L^2(\mathcal{O})} &= \sum_{i=1}^n (\Pi_N u_i^0, \psi_i)_{L^2(\mathcal{O})} \\ &\quad - \sum_{i=1}^n\int_0^t\int_{\mathcal{O}}\big(\delta\nabla v_i + v_i^+\nabla p_i(v)\big)\cdot\nabla\psi_i\,dx\,ds \\ &\quad + \sum_{i=1}^n\Big(\sum_{j=1}^n\int_0^t \sigma_{ij}(v(s))\,dW_j(s), \psi_i\Big)_{L^2(\mathcal{O})} \quad \mathbb{P}\text{-a.s.},\ t\in[0,T].\end{aligned}$$

The aim is to show that S is a contraction on $M_{T\wedge\tau_R}$ for sufficiently small $T > 0$.

First, we verify that S is a self-mapping. Let $v \in M_T$, $\psi \in H_N^n$, and $T_R := T\wedge\tau_R$. Then

$$\begin{aligned}&\|(S(v), \psi)_{L^2(\mathcal{O})}\|^2_{L^2(\Omega; L^\infty(0,T_R))} \le I_1 + \cdots + I_4, \quad \text{where} \quad (8)\\ &I_1 = \mathbb{E}\|u^0\|^2_{L^2(\mathcal{O})}\|\psi\|^2_{L^2(\mathcal{O})}, \\ &I_2 = \mathbb{E}\Big(\sup_{0<t<T_R}\Big|\sum_{i=1}^n\int_0^t\int_{\mathcal{O}}\delta\nabla v_i(s)\cdot\nabla\psi_i\,dx\,ds\Big|\Big)^2,\end{aligned}$$

$$I_3 = \mathbb{E}\bigg(\sup_{0<t<T_R}\bigg|\sum_{i,j=1}^n \int_0^t\int_{\mathcal{O}} a_{ij}v_i^+(s)\nabla v_j(s)\cdot\nabla\psi_i\,\mathrm{d}x\,\mathrm{d}s\bigg|\bigg)^2,$$

$$I_4 = \mathbb{E}\bigg(\sup_{0<t<T_R}\bigg|\sum_{i=1}^n\bigg(\sum_{j=1}^n\int_0^t \sigma_{ij}(v)\,\mathrm{d}W_j(s),\psi_i\bigg)_{L^2(\mathcal{O})}\bigg|\bigg)^2.$$

We estimate the term I_2, using the equivalence of the norms in H_N:

$$\begin{aligned}
I_2 &\le \delta^2\,T\mathbb{E}\bigg(\sup_{0<t<T_R}\sum_{i=1}^n\int_0^t\bigg|\int_{\mathcal{O}}\nabla v_i(s)\cdot\nabla\psi_i\,\mathrm{d}x\bigg|^2\mathrm{d}s\bigg)\\
&\le \delta^2\,TC\mathbb{E}\bigg(\int_0^{T_R}\|\nabla v(s)\|_{L^2(\mathcal{O})}^2\,\mathrm{d}s\bigg)\|\nabla\psi\|_{L^2(\mathcal{O})}^2\\
&\le C(N)\delta^2\,T^2\mathbb{E}\bigg(\sup_{0<s<T_R}\|v(s)\|_{L^2(\mathcal{O})}^2\bigg)\|\psi\|_{L^2(\mathcal{O})}^2\\
&= C(N)\delta^2\,T^2\mathbb{E}\|v\|_{M_{T_R}}^2\|\psi\|_{L^2(\mathcal{O})}^2,
\end{aligned}$$

where here and in the following $C>0$, $C_i>0$, etc. are constants independent of the solution, with values changing from line to line. The term I_3 is estimated in a similar way, taking into account that $\|\nabla v(s)\|_{L^2(\mathcal{O})}\le R$ for $s<T_R$:

$$\begin{aligned}
I_3 &\le CT\mathbb{E}\bigg(\int_0^{T_R}\|v(s)\|_{L^2(\mathcal{O})}^2\|\nabla v(s)\|_{L^2(\mathcal{O})}^2\,\mathrm{d}s\bigg)\|\nabla\psi\|_{L^\infty(\mathcal{O})}^2\\
&\le C(N)R^2T^2\mathbb{E}\bigg(\sup_{0<s<T_R}\|v(s)\|_{L^2(\mathcal{O})}^2\bigg)\|\psi\|_{L^2(\mathcal{O})}^2,
\end{aligned}$$

Finally, by the Burkholder–Davis–Gundy inequality [16, Theorem 1.1.7] and (H4),

$$\begin{aligned}
I_4 &\le C\mathbb{E}\bigg(\int_0^{T_R}\|\sigma(v(s))\|_{\mathcal{L}_2(U;L^2(\mathcal{O}))}^2\,\mathrm{d}s\bigg)\|\psi\|_{L^2(\mathcal{O})}^2\\
&\le C_\sigma^2 C\mathbb{E}\bigg(\int_0^{T_R}\|v(s)\|_{L^2(\mathcal{O})}^2\,\mathrm{d}s\bigg)\|\psi\|_{L^2(\mathcal{O})}^2 \le C_\sigma^2\,C(N)T\|v\|_{M_{T_R}}^2\|\psi\|_{L^2(\mathcal{O})}^2.
\end{aligned}$$

Inserting these estimates into (8) leads to

$$\|S(v)\|_{M_{T_R}}^2 \le \mathbb{E}\|u^0\|_{L^2(\mathcal{O})}^2 + C(N,R)(T^2+T)\|v\|_{M_{T_R}}^2.$$

This shows that S is a self-mapping. Next, we prove that S is a contraction if $T>0$ is sufficiently small. The estimations are similar as above with the exception of the nonlinear diffusion part. Let $u, v\in M_T$ and $\psi\in H_N^n$. Then

$$\|(S(u)-S(v),\psi)_{L^2(\mathcal{O})}\|_{L^2(\Omega;L^\infty(0,T_R))}^2 \le I_5+I_6+I_7, \quad \text{where} \tag{9}$$

$$I_5 = \mathbb{E}\bigg(\sup_{0<t<T_R} \bigg| \sum_{i=1}^n \int_0^t \int_{\mathcal{O}} \delta \nabla (u_i - v_i)(s) \cdot \nabla \psi_i \,\mathrm{d}x \,\mathrm{d}s \bigg| \bigg)^2,$$

$$I_6 = \mathbb{E}\bigg(\sup_{0<t<T_R} \bigg| \sum_{i,j=1}^n \int_0^t \int_{\mathcal{O}} a_{ij} \big(u_i^+(s) \nabla u_j(s) - v_i^+(s) \nabla v_j(s)\big) \cdot \nabla \psi_i \,\mathrm{d}x \,\mathrm{d}s \bigg| \bigg)^2,$$

$$I_7 = \mathbb{E}\bigg(\sup_{0<t<T_R} \bigg| \sum_{i=1}^n \bigg(\int_0^t \big(\sigma_{ij}(u(s)) - \sigma_{ij}(v(s))\big) \,\mathrm{d}W_j(s), \psi_i \bigg)_{L^2(\mathcal{O})} \bigg| \bigg)^2.$$

The terms I_5 and I_7 are estimated as I_2 and I_4, respectively, giving

$$I_5 \le C(N)\delta^2 T^2 \|u - v\|_{M_{T_R}}^2 \|\psi\|_{L^2(\mathcal{O})}^2, \quad I_7 \le C(N) T \|u - v\|_{M_{T_R}}^2 \|\psi\|_{L^2(\mathcal{O})}^2.$$

Writing $u_i^+ \nabla u_j - v_i^+ \nabla v_j = u_i^+ \nabla (u_j - v_j) + (u_i^+ - v_i^+) \nabla v_j$ and using the Lipschitz continuity of $z \mapsto z^+$, the remaining term becomes

$$\begin{aligned} I_6 &\le C_A T \mathbb{E}\bigg(\int_0^{T_R} \|u(s)\|_{L^2(\mathcal{O})}^2 \|\nabla (u - v)(s)\|_{L^2(\mathcal{O})}^2 \,\mathrm{d}s \bigg) \|\nabla \psi\|_{L^\infty(\mathcal{O})}^2 \\ &\quad + C_A T \mathbb{E}\bigg(\int_0^{T_R} \|(u - v)(s)\|_{L^2(\mathcal{O})}^2 \|\nabla v(s)\|_{L^2(\mathcal{O})}^2 \,\mathrm{d}s \bigg) \|\nabla \psi\|_{L^\infty(\mathcal{O})}^2 \\ &\le C_A C(N) R^2 T^2 \mathbb{E}\bigg(\sup_{0<s<T_R} \|(u - v)(s)\|_{L^2(\mathcal{O})}^2 \bigg) \|\psi\|_{L^2(\mathcal{O})}^2 \\ &= C(N, R) T^2 \|u - v\|_{M_{T_R}}^2 \|\psi\|_{L^2(\mathcal{O})}^2. \end{aligned}$$

We conclude from (9) that

$$\|S(u) - S(v)\|_{M_{T_R}}^2 \le C(N, R)(T^2 + T) \|u - v\|_{M_{T_R}}^2.$$

Then, choosing $T > 0$ such that $C(N, R)(T^2 + T) < 1$, the mapping $S : M_{T_R} \to M_{T_R}$ is a contraction. By Banach's fixed-point theorem, there exists a unique fixed point $u^N \in M_{T_R}$ that solves (6)–(7) in $[0, T_R]$. □

3.2 Uniform Estimates

We show that the solution u^N to (6)–(7) satisfies suitable uniform estimates. The following lemma is the key result of this subsection.

Lemma 2 *(Entropy inequality for u^N) Let $T > 0$ and let u^N be the pathwise unique strong solution to (6)–(7) on $[0, T \wedge \tau_R]$, constructed in Lemma 1. Then there exists a constant $C(\lambda, T)$, depending on λ and T, such that*

$$\mathbb{E}\bigg(\sup_{0<t<T} H(u^N(t))\bigg) + 2\delta\lambda\mathbb{E}\bigg(\int_0^T \int_{\mathcal{O}} |\nabla u^N(s)|^2\, dx\, ds\bigg) \\ + \sum_{i=1}^n \mathbb{E}\bigg(\int_0^T \int_{\mathcal{O}} (u_i^N)^+(s)|\nabla p_i(u^N(s))|^2\, dx\, ds\bigg) \le \mathbb{E}H(u^N(0))C(\lambda, T),$$

where $\lambda > 0$ is the smallest eigenvalue of PA.

Proof Since PA is symmetric positive definite, there exists $(PA)^{1/2}$ and the process $Y(t) = (PA)^{1/2}u^N(t)$ is well defined. Hence, we can write the entropy (3) as

$$H(u^N) = \frac{1}{2}\sum_{i,j=1}^n \int_{\mathcal{O}} \pi_i a_{ij} u_i^N u_j^N\, \mathrm{d}x = \frac{1}{2}\int_{\mathcal{O}} (u^N)^T (PA) u^N\, \mathrm{d}x \\ = \frac{1}{2}\|(PA)^{1/2}u^N\|^2_{L^2(\mathcal{O})},$$

Let $t_R = t \wedge \tau_R$ and $T_R = T \wedge \tau_R$. We apply Itô's lemma [15, Theorem 4.2.5] to $Y(t)$:

$$H(u^N(t_R)) = H(u^N(0)) + J_1 + \cdots + J_4, \quad \text{where} \tag{10}$$

$$J_1 = -\delta \sum_{i,j=1}^n \pi_i a_{ij} \int_0^{t_R} \int_{\mathcal{O}} \nabla u_i^N(s) \cdot \nabla u_j^N(s)\, \mathrm{d}x\, \mathrm{d}s \le -\delta\lambda \int_0^{t_R} \int_{\mathcal{O}} |\nabla u^N|^2\, \mathrm{d}x\, \mathrm{d}s,$$

$$J_2 = -\sum_{i=1}^n \int_0^{t_R} \int_{\mathcal{O}} (u_i^N)^+(s)|\nabla p_i(u^N(s))|^2\, \mathrm{d}x\, \mathrm{d}s \le 0,$$

$$J_3 = \frac{1}{2}\int_0^{t_R} \int_{\mathcal{O}} \mathrm{Tr}\left[\sigma(u^N(s))\mathrm{D}^2 h(u^N(s))\sigma(u^N(s))^*\right] \mathrm{d}x\, \mathrm{d}s,$$

$$J_4 = \sum_{i,j=1}^n \pi_i a_{ij} \int_{\mathcal{O}} \bigg(\int_0^{t_R} u_i^N(s)\sigma_{ij}(u^N(s))\, \mathrm{d}W_j(s)\bigg)\, \mathrm{d}x.$$

Using $\mathrm{D}^2 h(u^N) = PA$ and Hypothesis (H4),

$$J_3 = \frac{1}{2}\int_0^{t_R} \|(PA)^{1/2}\sigma(u^N(s))\|^2_{\mathcal{L}_2(U;L^2(\mathcal{O}))}\, \mathrm{d}s \tag{11} \\ \le C\int_0^{t_R} \|\sigma(u^N(s))\|^2_{\mathcal{L}_2(U;L^2(\mathcal{O}))}\, \mathrm{d}s \\ \le CC_\sigma^2 \int_0^{t_R} \|u^N(s)\|^2_{L^2(\mathcal{O})}\, \mathrm{d}s \le C(\lambda)\int_0^{t_R} H(u^N(s))\, \mathrm{d}s.$$

Inserting this estimate into (10), taking the supremum over $[0, T_R]$ and then the expectation, we find that

$$\mathbb{E}\Big(\sup_{0<t<T_R} H(u^N(t))\Big) \le \mathbb{E}H(u^N(0)) - \delta\lambda\mathbb{E}\Big(\sup_{0<t<T_R}\int_0^t\int_{\mathcal{O}} |\nabla u^N(s)|^2\, \mathrm{d}x\, \mathrm{d}s\Big) \tag{12}$$

$$- \mathbb{E}\Big(\sup_{0<t<T_R}\sum_{i=1}^n\int_0^t\int_{\mathcal{O}} (u_i^N)^+(s)|\nabla p_i(u^N(s))|^2\, \mathrm{d}x\, \mathrm{d}s\Big)$$

$$+ C(\lambda)\mathbb{E}\Big(\sup_{0<t<T_R}\int_0^t H(u^N(s))\, \mathrm{d}s\Big) + \mathbb{E}\Big(\sup_{0<t<T_R} J_4(t)\Big).$$

We deduce from the Burkholder–Davis–Gundy inequality that

$$\mathbb{E}\Big(\sup_{0<t<T_R} J_4(t)\Big) \le C\sum_{i,j,\ell=1}^n \mathbb{E}\bigg\{\sup_{0<t<T_R}\int_0^t\sum_{k=1}^\infty\Big(\int_{\mathcal{O}} u_j^N(s)\sigma_{i\ell}(u^N(s))\eta_k\, \mathrm{d}x\Big)^2 \mathrm{d}s\bigg\}^{1/2}$$

$$\le C\sum_{i,j,\ell=1}^n \mathbb{E}\bigg\{\sup_{0<t<T_R}\int_0^t\sum_{k=1}^\infty \|u_j^N(s)\|^2_{L^2(\mathcal{O})}\|\sigma_{i\ell}(u^N(s))\eta_k\|^2_{L^2(\mathcal{O})}\, \mathrm{d}s\bigg\}^{1/2}$$

$$\le C\mathbb{E}\Big(\int_0^{T_R}\|u^N(s)\|^2_{L^2(\mathcal{O})}\|\sigma(u^N(s))\|^2_{\mathcal{L}_2(U;L^2(\mathcal{O}))}\, \mathrm{d}s\Big)^{1/2}.$$

Therefore, using Hypothesis (H4) and Young's inequality,

$$\mathbb{E}\Big(\sup_{0<t<T_R} J_4(t)\Big) \le C_\sigma C\mathbb{E}\bigg\{\sup_{0<t<T_R}\|u^N(s)\|_{L^2(\mathcal{O})}\Big(\int_0^{T_R}\|u^N(s)\|^2_{L^2(\mathcal{O})}\, \mathrm{d}s\Big)^{1/2}\bigg\}$$

$$\le C(\lambda)\mathbb{E}\bigg\{\sup_{0<t<T_R}\|(PA)^{1/2}u^N(s)\|_{L^2(\mathcal{O})}\Big(\int_0^{T_R}\|(PA)^{1/2}u^N(s)\|^2_{L^2(\mathcal{O})}\, \mathrm{d}s\Big)^{1/2}\bigg\}$$

$$\le \frac{1}{2}\mathbb{E}\Big(\sup_{0<t<T_R} H(u^N(t))\Big) + C(\lambda)\int_0^{T_R}\mathbb{E}\Big(\sup_{0<s<t} H(u^N(s))\Big)\, \mathrm{d}t.$$

We insert this estimate into (12),

$$\frac{1}{2}\mathbb{E}\Big(\sup_{0<t<T_R} H(u^N(t))\Big) \le \mathbb{E}H(u^N(0)) - \delta\lambda\mathbb{E}\int_0^{T_R}\int_{\mathcal{O}} |\nabla u^N(s)|^2\, \mathrm{d}x\, \mathrm{d}s \tag{13}$$

$$- \sum_{i=1}^n \mathbb{E}\int_0^{T_R}\int_{\mathcal{O}} (u_i^N)^+(s)|\nabla p_i(u^N(s))|^2\, \mathrm{d}x\, \mathrm{d}s$$

$$+ C(\lambda)\int_0^{T_R}\mathbb{E}\Big(\sup_{0<s<t} H(u^N(s))\Big)\, \mathrm{d}t,$$

and apply Gronwall's lemma to obtain

$$\mathbb{E}\Big(\sup_{0<t<T_R} H(u^N(t))\Big) \le \mathbb{E}H(u^N(0))e^{2C(\lambda)T}.$$

Hence, the right-hand side of (13) does not depend on the chosen sequence of stopping times τ_R, and we can pass to the limit $R \to \infty$, finishing the proof. □

Corollary 1 *(Uniform estimates) Let $T > 0$ and let u^N be the pathwise unique strong solution to* (6)–(7) *on* $[0, T \wedge \tau_R]$*, constructed in Lemma 1. Then there exists a constant $C > 0$, which is independent of N, such that for all $i = 1, \ldots, n$,*

$$\mathbb{E}\Big(\sup_{0<t<T} \|u_i^N(t)\|_{L^2(\mathcal{O})}^2\Big) + \delta\lambda\mathbb{E}\Big(\int_0^T \|\nabla u_i^N(t)\|_{L^2(\mathcal{O})}^2\, dt\Big) \le C.$$

We also need higher-order moment estimates.

Lemma 3 *(Higher-order moment estimates) Let $T > 0$, $p > 2$, and let u^N be the pathwise unique strong solution to* (6)–(7) *on* $[0, T \wedge \tau_R]$*, constructed in Lemma 1. Then there exists a constant $C > 0$, which is independent of N, such that*

$$\mathbb{E}\Big(\sup_{0<t<T} \|u^N(t)\|_{L^2(\mathcal{O})}^p\Big) \le C.$$

Proof Let $T_R = T \wedge \tau_R$. We start from (10), neglect the terms J_1 and J_2, take the supremum over $(0, T_R)$, raise the power $p/2$ on both sides, take the expectation, and use Hölder's inequality:

$$\mathbb{E}\Big(\sup_{0<t<T_R} H(u^N(t))^{p/2}\Big) \le C\mathbb{E}H(u^N(0))^{p/2} + J_5 + J_6, \quad \text{where} \tag{14}$$

$$J_5 = CT^{p/2-1}\mathbb{E}\int_0^{T_R} \|(PA)^{1/2}\sigma(u^N(s))\|_{\mathcal{L}_2(U;L^2(\mathcal{O}))}^p\, \mathrm{d}s,$$

$$J_6 = \mathbb{E}\Big(\sup_{0<t<T_R}\Big|\sum_{i,j,k=1}^n \pi_i a_{ij}\int_{\mathcal{O}}\int_0^t u_j^N(s)\sigma_{ik}(u^N(s))\,\mathrm{d}W_k(s)\,\mathrm{d}x\Big|^{p/2}\Big)$$

The term J_5 can be estimated as in (11):

$$J_5 \le C(\lambda)\mathbb{E}\int_0^{T_R} H(u^N(s))^{p/2}\,\mathrm{d}s \le C(\lambda)\int_0^{T_R}\mathbb{E}\Big(\sup_{0<s<t} H(u^N(s))^{p/2}\Big)\,\mathrm{d}t.$$

Similarly as in Lemma 2, we apply the Burkholder–Davis–Gundy inequality and Young's inequality to find that

$$\begin{aligned} J_6 &\le C\mathbb{E}\bigg(\int_0^{T_R} \|u^N(s)\|_{L^2(\mathcal{O})}^2 \|\sigma(u^N(s))\|_{\mathcal{L}_2(U;L^2(\mathcal{O}))}^2 \,ds\bigg)^{p/4} \\ &\le CT^{1-p/2}\mathbb{E}\bigg\{\bigg(\sup_{0<s<T_R} \|u^N(s)\|_{L^2(\mathcal{O})}^p\bigg)^{1/2}\bigg(\int_0^{T_R} \|u^N(s)\|_{L^2(\mathcal{O})}^p \,ds\bigg)^{1/2}\bigg\} \\ &\le \frac{1}{2}\mathbb{E}\bigg(\sup_{0<s<T_R} H(u^N(s))^{p/2}\bigg) + C(T,\lambda)\int_0^{T_R} \mathbb{E}\bigg(\sup_{0<s<t} H(u^N(s))^{p/2}\bigg)\,dt. \end{aligned}$$

We insert these estimates into (14) and apply Gronwall's lemma to conclude that

$$\mathbb{E}\bigg(\sup_{0<t<T_R} H(u^N(t))^{p/2}\bigg) \le C(\lambda,T)\mathbb{E}H(u^N(0))^{p/2},$$

and the positive definiteness of $(PA)^{1/2}$ finishes the proof. □

3.3 Tightness

The tightness of the sequence of laws of u^N on a suitable subspace is proved similarly as in [12, Sect. 2.4]. We consider the following spaces:

- $C^0([0,T]; H^3(\mathcal{O})')$ with the topology $\mathcal{T}_1$, induced by its canonical norm.
- $L^2_w(0,T; H^1(\mathcal{O}))$ is the space $L^2(0,T;H^1(\mathcal{O}))$ with the weak topology $\mathcal{T}_2$.
- $L^2(0,T; L^2(\mathcal{O}))$ with the topology $\mathcal{T}_3$ induced by its canonical norm.
- $C^0([0,T]; L^2_w(\mathcal{O}))$ is the space of weakly continuous functions $v : [0,T] \to L^2(\mathcal{O})$ endowed with the weakest topology $\mathcal{T}_4$ such that for all $h \in L^2(\mathcal{O})$, the mappings

  $$C^0([0,T]; L^2_w(\mathcal{O})) \to C^0([0,T];\mathbb{R}), \quad u \mapsto (u(\cdot), h)_{L^2(\mathcal{O})},$$

  are continuous.

We define the following space:

$$\begin{aligned} Z_T := C^0([0,T]; H^3(\mathcal{O})') &\cap L^2_w(0,T; H^1(\mathcal{O})) \\ &\cap L^2(0,T; L^2(\mathcal{O})) \cap C^0([0,T]; L^2_w(\mathcal{O})), \end{aligned} \tag{15}$$

endowed with the topology $\mathcal{T}$, which is the maximum of the topologies $\mathcal{T}_1, \ldots, \mathcal{T}_4$ of the corresponding spaces, i.e. the smallest topology containing $\cap_{i=1}^4 \mathcal{T}_i$.

Lemma 4 *(Tightness) Let u^N be the pathwise unique strong solution to* (6)–(7), *constructed in Lemma 1. Then the set of laws of (u^N) is tight on $(Z_T, \mathcal{T})$.*

Proof We apply [6, Corollary 3.9] with $U = H^3(\mathcal{O})$, $V = H^1(\mathcal{O})$, and $H = L^2(\mathcal{O})$. Since $V \hookrightarrow H$ is compact, Corollary 1 shows that conditions (a) and (b) of [6, Corollary 3.9] are satisfied. It remains to verify that (u^N) satisfies the Aldous condition in $H^3(\mathcal{O})'$. For this, let $(\tau_N)_{N\in\mathbb{N}}$ be a sequence of $\mathbb{F}$-stopping times such that $0 \le \tau_N \le T$. Let $t \in [0, T]$, $i \in \{1, \dots, n\}$, and $\phi \in H^3(\mathcal{O})$. We write (6) as

$$\langle u_i^N(t), \phi\rangle = \langle K_1^N + K_2^N(t) + K_3^N(t), \phi\rangle, \quad \text{where} \tag{16}$$
$$\langle K_1^N, \phi\rangle = (\Pi_N(u_i^0), \phi)_{L^2(\mathcal{O})},$$
$$\langle K_2^N(t), \phi\rangle = -\int_0^t \int_{\mathcal{O}} \Pi_N\Big(\delta\nabla u_i^N(s) + \sum_{j=1}^n a_{ij}(u_i^N)^+(s)\nabla u_j^N(s)\Big)\cdot\nabla\phi \, dx \, ds,$$
$$\langle K_3^N(t), \phi\rangle = \sum_{j=1}^n \Big(\int_0^t \Pi_N\big(\sigma_{ij}(u^N(s))\big)\, dW_j(s), \phi\Big)_{L^2(\mathcal{O})},$$

and $\langle\cdot,\cdot\rangle$ denotes the dual pairing between $H^3(\mathcal{O})'$ and $H^3(\mathcal{O})$. (If $d \ge 1$ is arbitrary, we replace $H^3(\mathcal{O})$ by $H^m(\mathcal{O})$, where $m \ge 1 + d/2$, $m \in \mathbb{N}$.) Let $\theta > 0$. Then, using the embedding $H^3(\mathcal{O}) \hookrightarrow W^{1,\infty}(\mathcal{O})$ for $d \le 3$ and the Cauchy–Schwarz inequality,

$$\begin{aligned}
\mathbb{E}|\langle K_2^N(\tau_N+\theta) - K_2^N(\tau_N), \phi\rangle| &\le \mathbb{E}\int_{\tau_N}^{\tau_N+\theta}\Big(\delta\|\nabla u_i(t)\|_{L^2(\mathcal{O})}\|\nabla\phi\|_{L^2(\mathcal{O})} \\
&\quad + \sum_{j=1}^n a_{ij}\|(u_i^N)^+(s)\|_{L^2(\mathcal{O})}\|\nabla u_j^N\|_{L^2(\mathcal{O})}\|\nabla\phi\|_{L^\infty(\mathcal{O})}\Big)\, dt \\
&\le \delta\theta^{1/2}\Big(\mathbb{E}\int_{\tau_N}^{\tau_N+\theta}\|\nabla u_i(t)\|_{L^2(\mathcal{O})}^2\, dt\Big)^{1/2}\|\phi\|_{H^1(\mathcal{O})} \\
&\quad + C_A^{1/2}\theta^{1/2}\Big\{\mathbb{E}\Big(\sup_{0<t<T}\|u_i^N(t)\|_{L^2(\mathcal{O})}^2\Big)\Big\}^{1/2} \\
&\quad \times\Big\{\mathbb{E}\int_0^T\|\nabla u(t)\|_{L^2(\mathcal{O})}^2\, dt\Big\}^{1/2}\|\phi\|_{H^3(\mathcal{O})}.
\end{aligned}$$

Therefore, in view of the estimates of Corollary 1,

$$\mathbb{E}|\langle K_2^N(\tau_N+\theta) - K_2^N(\tau_N), \phi\rangle| \le C\theta^{1/2}\|\phi\|_{H^3(\mathcal{O})}. \tag{17}$$

For the stochastic term, we use the Itô isometry and Hypothesis (H4):

$$\begin{aligned}
&\mathbb{E}|\langle K_3^N(\tau_N+\theta) - K_3^N(\tau_N), \phi\rangle|^2 \\
&\quad\le \mathbb{E}\Big(\int_{\tau_N}^{\tau_N+\theta}\|\sigma(u^N(t))\|_{\mathcal{L}_2(U;L^2(\mathcal{O}))}^2\, dt\Big)\|\phi\|_{L^2(\mathcal{O})}^2 \\
&\quad\le C_\sigma\theta\mathbb{E}\Big(\sup_{0<t<T}\|u^N(t)\|_{L^2(\mathcal{O})}^2\Big)\|\phi\|_{L^2(\mathcal{O})}^2 \le C\theta\|\phi\|_{L^2(\mathcal{O})}^2.
\end{aligned} \tag{18}$$

Next, let $\zeta, \varepsilon > 0$. We conclude from the Chebyshev inequality and (17) that

$$\begin{aligned}\mathbb{P}\big(\|K_2^N(\tau_N+\theta) - K_2^N(\tau_N)\|_{H^3(\mathcal{O})'} \geq \zeta\big) \\ \leq \frac{1}{\zeta}\mathbb{E}\|K_2^N(\tau_N+\theta) - K_2^N(\tau_N)\|_{H^3(\mathcal{O})'} \leq \frac{C\theta^{1/2}}{\zeta}.\end{aligned}$$

Choosing $\theta_1 := (\zeta\varepsilon/C)^2$, we infer that

$$\sup_{N\in\mathbb{N}}\sup_{0<\theta<\theta_1}\mathbb{P}\big(\|K_2^N(\tau_N+\theta) - K_2^N(\tau_N)\|_{H^3(\mathcal{O})'} \geq \zeta\big) \leq \varepsilon.$$

In a similar way, it follows from the Chebyshev inequality and (18) that

$$\sup_{N\in\mathbb{N}}\sup_{0<\theta<\theta_2}\mathbb{P}\big(\|K_3^N(\tau_N+\theta) - K_3^N(\tau_N)\|_{L^2(\mathcal{O})} \geq \zeta\big) \leq \varepsilon.$$

We infer from (16) that

$$\sup_{N\in\mathbb{N}}\sup_{0<\theta<\min\{\theta_1,\theta_2\}}\mathbb{P}\big(\|u_i^N(\tau_N+\theta) - u_i^N(\tau_N)\|_{H^3(\mathcal{O})'} \geq \zeta\big) \leq 2\varepsilon.$$

This shows that (u_i^N) satisfies the Aldous condition in $H^3(\mathcal{O})'$. □

3.4 Convergence of the Approximate Solutions

According to [12, Theorem 23], the topological space $(Z_T, \mathcal{T})$ possesses a sequence of continuous funtions $Z_T \to \mathbb{R}$ that separates points of Z_T. We deduce from the theorem of Skorokhod–Jakubowski and the tightness of the laws of (u^N) on $(Z_T, \mathcal{T})$ that there exists a subsequence of (u^N), which is not relabeled, a probability space $(\widetilde{\Omega}, \widetilde{\mathcal{F}}, \widetilde{\mathbb{P}})$, and, on this space, $Z_T \times C^0([0,T]; U_0)$-valued random variables $(\widetilde{u}, \widetilde{W})$, $(\widetilde{u}^N, \widetilde{W}^N)$ with $N \in \mathbb{N}$ such that $(\widetilde{u}^N, \widetilde{W}^N)$ has the same law as (u^N, W) on $\mathcal{B}(Z_T \times C^0([0,T]; U_0)$ and

$$(\widetilde{u}^N, \widetilde{W}^N) \to (\widetilde{w}, \widetilde{W}) \quad \text{in } Z_T \times C^0([0,T]; U_0)\ \widetilde{\mathbb{P}}\text{-a.s. as } N \to \infty.$$

We deduce from the definition of Z_T that

$$\begin{aligned}\widetilde{u}^N &\to \widetilde{u} \quad \text{in } C^0([0,T]; H^3(\mathcal{O})') \text{ and } L^2(0,T; L^2(\mathcal{O})), \\ \widetilde{u}^N &\rightharpoonup \widetilde{u} \quad \text{weakly in } L^2(0,T; H^1(\mathcal{O})), \\ \widetilde{u}^N &\to \widetilde{u} \quad \text{in } C^0([0,T]; L^2_w(\mathcal{O})), \\ \widetilde{W}^N &\to \widetilde{W} \quad \text{in } C^0([0,T]; U_0).\end{aligned}$$

The strong convergence of $(\widetilde{u}_i^N)$ implies that $(\widetilde{u}_i^N)^+ \to \widetilde{u}_i^+$ in $L^2(0,T;L^2(\mathcal{O}))$ $\widetilde{\mathbb{P}}$-a.s. Using Corollary 1 and Lemma 3 and arguing as in [12, Section 2.5], we can prove that for $p > 2$,

$$\widetilde{\mathbb{E}}\bigg(\int_0^T \|\widetilde{u}(t)\|_{H^1(\mathcal{O})}^2 \,dt\bigg) \le C, \quad \widetilde{\mathbb{E}}\bigg(\sup_{0<t<T} \|\widetilde{u}(t)\|_{L^2(\mathcal{O})}^p\bigg) \le C.$$

Let $\widetilde{\mathbb{F}}$ and $\widetilde{\mathbb{F}}^N$ be the filtrations generated by $(\widetilde{u}, \widetilde{W})$ and $(\widetilde{u}^N, \widetilde{W}^N)$, respectively. The arguments of [7, Prop. B4] allow us to show that these new random variables are actually stochastic processes. The progressive measurability of $\widetilde{u}^N$ is a consequence of [5, Appendix B]. Set $\widetilde{W}_j^{N,k} = (\widetilde{W}_j^N(t), \eta_k)_{L^2(\mathcal{O})}$. Then $\widetilde{W}_j^{N,k}(t)$ are independent, standard $\widetilde{\mathcal{F}}_t$-Wiener processes. Passing to the limit $N \to \infty$ in the characteristic function, we infer that $\widetilde{W}(t)$ are $\widetilde{F}_t$-martingales with the correct marginal distributions. Then, by Lévy's charaterization theorem, $\widetilde{W}(t)$ is a cylindrical Wiener process. For the following lemma, we introduce

$$H_\nu^3(\mathcal{O}) = \big\{v \in H^3(\mathcal{O}) : \nabla v \cdot \nu = 0 \text{ on } \partial\Omega\big\}.$$

Lemma 5 *Let $\phi \in L^2(\mathcal{O};\mathbb{R}^n)$ and $\psi \in H_\nu^3(\mathcal{O};\mathbb{R}^n)$. Then it holds for $i = 1, \ldots, n$ that*

$$\lim_{N\to\infty} \widetilde{\mathbb{E}} \int_0^T \big(\widetilde{u}^N(t) - \widetilde{u}(t), \phi\big)_{L^2(\mathcal{O})} dt = 0, \quad \lim_{N\to\infty} \widetilde{\mathbb{E}}\big(\widetilde{u}^N(0) - \widetilde{u}(0), \phi\big)_{L^2(\mathcal{O})}^2 = 0,$$

$$\lim_{N\to\infty} \widetilde{\mathbb{E}} \int_0^T \bigg|\int_0^t \int_{\mathcal{O}} \bigg(\delta\nabla\widetilde{u}^N(s) + \sum_{j=1}^n a_{ij}(\widetilde{u}_i^N)^+(s)\nabla\widetilde{u}_j^N(s)$$
$$- \delta\nabla\widetilde{u}(s) - \sum_{j=1}^n a_{ij}\widetilde{u}_i^+(s)\nabla\widetilde{u}_j(s)\bigg) \cdot \nabla\psi_i \,dx\,ds\bigg| dt = 0,$$

$$\lim_{N\to\infty} \widetilde{\mathbb{E}} \int_0^T \bigg|\sum_{j=1}^n \int_0^t \big(\sigma_{ij}(\widetilde{u}^N(s))\,d\widetilde{W}_j^N(s) - \sigma_{ij}(\widetilde{u}(s))\,d\widetilde{W}_j(s), \phi_i\big)_{L^2(\mathcal{O})}\bigg|^2 dt = 0.$$

Proof The proof of this lemma is very similar to that one for [12, Lemma 16], using the previous convergence results and Vitali's convergence theorem. The only difference is the convergence in the nonlinear terms $(\widetilde{u}_i^N)^+\nabla\widetilde{u}_j^N$. In fact, the strong convergence of $(\widetilde{u}_i^N)^+$ in $L^2(0,T;L^2(\mathcal{O}))$ and the weak convergence of $\nabla\widetilde{u}_j^N$ in $L^2(0,T;H^1(\mathcal{O}))$ imply that

$$\bigg|\int_0^t \int_{\mathcal{O}} \big((\widetilde{u}_i^N)^+(s)\nabla\widetilde{u}_j^N(s) - \widetilde{u}_i^+(s)\nabla\widetilde{u}_j(s)\big) \cdot \nabla\psi_i \,\mathrm{d}x\,\mathrm{d}s\bigg|$$
$$\le \|(\widetilde{u}_i^N)^+ - \widetilde{u}_i^+\|_{L^2(0,t;L^2(\mathcal{O})}\|\widetilde{u}_j^N\|_{L^2(0,t;H^1(\mathcal{O}))}\|\psi_i\|_{W^{1,\infty}(\mathcal{O})}$$
$$+ \bigg|\int_0^t \int_{\mathcal{O}} (\widetilde{u}_i^N)^+(s)\nabla(\widetilde{u}_j^N - \widetilde{u}_i)(s) \cdot \nabla\psi_i \,\mathrm{d}x\,\mathrm{d}s\bigg| \to 0 \quad \widetilde{\mathbb{P}}\text{-a.s. as } N \to \infty.$$

Since $(\widetilde{u}_i^N)^+\nabla\widetilde{u}_j^N$ is uniformly bounded, Vitali's convergence theorem gives

$$\lim_{N\to\infty}\widetilde{\mathbb{E}}\int_0^T\left|\int_0^t\int_{\mathcal{O}}\big((\widetilde{u}_i^N)^+(s)\nabla\widetilde{u}_j^N(s)-\widetilde{u}_i^+(s)\nabla\widetilde{u}_j(s)\big)\cdot\nabla\psi_i\,\mathrm{d}x\,\mathrm{d}s\right|\mathrm{d}t=0.$$

This finishes the proof. □

We consider the maps $L_i^N,\ L_i : Z_T\times C^0([0,T];U_0)\times H^3_\nu(\mathcal{O})\to L^1(\widetilde{\Omega}\times(0,T))$ for $i=1,\dots,n$, defined by

$$\begin{aligned}L_i^N(\widetilde{u}^N,\widetilde{W}^N,\psi)(t)&=(\Pi_N(\widetilde{u}_i(0)),\psi)_{L^2(\mathcal{O})}\\&\quad-\int_0^t\int_{\mathcal{O}}\Pi_N\bigg(\delta\nabla\widetilde{u}_i^N(s)+\sum_{j=1}^n a_{ij}(\widetilde{u}_i^N)^+(s)\nabla\widetilde{u}_j^N(s)\bigg)\cdot\nabla\psi_i\,\mathrm{d}x\,\mathrm{d}s\\&\quad+\sum_{j=1}^n\bigg(\int_0^t\Pi_N\sigma_{ij}(\widetilde{u}^N(s))\,\mathrm{d}\widetilde{W}_j^N(s),\psi_i\bigg)_{L^2(\mathcal{O})},\end{aligned}$$

$$\begin{aligned}L_i(\widetilde{u},\widetilde{W},\psi)(t)&=(\widetilde{u}_i(0),\psi_i)_{L^2(\mathcal{O})}\\&\quad-\int_0^t\int_{\mathcal{O}}\bigg(\delta\nabla\widetilde{u}_i(s)+\sum_{j=1}^n a_{ij}\widetilde{u}_i^+(s)\nabla\widetilde{u}_j(s)\bigg)\cdot\nabla\psi_i\,\mathrm{d}x\,\mathrm{d}s\\&\quad+\sum_{j=1}^n\bigg(\int_0^t\sigma_{ij}(\widetilde{u}(s))\,\mathrm{d}\widetilde{W}_j(s),\psi_i\bigg)_{L^2(\mathcal{O})}.\end{aligned}$$

The previous lemma implies the following result.

Corollary 2 *It holds for any $\phi\in L^2(\mathcal{O};\mathbb{R}^n)$ and $\psi\in H^3_\nu(\mathcal{O};\mathbb{R}^n)$ that*

$$\begin{aligned}\lim_{N\to\infty}\|(\widetilde{u}^N,\phi)_{L^2(\mathcal{O})}-(\widetilde{u},\phi)_{L^2(\mathcal{O})}\|_{L^2(\widetilde{\Omega}\times(0,T))}&=0,\\\lim_{N\to\infty}\|L_i^N(\widetilde{u}^N,\widetilde{W}^N,\psi)-L_i(\widetilde{u},\widetilde{W},\psi)\|_{L^1(\widetilde{\Omega}\times(0,T))}&=0.\end{aligned}$$

3.5 *End of the Proof*

Since u^N is a strong solution to (6)–(7), we have

$$(u_i^N(t),\psi)_{L^2(\mathcal{O})}=L_i^N(u^N,W,\psi)(t)\quad\mathbb{P}\text{-a.s.}$$

for all $t\in[0,T]$, $\psi\in H^1(\mathcal{O})$, and $i=1,\dots,n$. In particular,

$$\mathbb{E}\int_0^T \big|(u_i^N(t),\psi)_{L^2(\mathcal{O})} - L_i^N(u^N, W, \psi)(t)\big|\,\mathrm{d}t = 0.$$

As the laws of (u^N, W) and $(\widetilde{u}^N, \widetilde{W}^N)$ are the same,

$$\widetilde{\mathbb{E}}\int_0^T \big|(\widetilde{u}_i^N(t),\psi)_{L^2(\mathcal{O})} - L_i^N(\widetilde{u}^N, \widetilde{W}^N, \psi)(t)\big|\,\mathrm{d}t = 0.$$

By Corollary 2, we can pass to the limit $N\to\infty$, yielding

$$\widetilde{\mathbb{E}}\int_0^T \big|(\widetilde{u}_i(t),\psi)_{L^2(\mathcal{O})} - L_i(\widetilde{u}, \widetilde{W}, \psi)(t)\big|\,\mathrm{d}t = 0$$

for all $\psi\in H_\nu^3(\mathcal{O})$. Since the space $H_\nu^3(\mathcal{O})$ is dense in $H^1(\mathcal{O})$, this identity also holds for all $\psi\in H^1(\mathcal{O})$. This means that $(\widetilde{u}_i(t),\psi)_{L^2(\mathcal{O})} = L_i(\widetilde{u},\widetilde{W},\psi)(t)$ $\mathbb{P}$-a.s. for a.e. $t\in(0,T)$, for all $\psi\in H^1(\mathcal{O})$ and $i=1,\ldots,n$. Thus, setting $\widetilde{V}=(\widetilde{\Omega},\widetilde{\mathcal{F}},\widetilde{\mathbb{F}},\widetilde{\mathbb{P}})$, we conclude that $(\widetilde{V},\widetilde{u},\widetilde{W})$ is a martingale solution to (1)–(2).

It remains to verify that $\widetilde{u}$ is nonnegative componentwise in $\mathcal{O}\times(0,T)$ $\mathbb{P}$-a.s. This follows from the stochastic Stampacchia truncation argument developed in [8]. The idea is to approximate the test function $g(z)=z^- := \max\{0,-z\}$ for $z\in\mathbb{R}$ by a smooth function g_ε with linear growth and to apply Itô's formula to $G_\varepsilon(v) = \int_{\mathcal{O}} g_\varepsilon(v(x))^2\,\mathrm{d}x$. It is proved in [12, Section 2.6] that the limit $N\to\infty$ and then $\varepsilon\to 0$ leads to

$$\begin{aligned}\mathbb{E}\|\widetilde{u}_i^-(t)\|_{L^2(\mathcal{O})}^2 &\le \mathbb{E}\|(u_i^0)^-\|_{L^2(\mathcal{O})}^2 + \sum_{j=1}^n \mathbb{E}\int_0^t \|\sigma_{ij}(-\widetilde{u}_i^-(s))\|_{\mathcal{L}_2(U;L^2(\mathcal{O}))}^2\,\mathrm{d}s \\ &\le C_\sigma \sum_{j=1}^n \mathbb{E}\int_0^t \|\widetilde{u}_i^-(s))\|_{L^2(\mathcal{O})}^2\,\mathrm{d}s,\end{aligned}$$

where we used that $\widetilde{u}_i^+\nabla\widetilde{u}_i^- = 0$. It follows from Gronwall's lemma that $\mathbb{E}\|\widetilde{u}_i^-(t)\|_{L^2(\mathcal{O})}^2 = 0$ for $t\in(0,T)$ and consequently $\widetilde{u}_i(t)\ge 0$ in $\mathcal{O}$ $\mathbb{P}$-a.s. for a.e. $t\in(0,T)$ and all $i=1,\ldots,n$.

4 Proof of Theorem 2

We apply Itô's lemma [15, Theorem 4.2.5] to the process $t\mapsto e^{\eta t/2}(PA)^{1/2}(u^N - \bar{u}^N)(t)$, where $\eta>0$ will be determined later and we set $\bar{u}^N(t) := |\mathcal{O}|^{-1}\int_{\mathcal{O}} u^N(x,t)\,\mathrm{d}x$. Recall that $H(u^N|\bar{u}^N) = \frac{1}{2}\|(PA)^{-1/2}(u^N-\bar{u}^N)\|_{L^2(\mathcal{O})}^2$. Then $\bar{u}^N$ solves

$$\mathrm{d}\bar{u}_i^N(t) = \sum_{j=1}^n \bar{\sigma}_{ij}(u^N(t))\,\mathrm{d}W_j(t),\quad \bar{\sigma}_{ij}(u^N) = \frac{1}{|\mathcal{O}|}\int_{\mathcal{O}} \sigma(u^N)\,\mathrm{d}x.$$

We compute

$$
\begin{aligned}
&e^{\eta t}H(u^N(t)|\bar{u}^N(t)) - H(u^N(0)|\bar{u}^N(0)) - \eta\int_0^t e^{\eta s}H(u(s)|\bar{u}^N(s))\,\mathrm{d}s \qquad (19)\\
&= -\delta\sum_{i,j=1}^n \pi_i a_{ij}\int_0^t e^{\eta s}\int_{\mathcal{O}}\nabla u_i^N(s)\cdot\nabla u_j^N(s)\,\mathrm{d}x\,\mathrm{d}s \\
&\quad - \sum_{i=1}^n \pi_i\int_0^t e^{\eta s}\int_{\mathcal{O}}(u_i^N)^+(s)|\nabla p_i(u^N(s))|^2\,\mathrm{d}x\,\mathrm{d}s \\
&\quad + \frac{1}{2}\int_0^t e^{\eta s}\int_{\mathcal{O}}\operatorname{Tr}\big[(\sigma(u^N)-\bar{\sigma}(u^N))\mathrm{D}_u^2 H(u^N|\bar{u}^N)(\sigma(u^N)-\bar{\sigma}(u^N))\big](s)\,\mathrm{d}x\,\mathrm{d}s \\
&\quad + \sum_{i,j,k=1}^n \pi_i a_{ij}\int_{\mathcal{O}}\int_0^t e^{\eta s}(u_j^N-\bar{u}_j^N)(s)\big(\sigma_{ik}(u^N(s))-\bar{\sigma}_{ik}(u^N(s))\big)\,\mathrm{d}W_k(s)\,\mathrm{d}x.
\end{aligned}
$$

With $\lambda > 0$ being the smallest eigenvalue of the positive definite matrix $(\pi_i a_{ij})$, the first term on the right-hand side is estimated according to

$$
-\delta\sum_{i,j=1}^n \pi_i a_{ij}\int_0^t e^{\eta s}\int_{\mathcal{O}}\nabla u_i^N(s)\cdot\nabla u_j^N(s)\,\mathrm{d}s\,\mathrm{d}s \geq -\delta\lambda\int_0^t e^{\eta s}\int_{\mathcal{O}}|\nabla u^N(s)|^2\,\mathrm{d}x\,\mathrm{d}s.
$$

Then, taking the expecation in (19), neglecting the second term on the right-hand side of (19), and observing that the stochastic integral vanishes, we obtain

$$
\begin{aligned}
&e^{\eta t}\mathbb{E}H(u(t)|\bar{u}(t)) - \mathbb{E}H(u(0)|\bar{u}(0)) - \eta\int_0^t e^{\eta s}\mathbb{E}H(u(s)|\bar{u}(s))\,\mathrm{d}s \\
&\quad \leq -\delta\lambda\int_0^t e^{\eta s}\int_{\mathcal{O}}|\nabla u(s)|^2\,\mathrm{d}x\,\mathrm{d}s \\
&\qquad + C\mathbb{E}\int_0^t e^{\eta s}\|(PA)^{1/2}(\sigma(u)-\bar{\sigma}(u))\|_{\mathcal{L}_2(U;L^2(\mathcal{O}))}^2\,\mathrm{d}s.
\end{aligned}
$$

We estimate the last term on the right-hand side:

$$
\begin{aligned}
\|(PA)^{1/2}(\sigma(u)-\bar{\sigma}(u))\|_{\mathcal{L}_2(U;L^2(\mathcal{O}))}^2 &= \sum_{k=1}^\infty \big\|(PA)^{1/2}\big(\sigma(u)-\bar{\sigma}(u)\big)\eta_k\big\|_{L^2(\mathcal{O})}^2 \\
&\leq C\sum_{i,j=1}^n\sum_{k=1}^\infty \|\sigma_{ij}(u(s))\eta_k - \bar{\sigma}_{ij}(u(s))\eta_k\|_{L^2(\mathcal{O})}^2 \\
&= C\sum_{i,j=1}^n\sum_{k=1}^\infty \int_{\mathcal{O}}\bigg(\sigma_{ij}(u(s))\eta_k - \frac{1}{|\mathcal{O}|}\int_{\mathcal{O}}\sigma_{ij}(u(s))\eta_k\,\mathrm{d}y\bigg)^2\,\mathrm{d}x
\end{aligned}
$$

$$= \frac{C}{|\mathcal{O}|^2}\sum_{i,j=1}^n\sum_{k=1}^\infty\int_\mathcal{O}\bigg(\int_\mathcal{O}\big(\sigma_{ij}(u(x,s))\eta_k - \sigma_{ij}(u(s,y))\eta_k\big)\,\mathrm{d}y\bigg)^2\,\mathrm{d}x$$
$$\le \frac{C}{|\mathcal{O}|}\sum_{i,j=1}^n\sum_{k=1}^\infty\int_\mathcal{O}\int_\mathcal{O}\big(\sigma_{ij}(u(x,s))\eta_k - \sigma_{ij}(u(s,y))\eta_k\big)^2\,\mathrm{d}y\,\mathrm{d}x,$$

where we used the Cauchy–Schwarz inequality in the last step. Next, we add and subtract the space-independent term $\sigma_{ij}(\bar{u}(s))\eta_k$ and apply the triangle inequality:

$$\|(PA)^{1/2}(\sigma(u) - \bar{\sigma}(\bar{u}))\|^2_{\mathcal{L}_2(U;L^2(\mathcal{O}))}$$
$$\le \frac{C}{|\mathcal{O}|}\sum_{i,j=1}^n\sum_{k=1}^\infty\int_\mathcal{O}\int_\mathcal{O}\big(\sigma_{ij}(u(x,s))\eta_k - \sigma_{ij}(\bar{u}(s))\eta_k\big)^2\,\mathrm{d}y\,\mathrm{d}x$$
$$+ \frac{C}{|\mathcal{O}|}\sum_{i,j=1}^n\sum_{k=1}^\infty\int_\mathcal{O}\int_\mathcal{O}\big(\sigma_{ij}(u(y,s))\eta_k - \sigma_{ij}(\bar{u}(s))\eta_k\big)^2\,\mathrm{d}y\,\mathrm{d}x$$
$$= 2C\sum_{i,j=1}^n\|\sigma_{ij}(u(s)) - \sigma_{ij}(\bar{u}(s))\|^2_{\mathcal{L}_2(U;L^2(\mathcal{O}))} \le 2C_1C_\sigma^2\|u(s) - \bar{u}(s)\|^2_{L^2(\mathcal{O})}.$$

Thus, by the Poincaré–Wirtinger inequality with constant $C_P > 0$,

$$e^{\eta t}\mathbb{E}H(u(t)|\bar{u}(t)) - \mathbb{E}H(u(0)|\bar{u}(0))$$
$$\le \bigg(C(\lambda)\eta + 2C_1C_\sigma^2 - \frac{\delta\lambda}{C_P^2}\bigg)\int_0^t e^{\eta s}\int_\mathcal{O}\|(u-\bar{u})(s)\|^2_{L^2(\mathcal{O})}\,\mathrm{d}s.$$

Without loss of generality, we may choose $C_1 \ge 1/2$. We set $c_0^2 := \delta\lambda/(2C_1C_P^2)$, choose $0 < C_\sigma < c_0$, and set $\eta := (c_0^2 - C_\sigma^2)/C(\lambda)$. Then $C(\lambda)\eta + 2C_1C_\sigma^2 - \delta\lambda/C_P^2 = (1-2C_1)(c_0^2 - C_\sigma^2) < 0$, and we end the proof.

5 Numerical Illustration

We discretize (1) for $n \ge 2$ species by a semi-implicit Euler–Maruyama scheme and centered finite differences in one space dimension. Let $\mathcal{O} = (0,1)$, $J \in \mathbb{N}$, $\Delta x = 1/J$, $\Delta t > 0$, $x_j = j\Delta x$ for $j = 0,\ldots,J$, and $t_k = k\Delta t$ for $k \in \mathbb{N}_0$. We approximate $u_i(x_j,t_k,\cdot)$ by u_{ij}^k, solving

$$u_{ij}^{k+1} = (I + \delta\Delta t A_\Delta)^{-1}\bigg(u_{ij}^k + \frac{\Delta t}{\Delta x}(F_{i,j+1/2}^k - F_{i,j-1/2}^k) + \sum_{j=1}^n\sigma_{ij}(u^k)\Delta W_j^k\bigg),$$

where the centered discrete fluxes are given by

$$F_{i,j+1/2}^k = \frac{1}{2\Delta x}(u_{i,j+1}^k + u_{ij}^k)\sum_{\ell=1}^{n} a_{i\ell}(u_{\ell,j+1}^k - u_{\ell,j}^k),$$

and we have set $u^k = (u_{ij}^k)_{i=1,2,\ j=0,\dots,J}$. The matrix $A_\Delta \in \mathbb{R}^{(J+1)\times(J+1)}$ is associated to the discrete Laplacian with Neumann conditions, and $\Delta W_j^k = W_j(t_{k+1}) - W_j(t_k)$.

First, we choose the initial data

$$u_1^0(x) = 1_{[0,1/2]}(x), \quad u_2^0(x) = 10x^2\left(\frac{1}{2} - \frac{x}{3}\right) \quad \text{for } x \in [0,1],$$

the diffusion parameter $\delta = 1$, the coefficients $a_{11} = a_{22} = 2$, $a_{12} = a_{21} = 1$, and the stochastic diffusion $\sigma_{ij}(u) = 0.001\sqrt{1+u_i}\delta_{ij}$ for $i, j = 1, 2$. Observe that the matrix (a_{ij}) is symmetric positive definite. The ℓ^2 errors of (u_1, u_2) versus time step size Δt (with $\Delta x = 0.02$ and 2^{11} samples) and space grid size Δx (with $\Delta t = 10^{-5}$ and 2^{11} samples) at time $t = 1$ are presented in Fig. 1. For the reference solution, we have chosen $\Delta t = 5 \cdot 10^{-5}$, $\Delta x = 0.02$ (left figure) and $\Delta t = 10^{-5}$ and $\Delta x = 1/128$ (right figure). As expected, we observe a half-order convergence in time and first-order convergence in space.

Next, we present the long-time behavior of the discrete ℓ^2 norms of the densities in the three-species case. The parameters are $\delta = 1$, $a_{12} = a_{23} = a_{31} = 1$ and $a_{ij} = 0$ else, $\sigma_{ij} = 0.1\sqrt{1+u_i}\delta_{ij}$ for $i, j = 1, 2, 3$. Notice that the matrix (a_{ij}) does not satisfy the detailed-balance condition such that, strictly speaking, Theorem 2 is not applicable. The time step is $\Delta t = 10^{-4}$ and the grid size $\Delta x = 0.02$. Figure 2 shows the time evolution of the ℓ^2 norms of u_i in the deterministic case, while the right figure illustrates the dynamics of the ℓ^2 norms in the stochastic situation (with 2^{11} samples). We observe that the ℓ^2 norms, which are equivalent to the Rao entropy, converge (in expectation) as $t \to \infty$, and the behavior is similar in both cases.

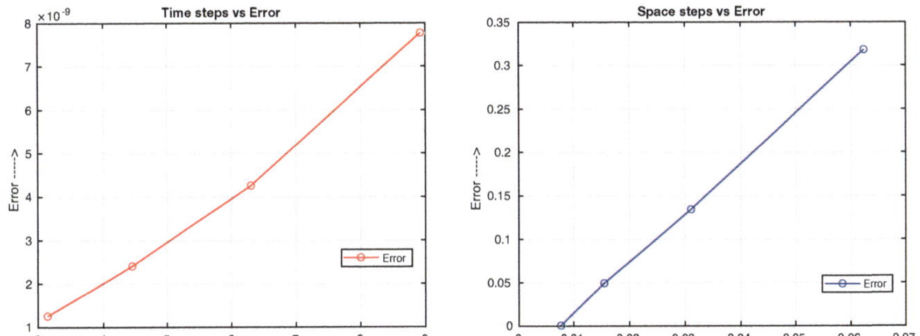

Fig. 1 Discrete ℓ^2 error of (u_1, u_2) versus time step size (left) and space grid size (right). Observe that the x-axis of the left figure is scaled with $\sqrt{\Delta t}$

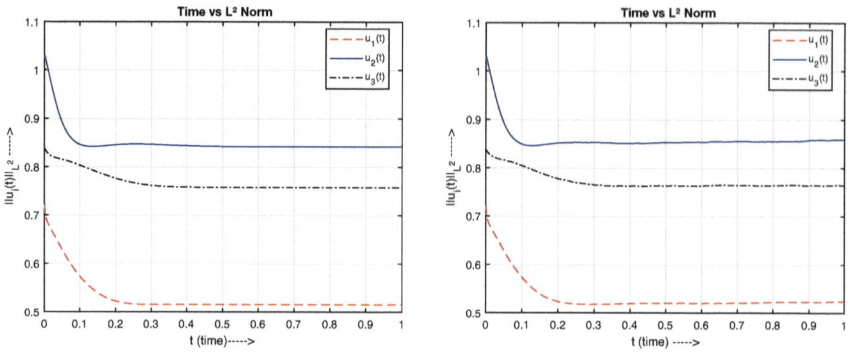

Fig. 2 Long-time behavior of the ℓ^2 norms of u_i in the deterministic (left) and stochastic (right) case

Acknowledgements The authors acknowledge partial support from the Austrian Science Fund (FWF), grants P33010, W1245, and F65. This work has received funding from the European Research Council (ERC) under the European Union's Horizon 2020 research and innovation programme, ERC Advanced Grant no. 101018153.

References

1. Amann, H.: Nonhomogeneous linear and quasilinear elliptic and parabolic boundary value problems. In: Schmeisser, H.J., Triebel, H. (eds.), Function Space Differentials Operator Theory for Nonlinear Analysis, Teubner, Wiesbaden, pp. 9–126 (1993)
2. Bendahmane, M., Karlsen, K.: Martingale solutions of stochastic nonlocal cross-diffusion systems. Netw. Heterog. **17**, 719–752 (2022)
3. Bertsch, M., Gurtin, M., Hilhorst, D., Peletier, L.L On interacting populations that disperse to avoid crowding: preservation of segregation. J. Math. Biol. **23**, 1–13 (1985)
4. Brzeźniak, Z., Goldys, B., Jegaraj, T.: Weak solutions of a stochastic Landau-Lifshitz-Gilbert equation. Appl. Math. Res. Express. **2013**, 1–33 (2013)
5. Brzeźniak, Z., Hausenblas, E., Motyl, E.: Uniqueness in law of the stochastic convolution process driven by Lévy noise. Electron. J. Probab. **18**, 1–15 (2013)
6. Brzeźniak, Z., Motyl, E.: The existence of martingale solutions to the stochastic Boussinesq equations. Global Stoch. Anal. **1**, 175–216 (2014)
7. Brzeźniak, Z., Ondreját, M.: Stochastic geometric wave equations with values in compact Riemannian homogeneous spaces. Ann. Probab. **41**, 1938–1977 (2013)
8. Chekroun, M., Park, E., Temam, R.: The Stampacchia maximum principle for stochastic partial differential equations and applications. J. Differ. Eqs. **260**, 2926–2972 (2016)
9. Chen, L., Daus, E., Jüngel, A.: Rigorous mean-field limit and cross-diffusion. Z. Angew. Math. Phys. **70**(122), 21 pages (2019)
10. Ciotir, I.: Stochastic porous media equations with divergence Itô noise. Evol. Eqs. Control Theory **9**, 375–398 (2020)
11. da Prato, G., Zabczyk, J.: Stochastic Equations in Infinite Dimensions, 2nd edn. Cambridge University Press, Cambridge (2014)
12. Dhariwal, G., Jüngel, A., Zamponi, N.: Global martingale solutions for a stochastic population cross-diffusion system. Stoch. Process. Appl. **129**, 3792–3820 (2019)

13. Fehrman, B., Gess, B.: Well-posedness of nonlinear diffusion equations with nonlinear, conservative noise. Arch. Ration. Mech. Anal. **233**, 249–322 (2019)
14. Kuehn, C., Neamţu, A.: Pathwise mild solutions for quasilinear stochastic partial differential equations. J. Differ. Eqs. **269**, 2185–2227 (2020)
15. Liu, W., Röckner, M.: Stochastic Partial Differential Equations. Springer, Cham (2015)
16. Lototsky, S., Rozovsky, B.: Stochastic Partial Differential Equations. Springer, Cham (2017)
17. Mayorcas, A., Tomašević, M.: Blow-up for a stochastic model of chemotaxis driven by conservative noise on $\mathbb{R}^2$. J. Evol. Equ. **23**, 57 (2023). https://doi.org/10.1007/s00028-023-00900-3
18. Rao, C.: Analysis of diversity: a unified approach. In: Gupta, S., Berger, J. (eds.) Statistical Decision Theory and Related Topics III, vol. 2, pp. 235–250. Academic Press, New York (1982)

MHD Casson Fluid Flow Over a Vertical Porous Surface in the Presence of Chemical Reaction and Radiation Effects

Sandhya Akuri, G. Venkata Ramana Reddy, and G. V. S. R. Deekshitulu

Abstract The article is examines the MHD Casson fluid flow under the influence of radiation, chemical reaction and magnetic field through a vertical porous surface. The nonlinear system of PDEs is converted to ODEs using similarity analysis. The resulting ODEs are then solved numerically using the 4th order R-K method. The attained outcomes are matched with the existing, good agreement with available standard results. The numerical outcomes are found for the various profiles like velocity, temperature and concentration, further, skin friction, Nusselt and rate of mass transfer coefficients for different values of critical constraints.

Keywords Porous medium · Casson fluid · Magnetic field · Non-Newtonian fluid

1 Introduction

A Newtonian fluid is a viscous fluid in which the stress at each point, resulting from its flow, is proportional to the rate of deformation over time. However, some fluids exhibit deviations from Newtonian behavior; these are known as non-Newtonian fluids. Unlike Newtonian fluids, whose viscosity remains constant, the viscosity of non-Newtonian fluids varies with the shear rate. Examples of non-Newtonian fluids include blood, shampoo, paints, starch suspensions, and custard.

Compared to Newtonian fluids, non-Newtonian fluids have wider, industrial applications. They are extensively used in foods and paper manufacturing, petroleum

S. Akuri (✉) · G. V. S. R. Deekshitulu
Department of Mathematics, JNTUK, Kakinada, AP, India
e-mail: akuri.sandhya@vrsiddhartha.ac.in

S. Akuri
Department of Mathematics, VR Siddhartha Engineering College (Deemed to be university), Kanur, Vijayawada, India

G. Venkata Ramana Reddy
Department of Mathematics, Koneru Lakshmaiah Education Foundation, Vaddeswaram, AP, India

S. Bhattacharyya and H. S. Mahato (eds.), *Modeling, Analysis and Simulations of Multiscale Transport Phenomena*, Springer Proceedings in Mathematics & Statistics 491, https://doi.org/10.1007/978-981-96-3098-1_2

drilling, certain separation processes, polymer engineering and few other industrial processes. Emmanuel et al. [1] focused their attention on analyzing the effect of permeability on the flow of Casson fluid over a vertical surface in presence of both chemical reaction and magnetic field. Arthur et al. [2] analyzed the combined impact of a magnetic field furthermore, suction on the flow of a stagnation point over the stretching surface. Bhattacharya et al. [3] observed that the cooling rate plays a crucial role in deportment on the properties of the final item during the procedures. Consequently, the final item quality relies upon the rate of transmission of heat as of the stretching surface. Nandy et al. [4] deliberated heat transference in the stagnation point area above the porous contracting sheet, considering boundary layer flow and the presence of radiation. Qasim et al. [5] researched the impact of mass transferal in the MHD Casson fluid flow in the occurrence of a chemical reaction over a porous stretching sheet. Alsaedi et al. [6] inquired investigated the impact of magnetic field on the boundary layer flow of a Casson fluid over a porous extending/contracting sheet. The mixed convection behaviour of a non-Newtonian stagnation-point Casson fluid flow built up by Alhothuali et al. [7]. Aziz et al. [8] examined the unsteady development of boundary layer and heat transferal Casson fluid flow with a parallel free stream over a flat moving plate. Yacob and Ishak et al. [9, 10] discussed the transmuted ODEs are resolved by shooting technique. Cui and Zhou et al. [11, 12] proposed some non-Newtonian fluid models depending on various physical characters. Salem et al. [13] analyzed the combined effects of Hall effects, mass diffusion and chemical species over a continuously stretching surface in the 1st and higher order reactions. Cortell and Hayat et al. [14, 15] discussed several industrial and manufacturing procedures like as hot rolling, fiber coating, metal spinning, production of glass-fiber and paper etc. Mehta et al. [16] considered a pipe filled with the Casson fluid flow with a same permeable medium. Reddy et al. [17] studied unsteady 2-D (dimensional) incompressible viscous flow conducting electrically fluid above an extending surface with the magnetic field effects.

A review of the literature clearly indicates that, although several studies have been conducted on Casson fluid flow, very little consideration is taken in to account for the flow of a Casson fluid through a vertical permeable surface with magnetic field and radiation impacts. The extensive industrial applications of certain non-Newtonian fluids have encouraged further investigation to know their behavior in numerous procedure of transport. Hence in this investigation the consistent progression of a Casson fluid and transmission of mass near a vertical porous, incompressible extending sheet are examined. The governing PDE are transformed into nonlinear systems of ODE by using the appropriate similitude changes. The outcomes are shown graphically and examined subjectively.

2 Mathematical Analysis

Consider a Casson fluid with a steady and incompressible flow of extending vertical permeable surface at $y' = 0$, keeping magnetic field in transverse direction, represented as Fig. 1.

The stretched sheet moves with a linear velocity expressed as $u_w = ax'$, a being positive constant. Along the continuous permeable extending sheet x' is directed and the y'–axis is normal to it. As well as it is taken that the flow of a fluid over $y' > 0$. The temperature and concentrations parameters are suitably kept at suggested constant values while, T'_w, C'_w, T'_∞ as well as C'_∞ are the surface and stable values they are away from boundary to a greater extent. It is also considered the 1st order homogeneous species chemical reaction with reaction rate constant γ.

For an isotropic flow of Casson fluid, the state of rheological equation is given by [1]

$$\tau_{ij} = \begin{cases} 2\left(\mu_B + \dfrac{P_{y'}}{\sqrt{2\pi}}\right)e_{ij}, \ \pi > \pi_{c,} \\ 2\left(\mu_B + \dfrac{P_{y'}}{\sqrt{2\pi_c}}\right)e_{ij}, \ \pi < \pi_{c,} \end{cases} \tag{1}$$

where $\pi = e_{ij}e_{ij}$ and e_{ij} represents the component of (i, j)th distortion rate, π being the resulted from distortion rate itself, the product π_c is a critical value reliant over non-Newtonian form, μ_B represents the non-Newtonian fluid viscosity relevant to a fluid of plastic dynamic and P_y gives the fluid yield pressure. The following governing equations are attained by taking the above assumption:

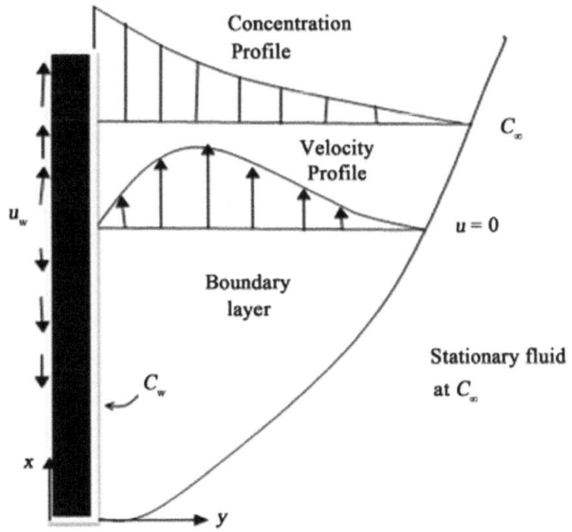

Fig. 1 Physical illustration of fluid model

$$\frac{\partial u'}{\partial x'} + \frac{\partial v'}{\partial y'} = 0 \tag{2}$$

$$u'\frac{\partial u'}{\partial x'} + v'\frac{\partial u'}{\partial y'} = \nu\left(1 + \frac{1}{\beta}\right)\frac{\partial^2 u'}{\partial y'^2} + g\beta_c(C' - C'_\infty)$$
$$+ g\beta_T(T' - T'_\infty) - \frac{\sigma B_0^2}{\rho}u' - \frac{\nu}{k'}u' \tag{3}$$

$$u'\frac{\partial T'}{\partial x'} + v'\frac{\partial T'}{\partial y'} = D_T\frac{\partial^2 T'}{\partial y'^2} - \frac{1}{\rho c_p}\frac{\partial q_r}{\partial y'} \tag{4}$$

$$u'\frac{\partial C'}{\partial x'} + v'\frac{\partial C'}{\partial y'} = D_m\frac{\partial^2 C'}{\partial y'^2} - \gamma(C' - C'_\infty) \tag{5}$$

Boundary conditions are of:

$$\begin{aligned} &u' = u_w,\ v' = -v'_0(x'),\ T' = T'_w,\ C' = C'_w \text{ at } y' = 0 \\ &u' = 0, T' = T'_\infty,\ C' = C'_\infty \text{ as } y' \to \infty \end{aligned} \tag{6}$$

Here, β represents the Casson parameter (non-Newtonian), ν being the viscosity of a kinematic, K permeable medium exhibiting porosity, D_m be the diffusion of mass, γ represents the rate of reaction, D_T thermal diffusion, $v_0(x')$ be the surface velocity of suction, k' represents medium permeability, C'_w and C'_∞ are indicative of surface concentration and concentration of the free flow respectively, β_c indicates expansion of solutal coefficient, β_T denotes thermal expansion coefficient, B_0 the strength of magnetic field applied, ρ fluid density, g is the acceleration due to gravity, σ indicates electrical conductivity and c_p is constant pressure specific heat.

By using Rosseland approximation [18, 19] Eq. (4) is expressed as

$$u'\frac{\partial T'}{\partial x'} + v'\frac{\partial T'}{\partial y'} = D_T\left(1 + \frac{16\sigma^* T_\infty^3}{3kk^*}\right)\frac{\partial^2 T'}{\partial y'^2} \tag{7}$$

Here σ^* represents the constant of Stefan-Boltzmann, k indicates thermal conductivity and k^* gives the coefficient of mean absorption.

Introducing dimensionless quantities:

$$\begin{aligned} &u' = u_w f'(\eta),\ v' = -\sqrt{\frac{\nu u_w}{x'}} f(\eta), \\ &\eta = y'\sqrt{\frac{u_w}{\nu x'}},\ \phi(\eta) = \frac{C' - C'_\infty}{C'_w - C'_\infty},\ \theta(\eta) = \frac{T' - T'_\infty}{T'_w - T'_\infty} \end{aligned} \tag{8}$$

Substituting the Eqs. (5) in (1–4) we get

$$\left(1+\frac{1}{\beta}\right)f''' + ff'' - f'^2 + Gc\phi + Gr\theta - (M+K)f' = 0 \tag{9}$$

$$\frac{1}{\Pr}\left(1+\frac{4}{3}R\right)\theta'' + f\theta' = 0 \tag{10}$$

$$\phi'' + Scf\phi' - ScKr\phi = 0 \tag{11}$$

respective boundary conditions are represented as

$$\begin{aligned} &f = f_w, f' = 1, \theta = 1, \phi = 1 \text{ at } \eta = 0 \\ &f' = 0, \theta = 0, \phi = 0 \text{ as } \eta \to \infty \end{aligned} \tag{12}$$

Here the symbol of prime represents differentiation w.r.to η. Where $M = \frac{\sigma B_0^2}{\rho a}$ the parameter of magnetic field is, $K = \frac{\nu}{k'a}$ is the permeability parameter, $Kr = \frac{\gamma}{a}$ is the parameter of chemical reaction, $fw = {v_0(x')}/{\sqrt{av}}$ is the parameter of suction, $Sc = {\nu}/{D_m}$ is the Schmidt number, $Gr = {g\beta_T(T'_w - T'_\infty)x'}/{u_w^2}$ is the Grashof number, $Gc = {g\beta_c(C'_w - C'_\infty)x'}/{u_w^2}$ represents the solutal Grashof number, $\Pr = \frac{\nu}{D_T}$ Prandtl number and $R = \frac{4\sigma^* T_\infty^3}{kk^*}$, Radiation parameter.

3 Numerical Solution

The Nonlinear DEs (9–11) along with the boundary conditions (12) can be resolved numerically using 4th and 5th R-K method along with shooting approach. The physical quantities i.e. skin friction coefficient, rate of heat and mass transformations are given as

$$C_f = \frac{\tau_w}{\rho(u_w)^2}, Nu = \frac{x' q_m}{\alpha_T(T'_w - T'_\infty)}, Sh = \frac{x' q_w}{D_m(C'_w - C'_\infty)} \tag{13}$$

Here $\tau_w = \left(\mu_B + \frac{P_{y'}}{\sqrt{2\pi_c}}\right)\left(\frac{\partial u'}{\partial y'}\right)_{y'=0}, q_w = -D_m\left(\frac{\partial C'}{\partial y'}\right)_{y'=0} q_m = -\alpha_T\left(\frac{\partial T'}{\partial y'}\right)_{y'=0}$ are surface shear stress, wall surface heat flux and mass flux parameters.

The non-dimensional terms for skin friction, rate of heat and mass transformations and Reynolds number are specified as

$$C_f \operatorname{Re}_x^{1/2} = \left[\left(1+\frac{1}{\beta}\right)f''(0)\right], \tag{14}$$

$$Nu \operatorname{Re}_x^{-1/2} = -\theta'(0), \tag{15}$$

Table 1 Comparison of values

Kr	Sc	Shehzad et al. [5]	Arthur et al. [1]	Present study
0.01	1.0	0.59136	0.59138	0.59140
0.10	1.0	0.66898	0.668983	0.668985
1.00	1.0	1.17650	1.176499	1.176498
10.00	1.0	3.23175	3.231228	3.231523

$$Sh\,\mathrm{Re}_x^{-1/2} = -\phi'(0), \tag{16}$$

$$\mathrm{Re}_x^{1/2} = \left(\frac{u_w^2}{\nu}\right)^{1/2}. \tag{17}$$

were calculated and obtainable in the tabular form.

The numerical estimations computed for critical values of physical parameters together with M, K, Gr, Gm, β, S, Pr, R, Sc and Kr to approve developed an algorithm in MATLAB software for the current technique, the computational outcomes for Sherwood number coefficient contrasted with the outcomes. Arthur et al. [1] and Shehzad et al. [5] are specified in Table 1, comparison exposed a closed agreement with them.

4 Result and Discussions

The numerical calculations are performed for the parameter values of $M = 1$, $K = 0.5$, $\beta = 0.5$, $Gr = 0.5$, $Gm = 2$, $S = 0.5$, $\Pr = 0.71$, $R = 1$, $Sc = 0.6$, $Kr = 0.5$. The velocity profiles for different flow parameters are illustrated in Figs. 2, 3, 4, 5, 6, and 7, while the temperature and concentration profiles are presented in Figs. 8, 9, 10, and 11. The effect of the magnetic parameter (M) profile shown in Fig. 2. The profile shows that enhance in the magnetic parameter field retard the velocity physically, due to Lorentz force, which increases thickness of a boundary layer. Consequently, the velocity decreases, weakening the boundary layer momentum, transversely the boundary. Figure 3 shows the effect of the permeability parameter (K) effects on the profiles of velocity. The results indicate that an increase in K enhances velocity. Notably, as the porosity parameter (K) increases, the improved velocity of fluid brought about the boundary layer thickness likewise increment. Figure 4 displays the Casson fluid parameter (β), for various velocity profiles. This figure shows that velocity reduced due to the rising viscosity of the fluid. In practice, enhancing of Casson fluid parameter (β), outcomes enhances in viscosity of plastic dynamic which creates an obstruction in the stream causes decline in the fluid velocity. Figures 5 and 6 represents Gr and Gc in the profiles of velocity. From the

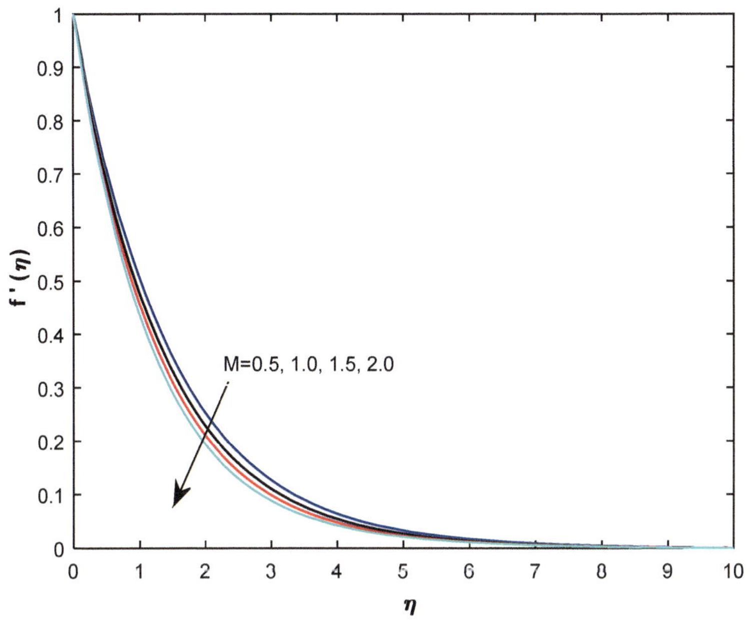

Fig. 2 Velocity profiles for dissimilar values of M

figures, it can be detected that the profiles of velocity enhance with the increases of Gr and Gc due to increased buoyancy force, i.e. enhancing buoyancy force will lead the improved stream kinematics. Figure 7 shows the profile of velocity at dissimilar suction parameter (S) values. It is observed that for an enhancing of S, the velocity stream field diminishes. Figure 8 indicates for distinct values of Prandtl number (Pr) the temperature distribution. An observation drawn with regard to Pr is that with increase in Pr, the temperature diminutions. Physically, the increase of Prandtl number (Pr) means the decrease of thermal conductivity of the fluid. The temperature profile for the radiation parameter (R) shown in Fig. 9. From the figure, it can be observed that with the R intensifications, growth in temperature profiles is observed. The modification in concentration profiles versus thermo physical is assessed through Figs. 10 and 11. The influence of the Sc and Kr are increased the concentration profiles decreases. The Kr raises the rate of interfacial mass transfer and decreases concentration. This leads to decrease in concentration gradient.

Table 2 shows the outcome of dissimilarities of physical constraints namely skin friction, Nusselt and rate of mass transfer coefficients for certain values of β, S, Pr, R, Sc, M, K, Gr, Gc and Kr. Due to the variation in several physical parameters the skin friction, Nusselt and rate of mass transfer coefficients are enhances.

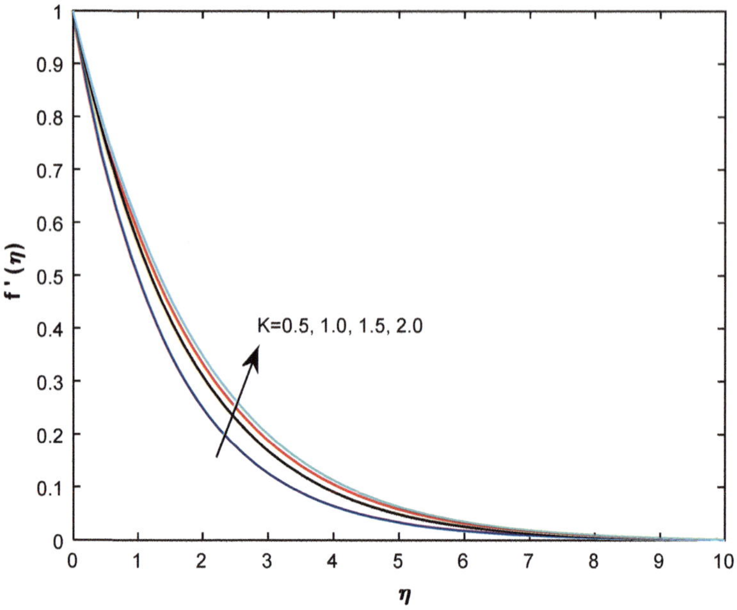

Fig. 3 Velocity profiles for dissimilar values of the Permeability parameter (K)

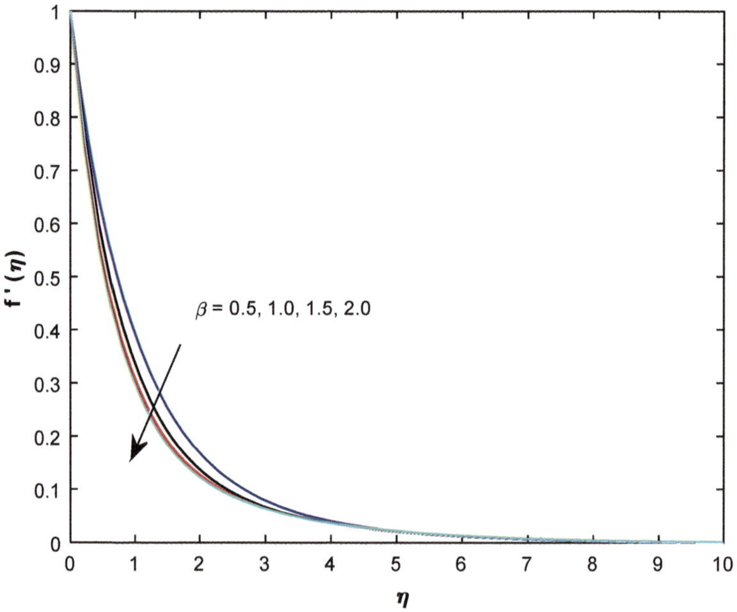

Fig. 4 Velocity profiles for different values of Casson fluid parameter (β)

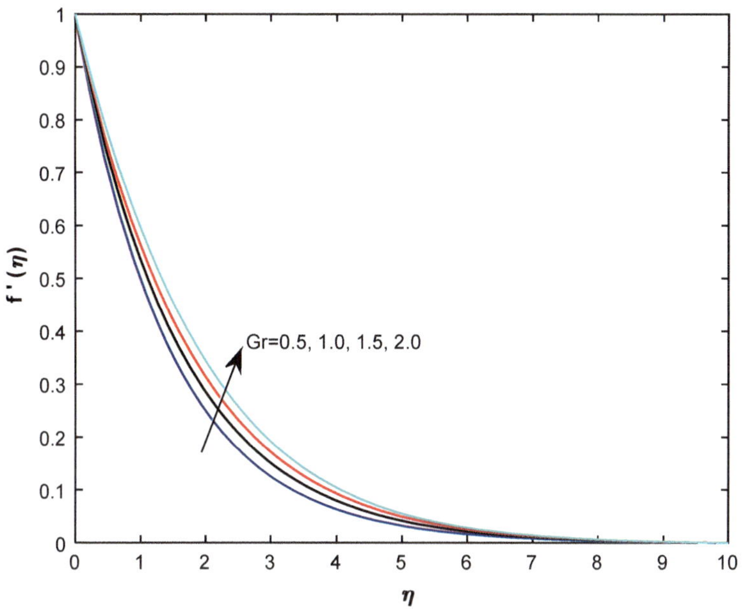

Fig. 5 Velocity profiles for different values of Grashof number (Gr)

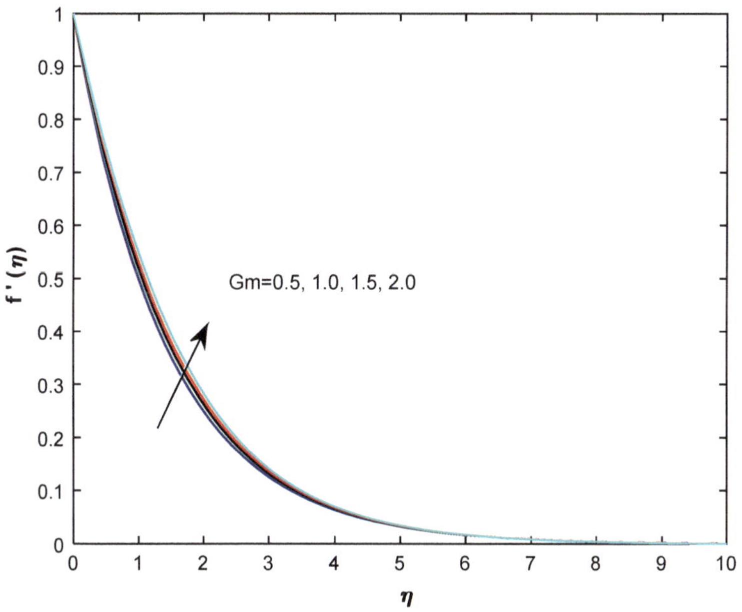

Fig. 6 Velocity profiles for different values of Gc

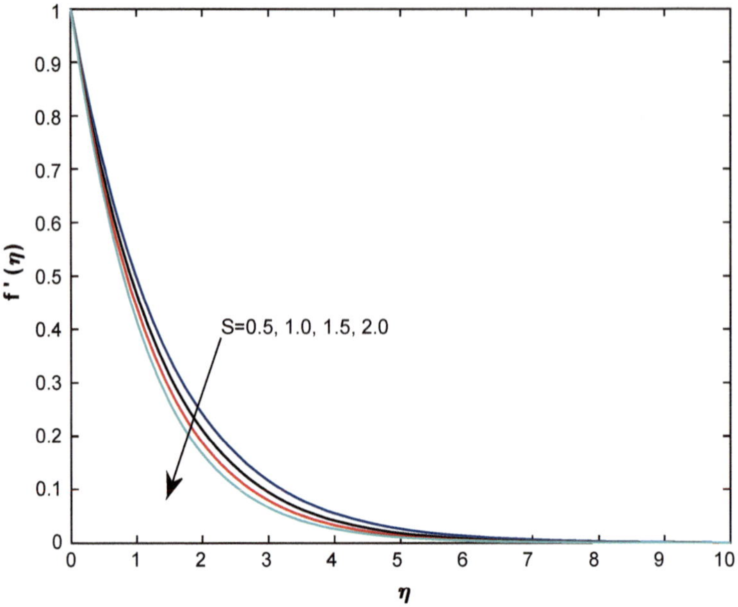

Fig. 7 Velocity profiles for distinct values of Suction parameter (S)

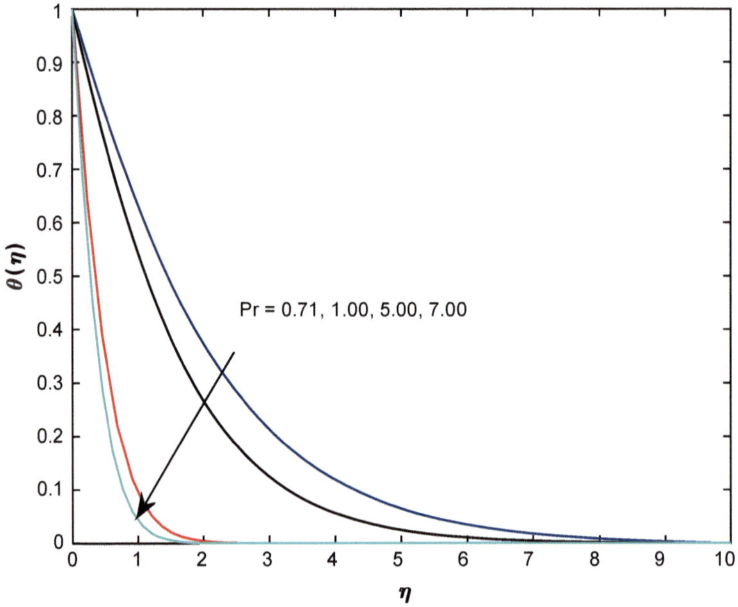

Fig. 8 Temperature profiles for distinct values of Prandtl number (Pr)

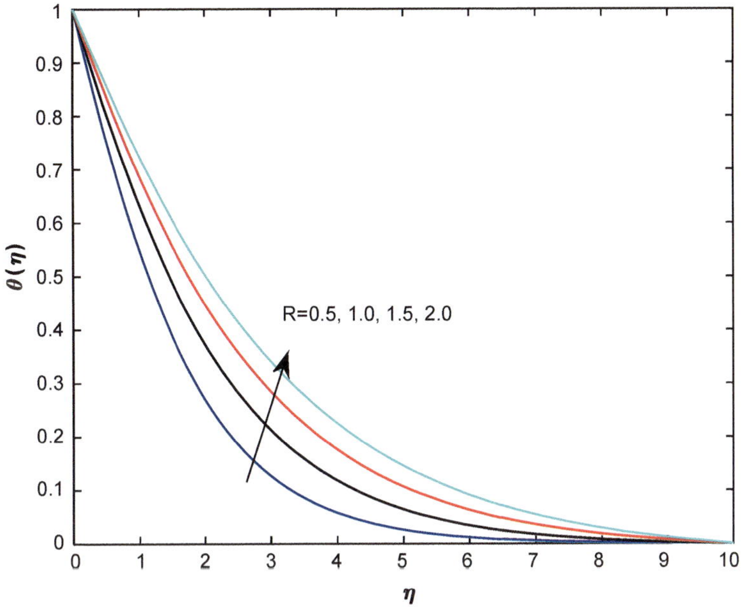

Fig. 9 Temperature profiles for dissimilar values of Radiation parameter (R)

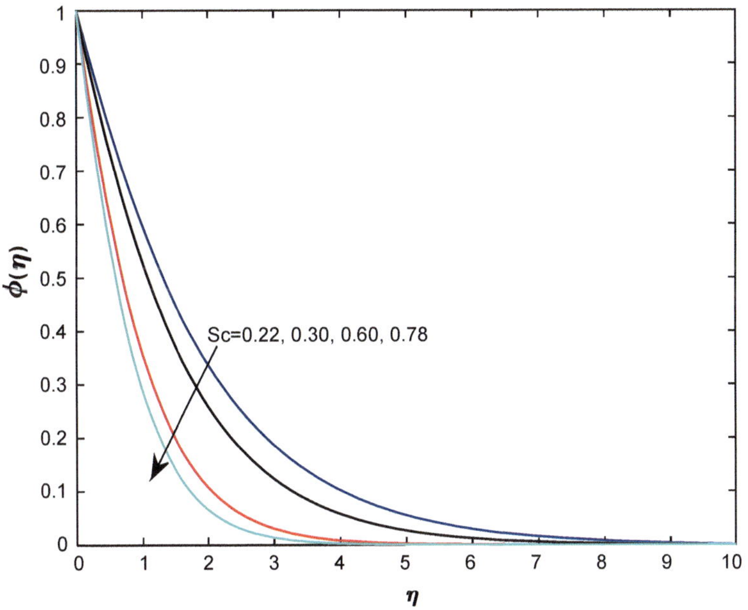

Fig. 10 Concentration profiles for dissimilar values of Schmidt number (Sc)

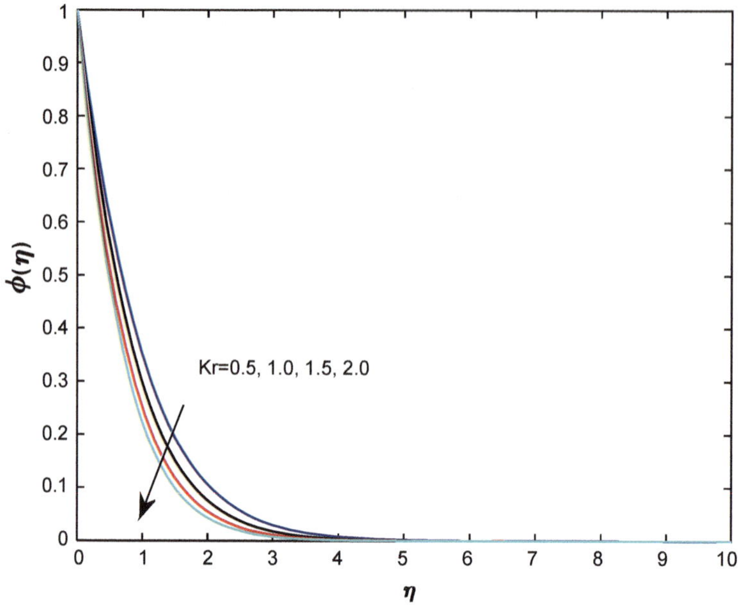

Fig. 11 Concentration profiles for dissimilar values of Chemical reaction parameter (Kr)

5 Conclusions

The following conclusions are drawn from the present analysis.

(1) The velocity is increases with the rise of the Permeability parameter (K), Grashof number (Gr) and solutal Grashof number (Gm), and while it decreases with the increasing values of the parameters like a magnetic parameter (M), Casson fluid parameter (β) and suction parameter (S).
(2) The temperature decreases with an increase in the Prandtl number (Pr) and increases with an increase in the radiation parameter (R).
(3) The boundary layer concentration decreases with the increasing Schmidt number (Sc) values and chemical reaction parameter (Kr) values.
(4) The skin friction coefficient, Nusselt number and Sherwood number are increased under the influence of several parameters like M, K, β, Gr, Gm, Pr, R, Sc, Kr and S.

Table 2 Numerical results

M	K	B	Gr	Gc	Pr	R	Sc	Kr	S	SF	Nu	Sh
0.5	0.5	0.2	0.5	0.5	0.71	1	0.5	0.5	0.5	0.689877	0.394290	0.910098
1										0.745572	0.400405	0.914028
1.5										0.797853	0.407123	0.918305
2										0.847259	0.414557	0.922995
0.5	0.5	0.2	0.5	0.5	0.71	1	0.5	0.5	0.5	0.493448	0.414557	0.922995
	1									0.518178	0.432219	0.934003
	1.5									0.565065	0.439188	0.938312
	2									0.689877	0.442929	0.940622
0.5	0.5	0.5	0.5	0.5	0.71	1	0.5	0.5	0.5	0.963553	0.356310	0.880902
		1								1.177024	0.360155	0.884160
		1.5								1.289815	0.367107	0.889828
		2								1.360527	0.383768	0.902416
0.5	0.5	0.5	0.5	0.5	0.71	1	0.5	0.5	0.5	0.493896	0.414557	0.922995
			1							0.557528	0.424710	0.929170
			1.5							0.622750	0.433761	0.934926
			2							0.689877	0.441961	0.940340
0.5	0.5	0.5	0.5	0.5	0.71	1	0.5	0.5	0.5	0.553778	0.414557	0.922995
				1						0.598764	0.418996	0.926202
				1.5						0.644124	0.423276	0.929333
				2						0.689877	0.427411	0.932391
0.5	0.5	0.5	0.5	0.5	0.71	1	0.5	0.5	0.5	0.689877	0.414557	0.917214
					1					0.697972	0.535972	0.917626
					5					0.731142	1.806330	0.921737
					7					0.736166	2.345036	0.922995
0.5	0.5	0.5	0.5	0.5	0.71	0.5	0.5	0.5	0.5	0.680101	0.298969	0.921758
						1				0.684244	0.344670	0.922995
						1.5				0.689877	0.414557	0.923905
						2				0.697828	0.533549	0.924589
0.5	0.5	0.5	0.5	0.5	0.71	0.5	0.22	0.5	0.5	0.678514	0.532442	0.490547
							0.3			0.684480	0.533549	0.594235
							0.6			0.697828	0.537264	0.921758
							0.78			0.702644	0.539195	1.092643
0.5	0.5	0.5	0.5	0.5	0.71	0.5	0.6	0.5	0.5	0.697828	0.531648	0.921758
								1		0.701869	0.532079	1.096699
								1.5		0.704774	0.532668	1.244106
								2		0.707026	0.533549	1.373883
0.5	0.5	0.5	0.5	0.5	0.71	0.5	0.6	0.5	0.5	0.697828	0.533549	0.921758

(continued)

Table 2 (continued)

M	K	B	Gr	Gc	Pr	R	Sc	Kr	S	SF	Nu	Sh
									1	0.752739	0.700341	1.132434
									1.5	0.808907	0.877645	1.361385
									2	0.866227	1.062449	1.604521

References

1. Arthur, E.M., Seini, I.Y., Bortteir, L.B.: Analysis of Casson fluid flow over a vertical porous surface with chemical reaction in the presence of magnetic field. J. Appl. Math. Phys. **3**, 713–723 (2015)
2. Arthur, E.M., Seini, Y.I.: MHD thermal stagnation point flow towards a stretching porous surface. Math. Theory Model. **4**, 163–169 (2014)
3. Bhattacharyya, K.: Boundary layer stagnation-point flow of Casson fluid and heat transfer towards a shrinking/stretching sheet. Front. Heat Mass Transf. (FHMT) **4**, 1–9 (2013)
4. Mahapatra, T.R., Nandy, S.K.: Stability of dual solutions in stagnation-point flow and heat transfer over a porous shrinking sheet with thermal radiation. Meccanica **48**, 23–32 (2013)
5. Shehzad, S.A., Hayat, T., Qasim, M., Asghar, S.: Effects of mass transfer on MHD flow of Casson fluid with chemical reaction and suction. Braz. J. Chem. Eng. **30**, 187–195 (2013)
6. Bhattacharyya, K., Hayat, T., Alsaedi, A.: Analytic solution for magnetohydrodynamic boundary layer flow of Casson fluid over a stretching/shrinking sheet with wall mass transfer. Chin. Phys. B **22**, 1–6 (2013)
7. Hayat, T., Shehzad, S.A., Alsaedi, A., Alhothuali, M.S.: Mixed convection stagnation point flow of Casson fluid with convective boundary conditions. Chin. Phys. Lett. **29**, 1–4 (2012)
8. Mustafa, M., Hayat, T., Pop, I., Aziz, A.: Unsteady boundary layer flow of a Casson fluid due to an impulsively started moving flat plate. Heat Transf.-Asian Res. **40**, 563–576 (2011)
9. Yacob, N.A., Ishak, A., Pop, I.: Melting heat transfer in boundary layer stagnation-point flow towards a stretching/shrinking sheet in a micropolar fluid. Comput. Fluids **47**, 16–21 (2011)
10. Ishak, A., Lok, Y.Y., Pop, I.: Stagnation-point flow over a shrinking sheet in a micropolar fluid. Chem. Eng. Commun. **197**, 1417–1427 (2010)
11. Cui, Z.W., Liu, J.X., Yao, G.J., Wang, K.X.: Borehole guided waves in a non-Newtonian (Maxwell) fluid-saturated porous medium. Chin. Phys. B **19**, 1–8 (2010)
12. Zhou, X.-F., Gao, L.: Effect of multipolar interaction on the effective thermal conductivity of nanofluids. Chin. Phys. B **16**, 2028–2032 (2007)
13. Salem, A.M., El-Aziz, M.A.: Effect of hall currents and chemical reaction on hydromagnetic flow of a stretching vertical surface with internal heat generation/absorption. Appl. Math. Model. **32**, 1236–1254 (2008)
14. Cortell, R.: Analysing flow and heat transfer of a viscoelastic fluid over a semi-infinite horizontal moving flat plate. Int. J. Non-Linear Mech. **43**, 772–778 (2008)
15. Hayat, T., Sajid, M., Pop, I.: Three-dimensional flow over a stretching surface in a viscoelastic fluid. Nonlinear Anal. Real World Appl. **9**, 1811–1822 (2008)
16. Dash, R.K., Mehta, K.N., Jayaraman, G.: Casson fluid flow in a pipe filled with a homogeneous porous medium. Int. J. Eng. Sci. **34**, 1145–1156 (1996)
17. Venkata Ramana Reddy, G., Hari Krishna, Y.: Numerical solutions of unsteady MHD flow heat transfer over a stretching surface with suction or injection. Fluid Dyn. Mater. Process. **14**, 213–222 (2018)
18. Brewster, M.Q.: Thermal Radiative Transfer and Properties. Wiley, New York (1992)
19. Babu, P.R., Rao, J.A., Shri, S.: Radiation effect on MHD heat and mass transfer flow over a shrinking sheet with mass suction. J. Appl. Fluid Mech. **7**, 641–650 (2014)

Paul-Painlevé Approach to Solve (3 + 1)-Dimensional Extended Sakovich Equation Arising in Fluid Dynamics

Shailendra Singh, B. Sagar, and S. Saha Ray

Abstract In this work, the (3 + 1)-dimensional extended Sakovich equation has been solved analytically by employing the Paul-Painlevé approach method. First, the governing equation has been converted into the nonlinear ODE by using the suitable wave transformation. Then, the Paul-Painlevé approach method has been applied to obtain the new abundant analytical solutions. The solutions yield distinct wave structures for different parametric values. Bell-shaped, anti bell-shaped, and complex periodic wave solutions are obtained and illustrated through graphical representations. The obtained solutions are useful in describing various nonlinear phenomena in the field of fluid dynamics.

Keywords Extended Sakovich equation · Paul-Painlevé method · Analytical solutions · Periodic wave solutions

1 Introduction

Nonlinear partial differential equations (NLPDEs) are widely used to model various nonlinear phenomena in the fields of applied science and engineering, such as wave propagation, fluid dynamics, ocean engineering, plasma physics, mechanics and many more. It is significant to obtain the new analytical solutions of these models to analyze the various wave characteristics and the nonlinear phenomenon. Several

S. Singh (✉)
Department of Mathematics, Institute of Technical Education and Research, Siksha 'O' Anusandhan (Deemed to be University), Bhubaneswar, India
e-mail: shailendra.chauhan012@gmail.com

B. Sagar
Department of Mathematics, University of Central Florida, Orlando FL. 32816, USA

S. S. Ray
Department of Mathematics, Indian Institute of Engineering Science and Technology, Howrah 711103, West Bengal, India

S. Bhattacharyya and H. S. Mahato (eds.), *Modeling, Analysis and Simulations of Multiscale Transport Phenomena*, Springer Proceedings in Mathematics & Statistics 491, https://doi.org/10.1007/978-981-96-3098-1_3

methods have been developed and executed to solve these NLPDEs, such as simplified Hirota's method, Painlevé analysis, auto-Bäcklund transformation, extended F-expansion method, Kudryashov method, tanh method, exp-function method, Lie symmetry analysis and many more [1–7].

This work aims to obtain the analytical solutions of (3 + 1)-dimensional extended Sakovich equation using the Paul-Painlevé approach (PPA) method. This method is used to find analytic solutions of both integrable as well as non-integrable NLPDEs.

In 1902, Paul-Painlevé [8] examined the 2nd order nonlinear ODE

$$y_{xx} + 3ay_x - 2y^3 + 2a^2y = 0, \tag{1}$$

and derived the general solution of the form

$$y(x) = -iaC_1 \exp(-ax) sn_{k^2=-1}(C_1 \exp(-ax) + C_2),$$

where C_1 and C_2 are any constants.

For solving Eq. 1, the composite function $y(x) = \phi(x)V(f)$, where $f = \psi(x)$ is substituted into the Eq. 1 and determined the functions $\phi(x)$ and $\psi(x)$. In 2019, Kudryashov [9] developed the PPA method from the above approach.

Now, consider the (3 + 1)-dimensional extended Sakovich equation as [10]

$$u_{xt} + u_{yy} + 2uu_{xy} + 6u^2u_{xx} + 2(u_{xx})^2 + u_{xx} + u_{xy} + u_{xz} + u_{yz} = 0. \tag{2}$$

This equation has remarkable applications in fluid dynamics. In [10], Wazwaz derived the integrability, multiple soliton solutions and multiple complex soliton solutions of Eq. 2. Recently, in [2], Singh and Saha Ray derived the Painlevé analysis, auto-Bäcklund transformation and new exact solutions of Eq. 2 with time dependent variable coefficients.

In this article, the PPA method has been employed to obtain the new abundant analytic solutions of Eq. 2. The surface solutions for various parametric values has been illustrated and the graphical representations depicts the bell-shaped, anti bell-shaped, and complex periodic wave solutions.

2 Brief Description of Paul-Painlevé Approach Method

The fundamental purpose of the Paul-Painlevé approach (PPA) method [11–13] is to derive numerous kinds of analytical solutions for NLPDEs. The following are the main steps of this method:

- Consider a NLPDE in Φ as a function with space-time variables as

$$W(\Phi, \Phi_{z_1}, \Phi_{z_2}, \Phi_t, \Phi_{z_1z_2}, \Phi_{tt}, \Phi_{z_1t}, \ldots) = 0, \tag{3}$$

where $z_1, z_2, \ldots$ are the space variables and t is the time variable.

- Let the transformation

$$\Phi(z_1, z_2, \ldots, t) = \Psi(\xi), \tag{4}$$

where $\xi = z_1 + z_2 + \ldots - C_0 t$. An ordinary differential equation (ODE) will occur by merging above transformation into Eq. 3 as

$$Q(\Phi', \Phi'', \Phi''', \ldots) = 0. \tag{5}$$

- The analytical solution of Eq. 5 will be stated in the form as per the PPA method

$$\Phi(\xi) = \sum_{i=0}^{m} A_i S^i(X) e^{-iN\xi}, \tag{6}$$

where m will be illustrated by using homogeneous balance method, $X = R(\xi) = C_1 - \frac{e^{-N\xi}}{N}$, and the Riccati-equation of the type $S_X - AV^2 = 0$ is satisfied by $S(X)$.

- The Riccati-equation is solved as

$$S(X) = \frac{1}{AX + X_0}, \tag{7}$$

where X_0 and A are the constants.

- By substituting Eq. 6 into Eq. 5 and setting the various powers of $S(X)e^{-N\xi}$ to zero, one may determine the values of constants A, N and A_i's for $i = 1, \ldots, m$.

3 Analytical Solutions for (3 + 1)-Dimensional Extended Sakovich Equation

The PPA method is applied to (3 + 1)-dimensional extended Sakovich equation to derive the different kinds of analytical solutions in this section. The (3 + 1)-dimensional extended Sakovich equation reads

$$u_{xt} + u_{yy} + 2uu_{xy} + 6u^2 u_{xx} + 2(u_{xx})^2 + u_{xx} + u_{xy} + u_{xz} + u_{yz} = 0. \tag{8}$$

Equation 8 will be converted into an ODE by substituting the following transformation

$$u(x, y, z, t) = \Phi(\xi), \qquad \xi = x + y + z - C_0 t. \tag{9}$$

Then the ODE occurs as

$$5\Phi''(\xi) - C_0\Phi''(\xi) + 2\Phi(\xi)\Phi''(\xi) + 6\Phi(\xi)^2\Phi''(\xi) + 2(\Phi''(\xi))^2 = 0, \tag{10}$$

where $' = \frac{d}{d\xi}$. The homogeneous balance principle is applied on Eq. 10 to balance $\Phi''(\xi)$ with $\Phi(\xi)^2$ such that $m+2=2m$ gives $m=2$. Then the solution given by Eq. 6 will be expanded up to $m=2$ as

$$\Phi(\xi) = A_0 + A_1 S(X)e^{-N\xi} + A_2 S^2(X)e^{-2N\xi}, \tag{11}$$

where the unknown constants A_0, A_1, A_2 and N have to be determined, and the Riccati-equation $S_X - AS^2 = 0$ is satisfied by $S(X)$, whose solution is given by Eq. 7, where $X = R(\xi) = C_1 - \frac{e^{-N\xi}}{N}$.

Now, $\Phi(\xi)^2$ and $\Phi''(\xi)$ will written as

$$\Phi(\xi)^2 = A_0^2 + A_2^2 S^4 e^{-4N\xi} + 2A_1A_2 s^3 e^{-3N\xi} + A_1^2 S^2 e^{-2N\xi} \\ +2A_0A_2S^2e^{-2N\xi} + 2A_0A_1Se^{-N\xi}, \tag{12}$$

$$\Phi''(\xi) = 6A^2A_2S^4e^{-4N\xi} + 2A^2A_1S^3e^{-3N\xi} + 10AA_2S^3e^{-3N\xi} \\ +3AA_1NS^2e^{-2N\xi} + 4A_2N^2S^2e^{-2N\xi} + A_1N^2Se^{-N\xi}. \tag{13}$$

The following system of equations is obtained using Eqs. 10–13 by setting the coefficients of numerous powers of $S(X)e^{-N\xi}$ to zero:

$$S^8e^{-8N\xi}: \quad 72A^4A_2^2 + 36A^2A_2^3 = 0. \tag{14}$$

$$S^7e^{-7N\xi}: \quad 48A^4A_1A_2 + 84A^2A_1A_2^2 + 240A^3A_2^2N + 60AA_2^3N = 0. \tag{15}$$

$$S^6e^{-6N\xi}: \quad 8A^4A_1^2 + 60A^2A_1^2A_2 + 12A^2A_2^2 + 72A^2A_0A_2^2 \\ +152A^3A_1A_2N + 138AA_1A_2^2N + 296A^2A_2^2N^2 + 24A_2^2N^2 = 0. \tag{16}$$

$$S^5e^{-5N\xi}: \quad 12A^2A_1^3 + 16A^2A_1A_2 + 96A^2A_0A_1A_2 + 24A^3A_1^2N \\ +96AA_1^2A_2N + 20AA_2^2N + 120AA_0A_2^2N + 176A^2A_1A_2N^2 \\ +54A_1A_2^2N^2 + 160AA_2^2N^3 = 0. \tag{17}$$

$$
\begin{aligned}
S^4e^{-4N\xi}: \quad & 4A^2A_1^2 + 24A^2A_0A_1^2 + 30A^2A_2 + 12A^2A_0A_2 + 36A^2A_0^2A_2 \\
& -6A^2A_2C_0 + 18AA_1^3N + 26AA_1A_2N + 156AA_0A_1A_2N + 26A^2A_1^2N^2 \\
& +36A_1^2A_2N^2 + 8A_2^2N^2 + 48A_0A_2^2N^2 + 88AA_1A_2N^3 + 32A_2^2N^4 = 0.
\end{aligned} \tag{18}
$$

$$
\begin{aligned}
S^3e^{-3N\xi}: \quad & 10A^2A_1 + 4A^2A_0A_1 + 12A^2A_0^2A_1 - 2A^2A_1C_0 + 6AA_1^2N \\
& +36AA_0A_1^2N + 50AA_2N + 20AA_0A_2N + 60AA_0^2A_2N \\
& -10AA_2C_0N + 6A_1^3N^2 + 10A_1A_2N^2 + 60A_0A_1A_2N^2 + 12AA_1^2N^3 \\
& +16A_1A_2N^4 = 0.
\end{aligned} \tag{19}
$$

$$
\begin{aligned}
S^2e^{-2N\xi}: \quad & 15AA_1N + 6AA_0A_1N + 18AA_0^2A_1N - 3AA_1C_0N \\
& +2A_1^2N^2 + 12A_0A_1^2N^2 + 20A_2N^2 + 8A_0A_2N^2 + 24A_0^2A_2N^2 \\
& -4A_2C_0N^2 + 2A_1^2N^4 = 0.
\end{aligned} \tag{20}
$$

$$
Se^{-N\xi}: \quad 5A_1N^2 + 2A_0A_1N^2 + 6A_0^2A_1N^2 - A_1C_0N^2 = 0. \tag{21}
$$

$$
(Se^{-N\xi})^0: \quad 0.
$$

Equation 14 is being solved for A_2. This gives

$$
A_2 = -2A^2.
$$

Now substituting the value of A_2 into Eqs. 15–20, then the system of Eqs. 15–21 yields the following set of solutions:

Set 1

$$
A_1 = 2iA(-29 + 6C_0)^{1/4}, \quad \text{where} \quad i = \sqrt{-1},
$$

$$
N = -i(-29 + 6C_0)^{1/4},
$$

$$
A_0 = \frac{1}{6}(-1 + \sqrt{-29 + 6C_0}).
$$

Therefore, Eq. 9 along with Eq. 11 gives the first solution of Eq. 8 as

$$\begin{aligned} u(x,y,z,t) =& \frac{1}{6}(-1+\sqrt{-29+6C_0}) \\ & - \frac{2A^2 e^{2i(-29+6C_0)^{1/4}(x+y+z-C_0t)}}{(A(C_1 - \frac{ie^{i(-29+6C_0)^{1/4}(x+y+z-C_0t)}}{(-29+6C_0)^{1/4}}) + X_0)^2} \\ & + \frac{2iA(-29+6C_0)^{1/4}e^{i(-29+6C_0)^{1/4}(x+y+z-C_0t)}}{A(C_1 - \frac{ie^{i(-29+6C_0)^{1/4}(x+y+z-C_0t)}}{(-29+6C_0)^{1/4}}) + X_0}, \end{aligned} \tag{22}$$

provided that $C_0 \neq \frac{29}{6}$.

Set 2

$$A_1 = -2iA(-29+6C_0)^{1/4}, \quad \text{where} \quad i = \sqrt{-1},$$

$$N = i(-29+6C_0)^{1/4},$$

$$A_0 = \frac{1}{6}(-1+\sqrt{-29+6C_0}).$$

Therefore, Eq. 9 along with Eq. 11 gives the second solution of Eq. 8 as

$$\begin{aligned} u(x,y,z,t) =& \frac{1}{6}(-1+\sqrt{-29+6C_0}) - \frac{2A^2 e^{-2i(-29+6C_0)^{1/4}(x+y+z-C_0t)}}{(A(C_1 + \frac{ie^{-i(-29+6C_0)^{1/4}(x+y+z-C_0t)}}{(-29+6C_0)^{1/4}}) + X_0)^2} \\ & - \frac{2iA(-29+6C_0)^{1/4}e^{-i(-29+6C_0)^{1/4}(x+y+z-C_0t)}}{A(C_1 + \frac{ie^{-i(-29+6C_0)^{1/4}(x+y+z-C_0t)}}{(-29+6C_0)^{1/4}}) + X_0}, \end{aligned} \tag{23}$$

provided that $C_0 \neq \frac{29}{6}$.

Set 3

$$A_1 = 2A(-29+6C_0)^{1/4},$$

$$N = -(-29+6C_0)^{1/4},$$

$$A_0 = \frac{1}{6}(-1-\sqrt{-29+6C_0}).$$

Therefore, Eq. 9 along with Eq. 11 gives the third solution of Eq. 8 as

$$u(x,y,z,t) = \frac{1}{6}(-1-\sqrt{-29+6C_0}) - \frac{2A^2 e^{2(-29+6C_0)^{1/4}(x+y+z-C_0 t)}}{(A(C_1+\frac{e^{(-29+6C_0)^{1/4}(x+y+z-C_0 t)}}{(-29+6C_0)^{1/4}})+X_0)^2} + \frac{2A(-29+6C_0)^{1/4} e^{(-29+6C_0)^{1/4}(x+y+z-C_0 t)}}{A(C_1+\frac{e^{(-29+6C_0)^{1/4}(x+y+z-C_0 t)}}{(-29+6C_0)^{1/4}})+X_0}, \tag{24}$$

provided that $C_0 \neq \frac{29}{6}$.

Set 4

$$A_1 = -2A(-29+6C_0)^{1/4},$$

$$N = (-29+6C_0)^{1/4},$$

$$A_0 = \frac{1}{6}(-1-\sqrt{-29+6C_0}).$$

Therefore, Eq. 9 along with Eq. 11 gives the fourth solution of Eq. 8 as

$$u(x,y,z,t) = \frac{1}{6}(-1-\sqrt{-29+6C_0}) - \frac{2A^2 e^{-2(-29+6C_0)^{1/4}(x+y+z-C_0 t)}}{(A(C_1-\frac{e^{-(-29+6C_0)^{1/4}(x+y+z-C_0 t)}}{(-29+6C_0)^{1/4}})+X_0)^2} - \frac{2A(-29+6C_0)^{1/4} e^{-(-29+6C_0)^{1/4}(x+y+z-C_0 t)}}{A(C_1-\frac{e^{-(-29+6C_0)^{1/4}(x+y+z-C_0 t)}}{(-29+6C_0)^{1/4}})+X_0}, \tag{25}$$

provided that $C_0 \neq \frac{29}{6}$.

4 Physical Interpretation of Solutions

This article examines the (3 + 1)-dimensional extended Sakovich Eq. 8. By utilizing the PPA method, four new analytical solutions have been derived given by Eqs. 22, 23, 24 and 25. Here, for the Eqs. 22 and 23, three-dimensional absolute graphs have been plotted, whereas the normal three-dimensional graphs have been plotted for the solutions given by Eqs. 24 and 25. These graphs represent the bell-shaped, anti bell-shaped, and complex periodic wave solutions by selecting specific values of free constants. Three-dimensional graphs have been drawn via taking y- and z-coordinates as 1. The description of the graphs are given as follows.

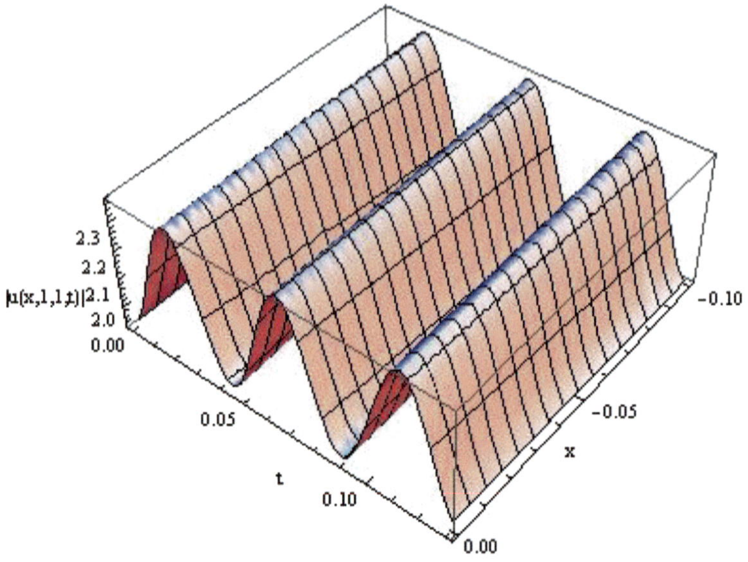

Fig. 1 Three-dimensional complex periodic wave solution surface for Eq. 22

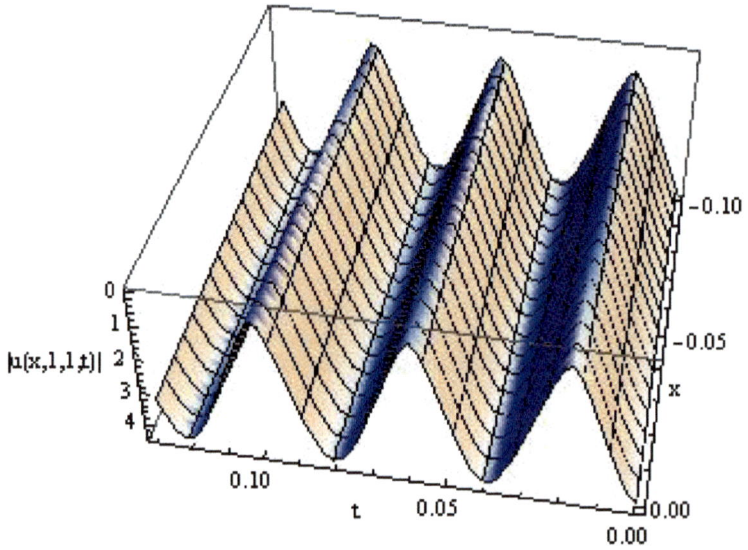

Fig. 2 Three-dimensional complex periodic wave solution surface for Eq. 23

The solution graph of Eq. 22 displays the complex periodic wave surface for the constants $A = -25.2$, $C_0 = 37$, $C_1 = 33.2$, and $X_0 = -11.3$ in Fig. 1, while Fig. 2 represents the same surface for the solution graph of Eq. 23 by taking the constants $A = 15.4$, $C_0 = 40$, $C_1 = -5.3$, and $X_0 = 20.5$. Figure 3 gives the anti

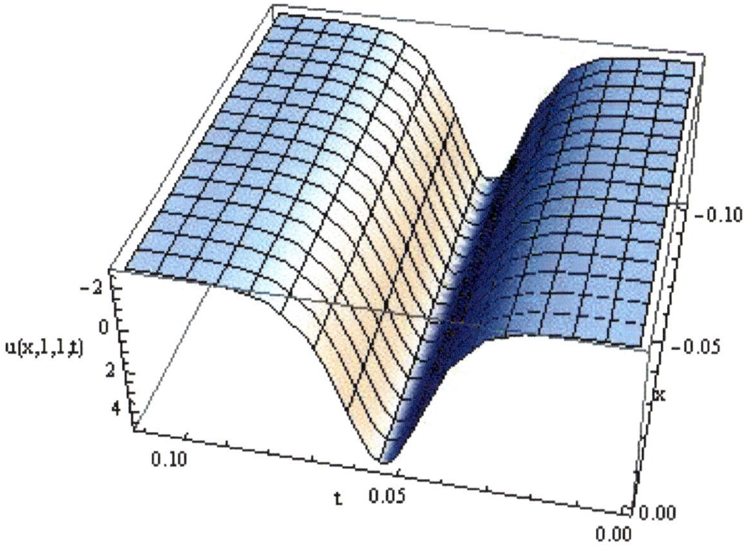

Fig. 3 Three-dimensional anti bell-shaped wave solution surface for Eq. 24

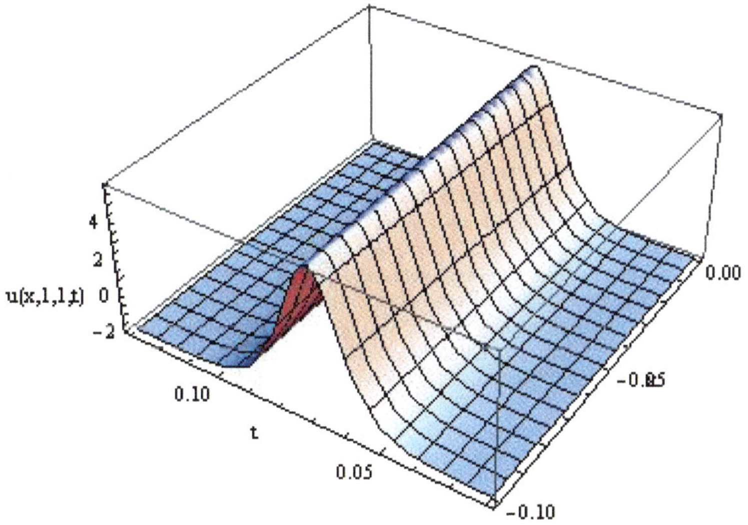

Fig. 4 Three-dimensional bell-shaped wave solution surface for Eq. 25

bell-shaped wave surface for the solution graph of Eq. 24 by opting the constants $A = 13.3$, $C_0 = 41.9$, $C_1 = 3.1$, and $X_0 = -40$. Finally, the solution graph of Eq. 25 represents the bell-shaped wave surface for the constants $A = 23$, $C_0 = 44.4$, $C_1 = -40$, and $X_0 = 20.5$ in Fig. 4.

5 Conclusion

This work makes the first attempt ever to implement the PPA method to solve the $(3 + 1)$-dimensional extended Sakovich equation analytically. First, the governing PDE equation has been converted into the nonlinear ODE by using the suitable wave transformation. Then, the PPA method has been employed to obtain the new abundant analytical solutions. The wave solutions are illustrated through graphical representations for different parametric values. The surface solutions representing the Bell-shaped, anti bell-shaped, and complex periodic wave solutions are obtained. The obtained wave solutions are useful in studying various nonlinear phenomena in the field of fluid dynamics.

Competing Interests The authors have no conflicts of interest to declare that are relevant to the content of this chapter.

References

1. Wazwaz, A.-M.: A two-mode Burgers equation of weak shock waves in a fluid: multiple kink solutions and other exact solutions. Int. J. Appl. Comput. Math. **3**, 3977–3985 (2017)
2. Singh, S., Saha Ray, S.: Painlevé analysis, auto-Bäcklund transformation and new exact solutions of $(2 + 1)$ and $(3 + 1)$-dimensional extended Sakovich equation with time dependent variable coefficients in ocean physics. J. Ocean Eng. Sci. (2022). https://doi.org/10.1016/j.joes.2022.01.008
3. Joseph, S.P.: Exact traveling eave doubly periodic solutions for generalized double Sine-Gordon equation. Int. J. Appl. Comput. Math. **8**, 42 (2022)
4. Sagar, B., Saha Ray, S.: Numerical and analytical investigation for solutions of fractional Oskolkov-Benjamin-Bona-Mahony-Burgers equation describing propagation of long surface waves. Int. J. Mod. Phys. B. **35**(32), 2150326 (2021)
5. Wang, K.J.: Abundant exact soliton solutions to the Fokas system. Optik. **249**, 168265 (2022)
6. Saha Ray, S.: Lie symmetries, exact solutions and conservation laws of the Oskolkov-Benjamin-Bona-Mahony-Burgers equation. Mod. Phys. Lett. B. **34**(01), 2050012 (2020)
7. Baleanu, D., Alshomrani, A.S., Ullah, M.Z.: A new fourth-order integrable nonlinear equation: breather, rogue waves, other lump interaction phenomena, and conservation laws. Adv. Differ. Equ. **2021**, 195 (2021)
8. Painlevé, P.: Sur les équations différentielles du second ordre et d'ordre supérieur dont l'intégrale générale est uniforme. Acta math. **25**(1), 1–85 (1902)
9. Kudryashov, N.A.: The Painlevé approach for finding solitary wave solutions of nonlinear nonintegrable differential equations. Optik. **183**, 642–649 (2019)
10. Wazwaz, A.-M.: Two new Painlevé-integrable extended Sakovich equations with (2+ 1) and (3+ 1) dimensions. Int. J. Numer. Methods Heat Fluid Flow. **30**(3), 1379–1387 (2020)
11. Bekir, A., Zahran, E.H.: Painlevé approach and its applications to get new exact solutions of three biological models instead of its numerical solutions. Int. J. Mod. Phys. B. **34**(29), 2050270 (2020)
12. Bekir, A., Shehata, M.S., Zahran, E.H.: Comparison between the exact solutions of three distinct shallow water equations using the Painlevé approach and its numerical solutions. Russ. J. Nonlinear Dyn. **16**(3), 463–477 (2020)
13. Bekir, A., Zahran, E.H.: Exact and numerical solutions for the nanosoliton of ionic waves propagating through microtubules in living cells. Pramana. **95**(4), 1–9 (2021)

Unsteady Electrophoresis of a Spherical Colloidal Particle: Time-Dependent Transient Henry Function

Hiroyuki Ohshima

Abstract The steady electrophoretic mobility of a spherical colloidal particle carrying low zeta potential in an electrolyte solution under an applied constant electric field is described with the Henry function. When a step electric field is applied to a spherical colloidal particle, the Henry function should be modified so as to be time-dependent. Huang and Keh presented a rigorous theory of the transient unsteady electrophoresis of a spherical particle. In a previous paper we have derived a simple analytic approximate expression for the time-dependent transient Henry function. The present paper reviews a study of the time-dependent transient Henry function.

Keywords Unsteady electrophoresis · Transient Henry function · Spherical particle

1 Introduction

The steady electrophoretic mobility of a spherical colloidal particle carrying low zeta potential in an electrolyte solution under an applied constant electric field is given by Henry's mobility formula Henry [1]. When, instead of a constant electric field, a step electric field is suddenly applied to a spherical colloidal particle, the particle starts to move with a time-dependent electrophoretic mobility, which tends to the steady mobility as $t \to \infty$. The electrophoretic motion of a charged colloidal particle under an applied step electric field is called unsteady transient electrophoresis [2–18]. The theory of transient electrophoresis was developed by Morison [2, 3] and later by Ivory [4, 5] and has been advanced significantly by Keh and his coworkers [6–8, 10, 11, 14–16]. For a weakly charged spherical particle, Huang and Keh [8] established the general theory of transient electrophoresis and derived an exact expression for the Laplace transform (with respect to time) of the transient electrophoretic mobility of a spherical particle. The transient electrophoretic

H. Ohshima (✉)
Faculty of Pharmaceutical Sciences, Tokyo University of Science, Chiba, Japan
e-mail: ohshima@rs.noda.tus.ac.jp

S. Bhattacharyya and H. S. Mahato (eds.), *Modeling, Analysis and Simulations of Multiscale Transport Phenomena*, Springer Proceedings in Mathematics & Statistics 491, https://doi.org/10.1007/978-981-96-3098-1_4

mobility of the particle can then be obtained by the numerical inverse Laplace transformation method. This method, however, which requires tedious numerical calculations, is not very convenient for practical uses. In a previous paper [18] we presented an approximation method to avoid numerical inverse Laplace transformations and derived a simple analytic mobility expression for a weakly charged spherical particle without involving the numerical inverse Laplace transformation. In the present paper, we review the theory of transient electrophoresis of Huang and Keh [8] and derive a simple analytic expression for the time dependent transient Henry function [18].

2 Theory

We briefly describe our theory of the time-dependent transient electrophoresis of a spherical particle [18], which is based on Huang and Keh's theory [8]. Let us consider a spherical colloidal particle of radius a, mass density ρ_p, and zeta potential ζ in an aqueous liquid of relative permittivity ε_r, mass density ρ_o, and viscosity η that contains a general electrolyte consisting of N ionic species of valence z_i, bulk concentration (number density) n_i^∞, and drag coefficient λ_i ($i = 1, 2, \ldots, N$). At time $t = 0$, a step electric field $\boldsymbol{E}(t)$ is applied to the particle, viz.,

$$\boldsymbol{E}(t) = \begin{cases} 0, t = 0 \\ \boldsymbol{E}_o, t > 0 \end{cases} \tag{1}$$

where $\boldsymbol{E}_o$ is constant and the particle starts to move with an electrophoretic velocity $\boldsymbol{U}(t)$in the direction parallel to $\boldsymbol{E}_o$. The unsteady transient electrophoretic mobility $\mu(t)$ of the particle is defined by $\boldsymbol{U}(t) = \mu(t)\boldsymbol{E}(t) = \mu(t)\boldsymbol{E}_o$. The origin of the spherical coordinate system (r, θ, ϕ) is held fixed at the center of the particle and the polar axis ($\theta = 0$) is set parallel to $\boldsymbol{E}_o$. We treat the case where the following conditions are satisfied: (i) The liquid can be regarded as incompressible. (ii) The applied electric field $\boldsymbol{E}(t)$ is weak so that the particle velocity $\boldsymbol{U}(t)$ is proportional to $\boldsymbol{E}(t)$ and terms involving the square of the liquid velocity in the Navier–Stokes equation can be neglected. (iii) The slipping plane (at which the liquid velocity $\boldsymbol{u}(\boldsymbol{r}, t)$ relative to the particle is zero) is located on the particle surface (at $r = a$). (iv) Electrolyte ions cannot penetrate the particle surface.

The fundamental electrokinetic equations for the liquid flow velocity $\boldsymbol{u}(\boldsymbol{r}, t)$ at position $\boldsymbol{r}(r, \theta, \phi)$ and time t and the velocity $\boldsymbol{v}_i(\boldsymbol{r}, t)$ of i th ionic species are given by [18]

$$\rho_o \frac{\partial}{\partial t}\{\boldsymbol{u}(\boldsymbol{r}, t) + \boldsymbol{U}(t)\} + \eta\nabla \times \nabla \times \boldsymbol{u}(\boldsymbol{r}, t) + \nabla p(\boldsymbol{r}, t) + \rho_{el}(\boldsymbol{r}, t)\nabla\psi(\boldsymbol{r}, t) = 0 \tag{2}$$

$$\nabla \cdot \boldsymbol{u}(\boldsymbol{r}, t) = 0 \tag{3}$$

$$\boldsymbol{v}_i(\boldsymbol{r}, t) = \boldsymbol{u}(\boldsymbol{r}, t) - \frac{1}{\lambda_i}\nabla\mu_i(\boldsymbol{r}, t) \tag{4}$$

$$\frac{\partial n_i(\boldsymbol{r}, t)}{\partial t} + \nabla \cdot \{n_i(\boldsymbol{r}, t)\boldsymbol{v}_i(\boldsymbol{r}, t)\} = 0 \tag{5}$$

with

$$\rho_{\rm el}(\boldsymbol{r}, t) = \sum_{i=1}^{N} z_i e n_i(\boldsymbol{r}, t) \tag{6}$$

$$\mu_i(\boldsymbol{r}, t) = \mu_i^{\rm o} + z_i e\psi(\boldsymbol{r}, t) + kT\ln[n_i(\boldsymbol{r}, t)] \tag{7}$$

$$\Delta\psi(\boldsymbol{r}, t) = -\frac{\rho_{\rm el}(\boldsymbol{r}, t)}{\varepsilon_{\rm r}\varepsilon_{\rm o}} \tag{8}$$

where e is the elementary electric charge, k is the Boltzmann constant, T is the absolute temperature, $\varepsilon_{\rm o}$ is the permittivity of a vacuum, $p(\boldsymbol{r}, t)$ is the pressure, $\rho_{\rm el}(\boldsymbol{r}, t)$ is the charge density, $\psi(\boldsymbol{r}, t)$ is the electric potential. Equations (2) and (3) are the Navier–Stokes equation and the equation of continuity for an incompressible flow (condition (i)). Equation (4) expresses that the flow $\boldsymbol{v}_i(\boldsymbol{r}, t)$ of the i th ionic species is caused by the liquid flow $\boldsymbol{u}(\boldsymbol{r}, t)$ and the gradient of the electrochemical potential $\mu_i(\boldsymbol{r}, t)$, given by Eq. (7), in which $\mu_i^{\rm o}$ is a constant term. Equation (5) is the continuity equation for the i th ionic species. Equation (8) is the Poisson equation.

Equations (2) and (5) are subject to the following boundary conditions at the particle surface (at $r = a$) and far from the particle ($r \to \infty$):

$$\boldsymbol{u}(\boldsymbol{r}, t) = 0 \text{ at } t = 0 \tag{9}$$

$$\boldsymbol{u}(\boldsymbol{r}, t) = 0 \text{ at } r = a \tag{10}$$

$$\boldsymbol{u}(\boldsymbol{r}, t) \to -\boldsymbol{U}(\boldsymbol{r}, t) + \frac{a^3(\rho_{\rm p} - \rho_{\rm o})}{3\rho_{\rm o}r^3}\left[\boldsymbol{U}(\boldsymbol{r}, t) - 3\{(\boldsymbol{U}(\boldsymbol{r}, t)\cdot\hat{\boldsymbol{r}})\hat{\boldsymbol{r}}\}\right] \text{ as } r \to \infty \text{ as } r \to \infty \tag{11}$$

$$\boldsymbol{v}_i(\boldsymbol{r}, t)\cdot\hat{\boldsymbol{n}} = 0 \text{ at } r = a \tag{12}$$

$$\psi(\boldsymbol{r}, t) = -\boldsymbol{E}(t)\cdot\boldsymbol{r} \text{ as } r \to \infty \tag{13}$$

where $\hat{\boldsymbol{r}} = \boldsymbol{r}/r$ ($r = |\boldsymbol{r}|$) and $\hat{\boldsymbol{n}}$ is the unit normal outward from the particle surface. Equation (10) states that slipping plane (at which $\boldsymbol{u}(\boldsymbol{r}, t) = \boldsymbol{0}$) is located on the particle surface (condition (iii)). Equation (11) can be derived from the equation of motion of the particle, as in the case of dynamic electrophoresis [19]. Equation (12) follows from condition (iv).

For a weak field $\boldsymbol{E}(t)$, the deviations of $n_i(\boldsymbol{r}, t)$, $\psi(\boldsymbol{r}, t)$, and $\mu_i(\boldsymbol{r}, t)$ from their equilibrium values due to the applied field $\boldsymbol{E}(t)$ are small and we may write

$$n_i(\boldsymbol{r}, t) = n_i^{(0)}(r) + \delta n_i(\boldsymbol{r}, t) \tag{14}$$

$$\psi(\boldsymbol{r}, t) = \psi^{(0)}(r) + \delta\psi(\boldsymbol{r}, t) \tag{15}$$

$$\mu_i(\boldsymbol{r}, t) = \mu_i^{(0)} + \delta\mu_i(\boldsymbol{r}, t) \tag{16}$$

where the quantities with superscript (0) refer to those at equilibrium and $\mu_i^{(0)}$ is a constant independent of r. We assume that the equilibrium concentration $n_i^{(0)}(r)$ obeys the Boltzmann distribution and the equilibrium electric potential satisfies the Poisson-Boltzmann equation, viz.,

$$n_i^{(0)}(r) = n_i^{\infty}(r)e^{-z_i y(r)} \tag{17}$$

$$\Delta y(r) = -\kappa^2 \frac{\sum_{i=1}^{N} z_i n_i^{\infty} e^{-z_i y(r)}}{\sum_{i=1}^{N} z_i^2 n_i^{\infty}} \tag{18}$$

with

$$y(r) = \frac{e\psi^{(0)}(r)}{kT} \tag{19}$$

$$\kappa = \sqrt{\frac{e^2}{\varepsilon_r \varepsilon_o kT} \sum_{i=1}^{N} z_i^2 n_i^{\infty}} \tag{20}$$

where $y(r)$ is the scaled equilibrium electric potential, and κ is the Debye-Hückel parameter ($1/\kappa$ is the Debye length).

Further, from symmetry considerations, we may write

$$\boldsymbol{u}(\boldsymbol{r}, t) = \left(-\frac{2}{r}h(r, t)E(t)\cos\theta, -\frac{1}{r}\frac{\partial}{\partial r}(rh(r, t))E(t)\sin\theta, 0\right) \tag{21}$$

$$\delta\mu_i(\boldsymbol{r}, t) = -z_i e\phi_i(r, t)E(t)\cos\theta \tag{22}$$

where $E(t)$ is the magnitude of $\boldsymbol{E}(t)$, and $h(r, t)$ and $\phi_i(r, t)$ are functions of r and t. By substituting Eqs. (14)–(16) into Eq. (2), and neglecting the products of the small quantities, we obtain the following equation for $h(r)$:

$$L\left[Lh(r,t) - \frac{1}{\nu}\frac{\partial h(r,t)}{\partial t}\right] = G(r,t) \tag{23}$$

with

$$L = \frac{\partial}{\partial r}\frac{1}{r^2}\frac{\partial}{\partial r}r^2 = \frac{\partial^2}{\partial r^2} + \frac{2}{r}\frac{\partial}{\partial r} - \frac{2}{r^2} \tag{24}$$

$$G(r,t) = -\frac{e}{\eta r}\frac{dy}{dr}\sum_{i=1}^{N} z_i^2 n_i^\infty e^{-z_i y}\phi_i(r,t) \tag{25}$$

$$\nu = \frac{\eta}{\rho_o} \tag{26}$$

where ν is the kinematic viscosity. It follows from Eqs. (11) and (21) that the transient electrophoretic mobility $\mu(t)$ can be obtained from

$$\mu(t) = \frac{U(t)}{E(t)} = \frac{U(t)}{E_o} = 2\lim_{r\to\infty}\frac{h(r,t)}{r} \tag{27}$$

Equation (23) for $h(r, t)$ is most easily solved by using the Laplace transformation with respect to time t. We introduce the Laplace transforms $\hat{h}(r,s)$, $\hat{G}(r,s)$, and $\hat{\mu}(s)$ of $h(r, t)$, $G(r, t)$, and $\mu(t)$, respectively, which are given by

$$\hat{h}(r,s) = \int_0^\infty h(r,t)e^{-st}dt \tag{28}$$

$$\hat{G}(r,s) = \int_0^\infty G(r,t)e^{-st}dt \tag{29}$$

$$\hat{\mu}(s) = \int_0^\infty \mu(t)e^{-st}dt \tag{30}$$

and the Laplace transform of Eq. (27) is

$$\hat{\mu}(s) = 2\lim_{r\to\infty}\frac{\hat{h}(r,s)}{r} \tag{31}$$

The Laplace transform of Eq. (23) gives

$$L\left[L\hat{h}(r,s) - \frac{s}{\nu}\hat{h}(r,s)\right] = \hat{G}(r,s) \tag{32}$$

We now consider the low ζ potential case, in which case it can be shown that

$$\phi_i(r,t) = r + \frac{a^3}{2r^2} \tag{33}$$

and Eq. (25) becomes

$$G(r,t) = -\frac{\varepsilon_r\varepsilon_o\kappa^2}{\eta}\left(1+\frac{a^3}{2r^3}\right)\frac{d\psi^{(0)}(r)}{dr} \tag{34}$$

where the equilibrium electric potential $\psi^{(0)}(r)$ for the low ζ potential case is given by

$$\psi^{(0)}(r) = \zeta\frac{a}{r}e^{-\kappa(r-a)} \tag{35}$$

Equation (35) is obtained from the linearized Poisson-Boltzmann equation $\Delta\psi^{(0)}(r) = \kappa^2\psi^{(0)}(r)$ (see Eq. (18)).

The Laplace transform $\widehat{G}(r,s)$ of $G(r,t)$ is given by

$$\widehat{G}(r,s) = \frac{G(r,t)}{s} = -\frac{\varepsilon_r\varepsilon_o\kappa^2}{\eta s}\left(1+\frac{a^3}{2r^3}\right)\frac{d\psi^{(0)}(r)}{dr} \tag{36}$$

By substituting Eq. (36) into Eq. (32) and solving Eq. (32), we obtain

$$\begin{aligned}\hat{\mu}(s) = &\frac{2\varepsilon_r\varepsilon_o\kappa^2\nu}{3\eta s^2\left\{1+a\sqrt{\frac{s}{\nu}}+\frac{a^2s}{3\nu}+\frac{2a^2(\rho_p-\rho_o)s}{9\nu\rho_o}\right\}}\\ &\times\int_a^\infty\left\{1+a\sqrt{\frac{s}{\nu}}+\frac{a^2s}{3\nu}-\left(1+\sqrt{\frac{s}{\nu}}r\right)\exp\left[-\sqrt{\frac{s}{\nu}}(r-a)\right]-\frac{sr^3}{3\nu a}\right\}\\ &\left(1+\frac{a^3}{2r^3}\right)\frac{d\psi^{(0)}(r)}{dr}dr\end{aligned} \tag{37}$$

which agrees with Huang and Keh's result (Eq. (31) in Ref. [8]). Huang ad Keh [8] obtained the transient electrophoretic mobility $\mu(t)$ by using the numerical inverse transform of Eq. (37). This method, however, involves tedious numerical calculations and is not very convenient for practical uses. The reason for this is that the integration in Eq. (37) cannot be carried out analytically, since it involves the factor $(1 + a^3/2r^3)$. In order to avoid this difficulty, we employ the same approximation method as used for the static electrophoresis problem [20]. That is, we approximately replace the r-dependent difficult factor $(1 + a^3/2r^3)$ with its constant value at $r = a + \delta/\kappa$, δ being a factor of order unity. In the static electrophoresis problem [20], we have found that the best approximation can be achieved if δ is chosen to be $2.5/\left(1+2e^{-\kappa a}\right)$. We use this choice of δ in the transient electrophoresis problem. By using this approximation, the integration in Eq. (37) can be carried out analytically to give

$$\widehat{\mu}(s) = \frac{1 + \left(1 + \frac{1}{\kappa a}\right) a\sqrt{\frac{s}{\nu}}}{\left\{1 + a\sqrt{\frac{s}{\nu}} + \frac{a^2 s}{3\nu} + \frac{2a^2(\rho_p - \rho_o)s}{9\nu\rho_o}\right\}\left(1 + \frac{1}{\kappa}\sqrt{\frac{s}{\nu}}\right)} \frac{\mu_{\mathrm{H}}}{s} \tag{38}$$

with

$$\mu_{\mathrm{H}} = \frac{\varepsilon_{\mathrm{r}}\varepsilon_{\mathrm{o}}\zeta}{\eta} f_{\mathrm{H}}(\kappa a) \tag{39}$$

and

$$f_{\mathrm{H}}(\kappa a) = \frac{2}{3}\left[1 + \frac{1}{2\left\{1 + \frac{2.5}{\kappa a(1 + 2e^{-\kappa a})}\right\}^3}\right] \tag{40}$$

where μ_{H} is the Henry function and Eq. (40) is an approximate expression for the Henry function $f_{\mathrm{H}}(\kappa a)$ for the steady electrophoresis of a spherical particle. Equation (40) has been found to be an excellent approximate expression for the exact expression of $f_{\mathrm{H}}(\kappa a)$ [20], viz.,

$$f_{\mathrm{H}}(\kappa a) = 1 - \left\{5e^{\kappa a}E_7(\kappa a) - 2e^{\kappa a}E_5(\kappa a)\right\} \tag{41}$$

where

$$E_n(\kappa a) = (\kappa a)^{n-1} \int_{\kappa a}^{\infty} \frac{e^{-x}}{x^n} dx \tag{42}$$

is the exponential integral of order n. Note that in the limit of $t \to \infty$, $\mu(t)$ approaches the static electrophoretic mobility μ_{H} as combined with Eq. (41). Equation (41) can be derived directly from Eq. (37), viz.,

$$\mu(\infty) = \lim_{s \to 0} s\widehat{\mu}(s) = -\frac{\varepsilon_r\varepsilon_o(\kappa a)^2}{9\eta} \int_a^{\infty} \left(1 - \frac{3r^2}{a^2} + \frac{2r^3}{a^3}\right)\left(1 + \frac{a^3}{2r^3}\right) \frac{d\psi^{(0)}(r)}{dr} dr \tag{43}$$

which yields Eq. (41), that is, $\mu(\infty) = \mu_{\mathrm{H}}$.

The transient electrophoretic mobility $\mu(t)$ can now be obtained analytically form Eq. (38) without recourse to the numerical inverse Laplace transformation. The result is

$$\mu(t) = \frac{\varepsilon_{\mathrm{r}}\varepsilon_{\mathrm{o}}\zeta}{\eta} f(\kappa a, t) \tag{44}$$

with

$$f(\kappa a, t) = \left\{ \frac{(\kappa a - q_1 - \kappa a q_1) q_2}{(\kappa a - q_1)(q_2 - q_1)} M\left(\frac{q_1\sqrt{\nu t}}{a}\right) - \frac{(\kappa a - q_2 - \kappa a q_2) q_1}{(\kappa a - q_2)(q_2 - q_1)} M\left(\frac{q_2\sqrt{\nu t}}{a}\right) \right.$$
$$\left. - \frac{\kappa a q_1 q_2 M\left(\kappa\sqrt{\nu t}\right)}{(\kappa a - q_1)(\kappa a - q_2)} \right\} f_{\mathrm{H}}(\kappa a) \quad (45)$$

where

$$q_1 = \frac{9}{2\left(1 + \frac{2\rho_{\mathrm{p}}}{\rho_{\mathrm{o}}}\right)}\left(1 + \frac{1}{3}\sqrt{5 - \frac{8\rho_{\mathrm{p}}}{\rho_{\mathrm{o}}}}\right) \quad (46)$$

$$q_2 = \frac{9}{2\left(1 + \frac{2\rho_{\mathrm{p}}}{\rho_{\mathrm{o}}}\right)}\left(1 - \frac{1}{3}\sqrt{5 - \frac{8\rho_{\mathrm{p}}}{\rho_{\mathrm{o}}}}\right) \quad (47)$$

$$M(z) = 1 - e^{z^2}\mathrm{erfc}(z) \quad (48)$$

and erfc(z) is the complementary error function, defined by

$$\mathrm{erfc}(z) = \frac{2}{\sqrt{\pi}}\int_z^{\infty} e^{-x^2} dx \quad (49)$$

Equation (44) as combined with Eq. (45) is found to be in excellent agreement with the exact numerical results obtained by Hang and Keh [8].

The function $f(\kappa a, t)$ is the time-dependent transient Henry function. At $t = 0$, $f(\kappa a, t)$ is zero, while as $t \to \infty$, $f(\kappa a, t)$ tends to the steady Henry function $f_{\mathrm{H}}(\kappa a)$ given by Eq. (41) (or Eq. (40) with negligible errors). In the limit of large κa ($\kappa a \gg 1$) (Smoluchowski limit), Eq. (45) tends to

$$f(\kappa a, t) = \frac{(1 - q_1) q_2}{q_2 - q_1} M\left(\frac{q_1\sqrt{\nu t}}{a}\right) - \frac{(1 - q_2) q_1}{q_2 - q_1} M\left(\frac{q_2\sqrt{\nu t}}{a}\right) \quad (50)$$

In the opposite limit of small κa ($\kappa a \ll 1$) (Hückel limit), Eq. (45) tends to

$$f(\kappa a, t) = \frac{2}{3}\left\{ \frac{q_2}{q_2 - q_1} M\left(\frac{q_1\sqrt{\nu t}}{a}\right) - \frac{q_1}{q_2 - q_1} M\left(\frac{q_2\sqrt{\nu t}}{a}\right) \right\} \quad (51)$$

Figure 1 shows the time-dependent transient Henry function $f(\kappa a, t)$ plotted as a function of κa and scaled time $\nu t/a^2$ at $\rho_{\mathrm{p}}/\rho_{\mathrm{o}} = 2$. The value of $f(\kappa a, t)$ at $t = \infty$, i.e., $f(\kappa a, \infty)$ is the Henry function $f_{\mathrm{H}}(\kappa a)$ for steady electrophoresis. It is seen how $f(\kappa a, t)$ increases with time from 0 to the steady Henry function $f_{\mathrm{H}}(\kappa a)$.

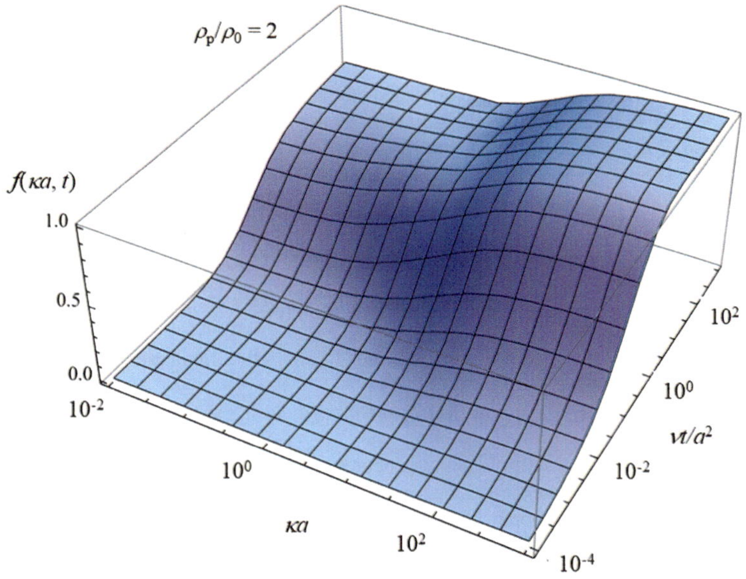

Fig. 1 A 3D plot for the time-dependent transient Henry function $f(\kappa a, t)$ of a spherical colloidal particle of radius a and mass density ρ_p in an electrolyte solution of mass density ρ_o, viscosity η, and Debye-Hückel parameter κ plotted as a function of κa and scaled time $\nu t/a^2$, ν being the kinematic viscosity ($\nu = \eta/\rho_o$) at $\rho_p/\rho_o = 2$. The value of $f(\kappa a, t)$ at $t = \infty$, i.e., $f(\kappa a, \infty)$ equals the Henry function $f_H(\kappa a)$ for steady electrophoresis

3 Conclusions

We have derived an approximate expression (Eq. (44) as combined with Eq. (45)) for the time-dependent transient electrophoretic mobility of a spherical colloidal particle with low zeta potential when a step electric field is applied. The results obtained with this expression is found to be in excellent agreement with the numerical exact results by Huang and Keh [8]. The function $f(\kappa a, t)$ given by Eq. (45) is the time-dependent transient Henry function, which is zero at $t = 0$ and tends to the steady Henry function $f_H(\kappa a)$ as $t \to \infty$.

Compliance with Ethical Standards

Conflict of Interest The author declares that he has no conflict of interest.

Source of Funding The author declares no sources of funding.

References

1. Henry, D.C.: The cataphoresis of suspended particles. Part I.—The equation of cataphoresis. Proc. R. Soc. Lond. A **133**, 106–129 (1931)
2. Morrison, F.A.: Transient electrophoresis of a dielectric sphere. J. Colloid Interface Sci. **29**, 687–691 (1969)
3. Morrison, F.A.: Transient electrophoresis of an arbitrarily oriented cylinder. J. Colloid Interface Sci. **36**, 139–145 (1971)
4. Ivory, C.F.: Transient electroosmosis: the momentum transfer coefficient. J. Colloid Interface Sci. **96**, 296–298 (1983)
5. Ivory, C.F.: Transient electroosmosis of a dielectric sphere. J. Colloid Interface Sci. **100**, 239–249 (1984)
6. Keh, H.J., Tseng, H.C.: Transient electrokinetic flow in fine capillaries. J. Colloid Interface Sci. **242**, 450–459 (2001)
7. Keh, H.J., Huang, Y.C.: Transient electrophoresis of dielectric spheres. J. Colloid Interface Sci. **291**, 282–291 (2005)
8. Huang, Y.C., Keh, H.J.: Transient electrophoresis of spherical particles at low potential and arbitrary double-layer thickness. Langmuir **21**, 11659–11665 (2005)
9. Khair, A.S.: Transient phoretic migration of a permselective colloidal particle. J. Colloid Interface Sci. **381**, 183–188 (2012)
10. Chiang, C.C., Keh, H.J.: Startup of electrophoresis in a suspension of colloidal spheres. Electrophoresis **36**, 3002–3008 (2015)
11. Chiang, C.C., Keh, H.J.: Transient electroosmosis in the transverse direction of a fibrous porous medium. Colloids Surf A: Physicochem Engin Asp **481**, 577–582 (2015)
12. Saad, E.J., Faltas, M.S.: Time-dependent electrophoresis of a dielectric spherical particle embedded in Brinkman medium. Z. Angew. Math. Phys. **69**, 43 (2018)
13. Saad, E.J.: Unsteady electrophoresis of a dielectric cylindrical particle suspended in porous medium. J. Mol. Liquid **289**, 111050 (2019)
14. Li, M.X., Keh, H.J.: Start-up electrophoresis of a cylindrical particle with arbitrary double layer thickness. J. Phys. Chem. B **124**(44), 9967–9973 (2020)
15. Lai, Y.C., Keh, H.J.: Transient electrophoresis of a charged porous particle. Electrophoresis **41**, 259–265 (2020)
16. Lai, Y.C., Keh, H.J.: Transient electrophoresis in a suspension of charged particles with arbitrary electric double layers. Electrophoresis **42**, 2126–2133 (2021)
17. Sherief, H.H., Faltas, M.S., Ragab, K.E.: Transient electrophoresis of a conducting spherical particle embedded in an electrolyte-saturated Brinkman medium. Electrophoresis **42**, 1636–1647 (2021)
18. Ohshima, H.: Approximate analytic expression for the time-dependent transient electrophoretic mobility of a spherical colloidal particle. Molecules **27**, 5108 (2022)
19. Ohshima, H.: Dynamic electrophoretic mobility of a spherical colloidal particle. J. Colloid Interface Sci. **179**, 431–438 (1996)
20. Ohshima, H.: A simple expression for Henry's function for the retardation effect in electrophoresis of spherical colloidal particles. J. Colloid Interface Sci. **168**, 269–271 (1994)

Reverse Transcription Polymerase Spin Chain Reaction (RT-PSCR)

Subhamoy Singha Roy and Mallika Aich

Abstract This paper investigates the chimera characteristics of DEOXYRIBONUCLEIC ACID molecules that may be tracked within the Heisenberg spin structure, as well as the decrease in the temperature-dependent development among the consequence by the use of the Quantum-Phase-Transition (QPT) at the cleansing instance. Backward carbon copy, a supplement that employs Ribonucleic acid molecules to create distinct-stranded DEOXYRIBONUCLEIC ACID molecules, aids in the proliferation of DEOXYRIBONUCLEIC ACID compounds by delaying the free rotation of molecules. The significant triple stress seen close to the backward copy point can be explained by this. We have accurately approximated the produced curve limit.

Keywords DEOXYRIBONUCLEIC ACID molecules · Heisenberg spin structure · QPT

1 Introduction

Nelson came to the conclusion in a groundbreaking work [1] to facilitate DEOXYRIBONUCLEIC ACID is a composite polymer and that it is physically crooked about length balance longer compared to its actual extent of around 50 nm. The ordinary bends prevent the bent molecule from dragging sideways across the surrounding material by resisting its ability to bend. Flexing creates resistance, which increases the cohesive force in the neighboring water. The enhancement in pull clarifies the lengthy torsional tension close to the purpose of sequencing. Nelson [1] made the assumption that repeal Ribonucleic acid is a heap of randomly sequenced, same-curved, non-identical subgroups. Although the bends are random, it has been thought that they do not signify zero on length scales. In contrast to polymerase, copy replication is what causes the axial rotation of duplicated RNA. The incurable tension will

S. S. Roy (✉) · M. Aich
Department of Physics, JIS College of Engineering (Autonomous), West Bengal University of Technology, Kalyani, Nadia, India
e-mail: subhomay.singharoy@jisacolege.ac.in

S. Bhattacharyya and H. S. Mahato (eds.), *Modeling, Analysis and Simulations of Multiscale Transport Phenomena*, Springer Proceedings in Mathematics & Statistics 491, https://doi.org/10.1007/978-981-96-3098-1_5

move down the DEOXYRIBONUCLEIC ACID whenever if there is any interference with the rotation, which will disrupt the integrity of the twofold coil configuration. A frictional resistance to axial rotation of an endless rod that is directly in a viscous medium is given by $\Gamma = (\mu L)$, where μ stands for the friction constant, the torque in Γ has an energy measurement and L indicates the length.

Although experimental observations suggest that there are significant diagonal strains in exposed reverse RNA in both living cells and in vitro, Liu and Wang [2] illustrious to facilitate frictional strength was negligible and But without a corporal secure, DEOXYRIBONUCLEIC ACID of the proper distance end to end could not experience any meaningful incurable stress. Coco and Monason [3] suggested the hypothesis of DEOXYRIBONUCLEIC ACID denaturation shifts ambitious through heat or influenced by external reflex rotational analogue of linear force. Indeed, a torque-induced conversion was seen in an testing by Streak et al. [4] in a largely discernible DEOXYRIBONUCLEIC ACID. A DEOXYRIBONUCLEIC ACID molecule's measurement is based on the stage of permanent-edge supercooling, represented by the number $b = \{(A \times w - A \times w_0)/A * w_0\}$, which recognizes the twist as a sign that the two strands of the double helix are coming together and serves as a model for an exposed linear molecule. The torque motivation test has been amended to state that the cruel molecule splits keen on unadulterated B-DEOXYRIBONUCLEIC ACID stage within $[b] = (b_c)$ and denatured regions through $b = (-1.0)$ outer surface of regions significant supercoiling $b_c = -0.0150$ and an accompanying significant rotational analogue of linear force $[\Gamma_c] = -\{0.05(ev/rad)\}$. The molecule was subjected to further curves enhance the comparative divisions with respect to the opposing RNA with B-DEOXYRIBONUCLEIC ACID. Denaturation was defined by Coco and Monason [3] as the first-order phase shift from B-DEOXYRIBONUCLEIC ACID to RIBONUCLEIC ACID, with temperature and external torque acting as control factors. According to single molecular experiment findings [4], it offers a plausible clarification regarding the sequence of B-DEOXYRIBONUCLEIC ACID and RIBONUCLEIC ACID events.

We stated in a previous work [5] that only if the molecules were spined on possible mapping of the axis to the Heisenberg antiferromagnetic spin structure is for a dual helix DEOXYRIBONUCLEIC ACID molecule. The two neighboring coils spin orientations should be facing each other as, every time you revolve, the 2-strands actually go in opposing orders down the alliance. This model provides a more comprehensive explanation of the investigational information than the Rod-Like-Chain (RLC) model since it shares some particular traits with it [6]. It is the courtesy that the thermodynamic entropy and the entropy concerned inside the spin sequence coincide. The quantum phase transition caused by temperature is started by the splitting that occurs when refinement occurs and the temperature effect merges [7]. Observations show that the axial force is observed that when Γ is the rotational analogue of linear force and the twist angle is γ, the torsional energy V is provided by $V = -(\Gamma\gamma)$. The molecule is under twisted by $\Gamma\langle 0$ and over twisted by $\Gamma\rangle 0$ due to torque. The temperature affects the twist angle γ started by merging. In actuality, experimental observations of $\left(\Delta\gamma / \Delta T\right) = -1.70 \times 10^{-4}$ rad/K at $(\Gamma) = 0$, , indicate that heat changes cause an under twisting. This model better explains the experimental results because it has some specific characteristics with the rod-like-chain (RLC) model [8]. The spin chain's entropy and the thermodynamic entropy agree in this manner. It is thought that the temperature impact during temperature refinement is a good representation of the quantum phase change that fusion initiates.

2 Theoretical Background

Two strands can be considered as being represented by two spins with opposing orientations, such as $+\frac{1}{2}$ and $-\frac{1}{2}$, when they are twisted in reciprocally opposed directions. Two face-to-face oriented spins that are inserted on the axis and arranged at two neighboring locations separated by a single helix period on the lattice, to form an antiferromagnetic spin system. When the entanglement entropy of the spin system disappears, denaturation transition takes place [9]. The steady rise in the proportion of d-DEOXYRIBONUCLEIC ACID relative to RNA will be examined in this note from the perspective of entropy release upon untwisting in the temperature range of $T = 298.0K$, or ambient temperature, with $\Gamma_c = (-1) \times 0.050(ev/rad)$ to $T = 350.0K$, or the melting temperature with $(\Gamma) = 0$.

It has been demonstrated that the topological, stretchy characteristics of the opposite DEOXYRIBONUCLEIC ACID may be translated into spin degrees of freedom on a Heisenberg antiferromagnetic sequence when a DEOXYRIBONUCLEIC ACID particle is mapped [10]. In fact, a spin may be represented similar to an $\{SU((2))\}$ gauge pack [11], and the vector among the $\{SU((2))\}$ gauge budding (E_μ), which has 3-set Indexes, is connected to the digression vector $\left[\vec{T}\right] = \{\partial_s\{\vec{r}(s)\}\}$, $\left[\vec{T}\right] = \{\partial_s\{\vec{r}(s)\}\}$ Whereas $\left[\vec{r}(s)\right]$ depicts a space curvature through arc length s as its parametrization. The $\{SU((2))\}$ gauge impending is given by

$$E_\mu = \left(\vec{a}_\mu \vec{b}\right) \tag{1}$$

In this instance, the Pauli matrix vector is $\vec{b}$. Apparently, $\left(\partial_s^2 \vec{r}\right)$ matches the vector $\vec{f}_{\mu\nu}$ connected to the gauge field strength of $\{SU(2)\}\{SU(2)\}$

$$F_{\mu\nu} = \left(\vec{f}_{\mu\nu}.\vec{b}\right) \tag{2}$$

It is therefore possible to express the internal energy associated with the curvature (bending) as.

$$\int ds(\partial_s^2 \vec{r})^2 = \int d^3x Tr F_{\mu\nu} F_{\mu\nu} \tag{3}$$

According to the Lagrangian, twisting stretchy energy related to torsion may be linked to the spin–spin interaction [12].

$$L = (m^2) \int \vec{R_\mu} \times R_\mu \vec{\times} d^3x \tag{4}$$

In where $\left(\vec{R_{(\mu)}}\right)$ stands for the topological current.

Where the topological current is denoted by $\left(\vec{R_\mu}\right)$.

$$\vec{R_\mu} = \left\{(\varepsilon^{\mu\nu\lambda b})\left((\vec{a}_\nu) \times \left(\vec{f}_{\lambda b}\right)\right)\right\} = (\varepsilon^{\mu\nu\lambda b})\left(\partial_\nu \vec{f}_{\lambda b}\right) \tag{5}$$

At this time, Tensor Levi-Cirita is represented by $\varepsilon^{(\mu\nu\lambda b)}$. The topological Lagrangian is linked to topological currents

$$L = [-1]\{\gamma \times [Tr\varepsilon^{(\mu\nu\lambda b)}](F_{\mu\nu} \times F_{\lambda b})\} \tag{6}$$

inside non-Abelion gauge assumption is equivalent to the γ term. The axial vector current connected to a chiral fermion $R^5_\mu = (\overline{\psi}\gamma_\mu\gamma_5\psi)$ is represented as the current $\vec{R_\mu}$, with (γ_μ, γ_5) representing the Dirac matrices and ψ representing the fermionic wave function.

The connection is satisfied by the three elements of $\vec{R_\mu}$ [13].

$$\partial_\mu R^2_\mu = -\{(1/2)\partial_\mu R^5_\mu\},$$

$$R^1_\mu = -(1/2)R^2_\mu,$$

$$R^3_\mu = (1/2)R^2_\mu \tag{7}$$

It appears that the chiral anomaly shown by the relation is connected to $\vec{R_\mu}$

$$\partial_\mu R^5_\mu = \left\{(1/8\pi^2)TrF^*_{\mu\nu}F_{\mu\nu}\right\} \tag{8}$$

the Hodge dual is represented by $F^{(*)}_{\mu\nu}$.

$$F^*_{\mu\nu} = \{(1/2)\varepsilon^{\mu\nu\lambda b}F_{\lambda b}\} \tag{9}$$

A few previous studies [14–16] have demonstrated that the topological (θ) component, when included into the gauge field theory's Lagrangian, as explained by

$$L = \{(0.250)Tr[F_{\mu\nu} \times F_{\mu\nu}]\} \tag{10}$$

L results in an Abelian gauge field, which is equivalent to a magnetic field inside the region of the gauge orbit connected to the non-Abelian gauge changes. Considering this, we observe that in a non-Abelian gauge field theory.

With closest neighbor interaction, the XXX model's (one-dimensional Heisenberg spin chain) Hamiltonian is provided by

$$H = \left\{ \sum_i \left(b_i^X b_{i+1}^X + b_i^Y b_{i+1}^Y + b_i^Z b_{i+1}^Z \right) + (\lambda) \sum_i b_i^2 \right\} \tag{11}$$

Representing the spin vector using revolution $\vec{S} = \left\{ (0.5)\vec{b} \right\}$ is given by Pauli matrix $b_i^{\{X(Y,Z)\}}$ at site i, and the parameter λ denotes the external magnetic field represents torsional stress energy in the circumstance of reverse DEOXYRIBONUCLEIC ACID [17]. The critical behavior in a spin system that experiences a quantum phase transition (QPT) is associated with a two-limit behavior. The system is representative of the ferromagnetic state at $\lambda - (2)$ and the antiferromagnetic state at $\lambda = 0$. The key region is the area that lies among $(\lambda) = zero$ and $(\lambda) = Two$. The entanglement entropy (EE) reaches its upper limit at $(\lambda) = zero$ and then diminishes at $(\lambda) = 2$ as λ grows. As the nonequilibrium effect in zero temperature QPT exhibits an amazing parallel connecting phase transition at fixed and nil temperature, it has been discovered that a quench-induced nil temperature QPT can acquire the critique of a limited temperature shift [18]. Actually, there are two methods to approach the concept of quantum criticality. We can see that the organize parameter λ reaches the crucial value $\lambda = [\lambda_c]$ as $(T \to 0)$ increases. Additionally, that may be seen at $(T = 0)$, $\lambda \to (\lambda_c)$ Both the temporal and spatial correlations of the order parameter fluctuations lengthen as the critical point gets closer. The association occasion, which differs close to critique, is the usual time scale for the fluctuations' decline. In a quench-induced QPT, the intersect from quantum to classical performance takes place when the connection time surpasses $\beta = 1/(k_B T)$. It occurs when the Kibble-Zurek (KZ) method [19, 20] is introduced into QPT while accounting for time-reliant direct specifications. This results in a nonequilibrium effect that is close to criticality. This suggests that when temperature influence is taken into account in the (quench-time) τ_q bearing the relative $T \sim (1/\tau_q)$, nil temperature QPT caused by a quench may be translated against a conventional restricted temperature phase change.

Once the Heisenberg spin chain experiences a quench in QPT, we are able to obtain

$$\lambda(t\langle 0) = \left\{ (2.0) - 2 \times \left(\frac{t}{\tau_q} \right) \right\} \tag{12}$$

such that $(\lambda) = 0$ is present at $[t] = (\tau_{(q)})$, and when t approaches zero beginning at τ_q, λ rises, resulting in $(t = Zero)$, $(\lambda = Two)$, where the entanglement entropy disappears. A recent research [21] has demonstrated that the λ dependency of the entanglement entropy Z in the Heisenberg spin sequence is given by $Z(\lambda)$.

$$Z(\lambda) = \{Z[(\lambda = Zero)] + \Delta Z\} \tag{13}$$

within

$$[\Delta Z] = (1/6)\{\log_2(1 - 0.50 \times \lambda)\} \tag{14}$$

In a spin sequence Heisenberg, this has been demonstrated in a recent article [22] that the entanglement entropy(EE) z, λ dependency is provided with

$$Z(\lambda) = \left\{[Z(\lambda = Zero)] + ({}^{1}/_{6}) \log_2(1.0 - {}^{\lambda}/_{2})\right\} \tag{15}$$

With this relationship in mind, we will now examine the phase transition linked to denaturation given that λ rises from zero to *Two* and the Entropy progressively drops until it vanishes at $\lambda = 2$.

We propose that the supercoil stricture $[b]$ and temp. $[T]$ may be related to the torsional energy give by $V = (-1) \times (\Gamma \times \gamma)$, where Γ is the rotational analogue of linear force and γ is the interweave angle, respectively, through their interaction with Γ and γ. Because of this, we may consider λ to be a function of both T and b. It is exposed to at $Temp = 298.0K$, the critical rotational analogue of linear force value $\Gamma_c = -0.050(ev/rad)$ and the corresponding supercoiling $b_c = -0.0150$ result in a torque-induced partial denaturation. Beyond this, as has been shown experimentally [23], the perverse molecule splits keen on unadulterated Ribonucleic Acid with regard to the B-DEOXYRIBONUCLEIC ACID phase through $[b] = ([b_c])$ and denatured portions by $[b] = (-1.0)$. When the melting temperature with $(\Gamma) = 0$ is at $T = 350.0K$, more turns given to the particle raise the comparative proportion of Ribonucleic Acid through deference to B-DEOXYRIBONUCLEIC ACID.

We can conclude that λ fulfills the linear relation in light of this.

$$\lambda = ([P \times T] + Q) \tag{16}$$

with

$$Zero = ([298 \times P] + Q) \tag{17}$$

$$Two = ([350 \times P] + Q) \tag{18}$$

At this point P and Q are 2-constabularies.

Resolving this we have $[P] = 0.03850$ and $\left[Q\right] = -11.470$.

We may write for σ in a similar fashion.

$$\lambda = \left(\left[\overline{P} \times b\right] + \overline{Q}\right) \tag{19}$$

with

$$0 = \left([-1 \times 0.0150]\overline{P} + \overline{Q}\right) \tag{20}$$

$$Two = \left([-1] \times \overline{P} + \overline{Q}\right) \tag{21}$$

Explaining this $\overline{P} = -(2.030)$ and $\overline{Q} = -(0.030)$.

Now the relationships allow us to calculate how the entropy changes as a function of temperature and untwisting.

We can now calculate the entropy at a particular temperature using Eqs. (13–15), which provides a measurement of the openness of stand pairs at a certain temp [24]. A common characteristic of quench-caused QPT is the creation of kinks and antikinks as well as the introduction of the standard coldness linking a kink and the next antikink, or the Kibble-Zurek (KZ) correlation length, is a new span size [25, 26]. In order for the KZ correlation length ξ to satisfy the connection $\xi \sim \left(\sqrt{\tau_q}\right)$, the density of defects ρ is dependent upon the quench time τ_q through the relation $\rho \sim \left(1/\sqrt{\tau_q}\right)$. We see that the KZ correlation extent ξ relies on the temperature through the connection $1/\left(\sqrt{k_B T}\right)$, using the previously described link between the quench time $\left(\tau_q\right)$ and temperature T during the next of kin $T \sim \left(1/\tau_q\right)$.

We now translate the conclusions we obtained before, which concerned the mapping of a twofold coil against the Heisenberg spin sequence, keen on an stretchy rod model. It has been observed that restricted denaturation linked to transcriptions induced by quench-caused QPT results in the formation of defects such as kinks and antikinks, which correspond to dynamical meandering of the stretchy rod and correspond to bends in the DEOXYRIBONUCLEIC ACID molecule's axis by deference to the reverse Ribonucleic Acid molecule. The bend length and the KZ correlation length match.

These bend are generated animatedly, and the corresponding curve length differs from the geometrical attribute of the rod, the structural persistent length F. Because of the way the flaws arise, every joint periodically bends in the opposite direction from one another, with each kink followed by an antikink. We may formulate the connection $P \sim \left(1/\sqrt{T}\right)$, curve length ξ, based on the relationship between the KZ correlation lengths $\xi \sim \sqrt{\tau_q}$ and $\tau_q \sim \left(T^{-1}\right)$.

Now

$$\left[(P^2 \div F)\right] = [(k)/(K_B \times T)] \tag{22}$$

where the curve rigidity with a measurement of *energy. length* is represented by k. Since this relates to a constant length level for the arrangement, Seeing that bend stiffness $k = 45.0\,\text{nm}(K_B T)$ and $F = 130.0\,\text{nm}$ [27] are seen experimentally, we now have $P = 76.50\,\text{nm}$. Comparing this with the investigational assessment 80.0 nm is necessary [27]. This gives us the overall persistence-length(PL), which is indicated by

$$A_{tot} = 1/(P^{-1} + F^{-1}) \tag{23}$$

From Eq. (23) we have $P_{tot} = 48.30$ nm, It has to be contrasted with the reference value 50.0 nm.

There, we discover that our findings align with previous trials.

It is discovered that [16] gives the entanglement entropy(EE) of a plectonomic supercoil.

$$Z = \left(P^2/B^2\right) + \left(P^2/(\pi F)^2\right) \tag{24}$$

where P is the winding stretchy continuous and B radius and F is pitch. The corresponding imprisonment $Freeenergy/unitlength$ at steady temperature, denoted as T, is therefore obtained and is provided by [16, 28].

3 Discussion

In order to avoid flexing, Nelson [1] planned using a amalgam model to pick a DEOXYRIBONUCLEIC ACID intersect length and achieving a amalgam movement that is less than the movement in fibers of the planned distance end to end. The velocity rises over the revolving (plumber's snake) at long length scales due to the large viscosity of the stiff (crankshaft) motion. A haphazardly twisted rod has a configuration according to the scale for each length, and until the crossover length is reached, the amount of torque pulled for each length rises without being limited. Since around [29], confined DEOXYRIBONUCLEIC ACID is thought to be the primary cause of the high Torjan pressure close to the spin and duplication sites protected on comparatively tiny sizes. In our present convention, we have observed that when the theoretical properties of a DEOXYRIBONUCLEIC ACID particle related to the Heisenberg spin arrangement are mapped to a representation of stretchy rod, bends are animatedly shaped throughout arrangement and portions flanked by 2-curves go from side to side a supple movement that keeps from litheness. In reality, we also know that the configuration of twists throughout restricted dissociation connected to the transcription procedure successfully merges flaws like kinks and antinociceptive, and these flaws are passed on to the transcription DEOXYRIBONUCLEIC ACID in molecules [30]. This is because we think of DEOXYRIBONUCLEIC ACID molecules as clean system that exceed from side to side a white-induced QPT. This explanation, terminal stress is the major observation close to the transcription point. We have seen that there is a lack of steadfastness around our criticism when a refinement spurs the quantum phase transitions. When conditions are met such that an RNA molecule splits into two ends immediately underneath the dangerous point, the spins start to slant. When the system reaches the dangerous point, a knot forms, causing all of the spins in an antikink to point in the identical track. This is known as the spin flip. The overturn procedure proceeds gradually from side to side the tube when a spin begins to move backwards to the crucial point located somewhat below this point. The structural characteristics of RNA molecules with DEOXYRIBONUCLEIC ACID molecules start to alter during this phase, and the strands start to open up more base pairs. The ternary stress around the reproduction position is substantial due to the vibrant arrangement of the curves, which fully stops the rotating velocity.

References

1. Liu, L., Wang, J.: Proc. Natl. Acad. Sci. USA **84**, 7024 (1987)
2. White, J.H.: Am. J. Math. **91**(3), 693 (1969)
3. Cocco, S., Monasson, R.: Phys. Rev. Lett. **83**, 5178 (1999)
4. Strick, T.R., Allemand, J.F., Bensimon, D., Lavery, R., Croquette, V.: Physica **263A**, 392 (1999)
5. Azbel, M.Y.A.: J. Chem. Phy. S. **62**, 3635 (1975)
6. Roy, S.S., Bandyopadhyay, P.: Phys. Lett. A, **382**, 1973 (2018)
7. Goswami, G., Bandyopadhyay, P.: J. Math. Phys. **34**(2), 749 (1995)
8. Bouchiat, C., Mezard, M.: Eur. Phys. J. E **2**, 377 (2000)
9. Roy, S.S., Bandyopadhyay, P.: Phys. Lett. A. **337**, 2884 (2013)
10. Vojta, M.: Rep. Prog. Phys. **66**, 2069 (2013)
11. Bandyopadhyay, P.: Proc. R. Soc. Lond. **A466**, 2917 (2010)
12. Bandyopadhyay, A., Chatterjee, P., Bandyopadhyay, P.: Gen. Rel. Grav. **18**, 1293 (1986)
13. Sen, K., Bandyopadhyay, P.: J. Math. Phys. **35**, 2270 (1994)
14. Roy, A., Bandyopadhyay, P.: J. Math. Phys. **30**, 2366 (1989)
15. Wn, Y.S., Zee, A.: Nucl. Phys. B **288**, 157 (1985)
16. Mondal, S.S., Ravishankar, V.: Phys. Rev. B **54**, 8688 (1996)
17. Roy, S.S.: Theor. Phys. **2**(3), 141–144 (2017)
18. Zurek, W.H.: London. Nature **327**, 505; Phys. Rep. **276**, 177 (1985)
19. Schutzhold, R.: Phys. Rev. Lertt. **95**, 133703 (2005)
20. Kibble, T.W.B.: J. Phys. A Math. Gen. **9**, 1387 (1976)
21. Basu, B., Bandyopadhyay, P., Majumder, P.: Phys. Rev. A **86**, 022303 (2012)
22. Hurst, C.A.: Ann. Phys. **50**, 51 (1968)
23. Roy, S.S., Bandyopadhyay. P.: Eur. Lett. **109**, 48002 (2015)
24. Basu, B., Bandyopadhyay. P.: Int. J. Mod. Phys. B **12**, 49 (1998)
25. Cincio, L., et al.: Phys. Rev. A **75**, 052321 (2007)
26. B-DEOXYRIBONUCLEIC ACID: J. Mol. Biol. **254**, 579 (1995)
27. Basu, B., Bandyopadhyay, P., Majumder, P.: Phys. Rev. A **83**, 03231 (2011)
28. Marko, J.F., Siggia, E.D.: Phys. Rev. E **52**, 2912 (1995)
29. Carmeli, M., Malin, S.: Ann. Phys. **103**, 208 (1977)
30. Bouchiat, C., Mezard, M.: Phys. Rev. Lett. **80**, 1556; Euro. Phys. J. E **2**, 377 (1997)

Numerical Study of Ion Transport and Convective Mixing in Micro Channel with Nozzle/Diffuser

A. K. Nayak

Abstract The elctroosmotic flow and mixing is numerically studied through a Nozzle/Diffuser placed between two large reservoirs connected through a micro channel. The analytical and numerical studies are performed to illustrate the effects of convection and diffusion mechanism due to different aspect ratios of the transformation. The hydrodynamics of the ion and flow components are analyzed through Poisson-Nernst-Plank based Navier Stokes model of Newtonian fluid subjected to specified system of forces. The numerical results for the reservoir effects on velocity, pressure, potential have found a significant variation when the transport mechanism is changing from nozzle to diffuser. Mixing efficiency is also evaluated for different geometric configurations and compared with the plane slit channel when the reservoir is connected. Various effects of Debye-Hückel parameter, conical angles or slopes and reservoir height/ width on mixing efficiency are discussed. A notable findings are obtained for the effective mixing in reservoir connected nozzle channel is very high without implementing any hurdles or heterogeneous zeta potential. The flow solutions observed a low torque when the equilibrium conditions are justified and is quite effective for mixing enhancement which is optimized in terms of aversion of frictional factor and viscous dissipation. Also, the findings manifest the species patterns with high accuracy and versatility, which could possibly help to handle the technical challenges associated with the design of pumpless actuated micro-fluidic devices.

Keywords Electrokinetic flow · Newtonian fluid · Micro-mixing · Nozzle/ Diffuser · Pressure drop · Reservoirs · End effects · Transformed geometry

A. K. Nayak (✉)
Department of Mathematics, Indian Institute of Technology Rorkee, Roorkee 247667, India
e-mail: ameeyakumar@gmail.com

S. Bhattacharyya and H. S. Mahato (eds.), *Modeling, Analysis and Simulations of Multiscale Transport Phenomena*, Springer Proceedings in Mathematics & Statistics 491, https://doi.org/10.1007/978-981-96-3098-1_6

Nomenclature

C	Dimensionless eluted species, C^*/C_{ref}
D_i	Diffusivity of ith ionic species, $[\mathrm{m^2s^{-1}}]$
D_s	Diffusivity of eluted species, $[\mathrm{m^2s^{-1}}]$
e	Charge of a electron or proton, $[C]$
E_0	External electric field, $[V\mathrm{m}^{-1}]$
F	Faraday's constant, $[C\ \mathrm{mol}^{-1}]$
h_i	Inlet height of the channel, [m]
h_o	Outlet height of the channel, [m]
k_B	Boltzmann constant, $[\mathrm{m^2kgs^{-2}}K^{-1}]$
L	Length of the channel, [m]
L_R	Length of the reservoir, [m]
H_R	Height of the reservoir, [m]
n_0	Ion density, $[ions\ \mathrm{m}^{-3}]$
$n_1,\ n_2$	Ionic species
p	Scaled pressure, $p^*/(mU_{HS}/H)$
Pe	Peclet number, $U_{HS}H/D_i$
Pe_s	Uncharged eluted species Peclet number, $U_{HS}H/D_s$
$\mathbf{q}$	Velocity vector, $[\mathrm{ms^{-1}}]$
R	Universal gas constant, $[J\mathrm{mol}^{-1}K^{-1}]$
Re	Reynolds number, $U_{HS}H/\nu$
Sc	Schmidt number, ν/D_i
t	Time, [s]
U_{HS}	Generalized Helmholtz-Smolouchowski velocity
$u,\ v$	Dimensionless velocity components
(x,y)	Cartesian coordinate, [m]
(X,Y)	Coordinates in transformed domain, [m]
z_i	Valence of the ith ionic species

Greek

ϵ	Electrical permittivity of the solution, $[CV^{-1}\mathrm{m}^{-1}]$
$\dot{\gamma}$	Scaled rate of strain tensor, $\dot{\gamma}^*/(U_{HS}/H)$
κ	Inverse Debye length, $[\mathrm{m}^{-1}]$
λ	EDL thickness, $[m]$
Λ	Scaled external electric field, E_0H/Φ_0
μ	Scaled apparent viscosity, $\mu^*/(U_{HS}H)$
ϕ	Induced potential, V
ψ	External potential, V
ρ_e	Net charge density, $[C\mathrm{m}^{-3}]$
ρ	Fluid density, $[\mathrm{kgm^{-3}}]$
ζ	Zeta potential, $[V]$

Subscripts

$*$ Dimensional

1 Introduction

Electrokinetic mechanism plays an important role in various chemical and biological developments such as sample separation, drug delivery, DNA hybridization, biochemical investigation. These techniques are used in Lab-on-a-chip devices, microelectrochemical systems (MEMS) etc. [1]. There are many advantages of using microfluidic devices as these devices require low amount of sample reagent as well as power cost is also very low. In general, mechanical pumping are used for large characteristic length channel flow, i.e. pressure driven flow but in micro/nano channel flow, the mechanical pumping cannot be used as it may cause various difficulties like mechanical failure or fabrication difficulty. To defeat such failures, electroosmotic flow (EOF) is used in micro-nano channel. In EOF, the opposite charged ions (counter ions) of electrolyte solution are attracted to the surface charge (zeta potential) [2] while same charged ions (co-ions) are reppled. These counter ions make a region near the surface wall and is called Electric Double Layer (EDL). By the help of externally applied electric field, the counter ions in the EDL can move and drags the surrounding liquid with it in motion, thus generates Electroosmotic flow [3].

Micro/nano nozzle and diffuser have a lot of application in EOF and mixing. Recently micro nozzle diffuser have obtained a lot of attention in different field of industry, medical uses and biological uses for their simple structure and higher efficiency. Micro-nozzles are first introduced by Lehnert et al. [4] in their work of realization of hollow SiO_2 micronozzles for electrical measurements on living cells. Chen et al. [5] introduced different polymers for the fabrication of micro-nozzle/diffusers for the fabrication of micro nozzle/diffusers. A piezoelectric micro-pump using MT135 type valves in low pressure gradient has been used to deliver insulin, drugs etc. [6]. An improvement of nozzle-diffuser micropump design has been established by Yang et al. [7]. It was found that the pressure drop increases in huge amount when two fin or obstacles are added with the microdiffuser. A two layer nozzle-diffuser electroosmotic micromixer was proposed by Ahmed and Kim [8]. Their numerical results concluded a far better mixing performance with such a short channel length. Yeh et al. [9] numerically investigated electroosmotic flow pumping and ion concentration polarization based on negatively charged nozzle shaped nanopore. Electroosmotic flow and mixing in different shaped microchannel has been investigated numerically by Lattice Boltzman method and optimized the geometry and zeta potential for maximum mixing flow rate [10].

The majority of earlier works on electrokinetic flow used the assumption that the flow was fully developed in their work. Therefore, the impact of reservoirs on channel flow has not yet been calculated. In a realistic setting, reservoirs at the channel's entrance and outflow impact flow, with effects that really travel the whole length of the channel. Mirbozorgi et al. [11] evaluated the impact of different non-uniform zeta potentials on electroosmotic flow in slit microchannels coupled with reservoirs. The influence of a limited reservoir size on electroosmotic flow in a rectangular microchannel was quantitatively examined by Yan et al. [12]. Because of the existence of reservoirs, they also discovered the backpressure gradient analytically. The impact of streaming potential in reservoir linked flat microchannel has been studied analytically and quantitatively by Mirbozorgi et al. [13]. They demonstrated how the zeta potential along the channel wall affects the streaming potential nonlinearly and how a particular zeta potential yields the highest potential. Yang et al. [14] conducted a numerical analysis of the impact of reservoirs on electroosmotic flow in microchannels. Additionally, their analytical analysis demonstrated the influence of the axial pressure gradient on velocity owing to the microchannel's end effect.

Mixing in microfluidic devices plays an important role due to potential application in various fields such as DNA hybridization, drug delivery, chemical synthesis and others biomedical devices. Rapid mixing in these devices is challenging due to low Reynolds number, thus the main mechanism for mixing in such devices is by diffusion. This process takes much time and the channel length should be long enough. For this reason micromixers is one of the essential components in microfluidic systems. As reported in previous literatures, micromixers are categorized in two types : Active micromixers and passive micromixers. Active micromixers depend on external energy sources to perform the mixing of fluids [15]. On the other hand, passive micromixers are based on different geometric fabrication and surface heterogeneity. They donot require external energy sources for fluid mixing [16]. Chang and Young [17] numerically investigated the EOF mixing in straight channel patterned with rectangular block and concluded that mixing efficiency can be increased as heterogeneous zeta potential introduced in the upper surface of rectangular blocks. Mixing efficiency may also be increased by introducing wavy surfaces along with heterogenous surface zeta potential on the channel wall [18].

In this study, electro-osmotic flow and mixing in reservoir connected two types of micro-channel (nozzle and diffuser) has been discussed. For this, Poisson-Nernst-Plank based Navier-Stokes equation for Newtonian fluid is considered. There is a huge effect of reservoir on the electro-osmotic flow when the reservoirs are included in computation. To the best of our knowledge, this work has not been done yet. Section 2 is following with the geometric configuration of two types of micro-channel (nozzle, diffuser), mathematical model and flow governing equations along with the boundary conditions. In Sect. 3, we have presented the numerical scheme and discretization techniques to solve nonlinear coupled set of equations. The effect of reservoirs on channel flow and mixing due to different physical parameters have been discussed in Sect. 4.

2 Mathematical Model

In this paper, the electroosmotic flow of an aqueous ionic solution is considered in a finite length channel positioned between two cylindrical reservoires. Here we have considered the channel with four varieties, i.e. Nozzle, Diffuser, Diffuser-Nozzle and Nozzle-Diffuser as shown in Fig. 1. Both the reservoirs are connected to electrodes which produce an electric field in the tangential direction to the EDL. The width of the channel is assumed to be much larger than the height of the channel, so the flow field can be considered as two dimensional.

The symmetrical channel (about x-axis) is considered as a linear equation of $x*$ such as (1),

$$y^*(x) = h^*(x) = \pm\left[h_i \pm \frac{h_o - h_i}{L}x^*\right] \tag{1}$$

where h_i and h_o are assumed as the half heights of channel inlet and outlet respectively, L is the length of the channel. The 1st '±' represents the upper wall (+) and lower wall (−) whereas the 2nd '±' represents the diffuser (+) and nozzle (−) channel respectively. Along with that H_R and L_R are the heights and lengths of both reservoirs respectively.

After scaling the parameters with $h = \frac{h*}{h_i}$ and $x = \frac{x*}{L}$, the non-dimensionalized equation becomes (2),

$$y(x) = h(x) = h = \pm[1 \pm mx] \tag{2}$$

where $m = \frac{h_o - h_i}{h_i}$ is the conical angle or slope of the channel.

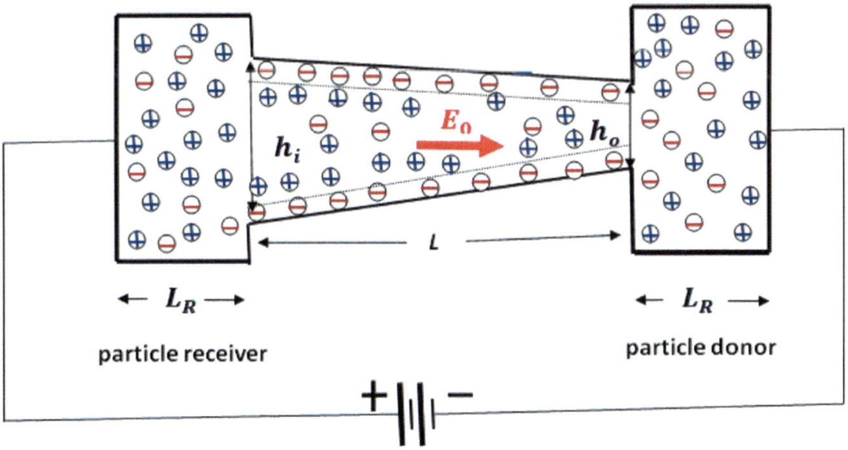

Fig. 1 Schematic diagram of reservoir connected nozzle microchannel

The physical domain of the channel is now transformed into a rectangular geometry with a transformation such as,

$$X = x, \quad Y = \frac{y}{h(x)}. \tag{3}$$

Using the above transformation, the external potential due to the externally applied electric field obeys the Laplace equation, i.e.,

$$\nabla^2 \psi = 0. \tag{4}$$

The wall charge density produces the induced potential which obeys the Poisson's equation (5),

$$\nabla^2 \phi = \frac{(\kappa h_i)^2}{2} \rho_e. \tag{5}$$

The transport of the charged species, i.e. cations and anions obeys the convection—diffusion—migration equation or the Nernst-Plank equation (6),

$$\rho_e(\mathbf{q}.\nabla n_i) = \nabla^2 n_i + z_i n_i.\nabla(\phi + \psi) - \frac{(\kappa h_i)^2}{2} z_i n_i \rho_e. \tag{6}$$

The non-dimensional number Pe is the Peclet number described as $Pe = \frac{h_i U_{HS}}{D}$. D is the diffusivity of both co-ions and counter ions, valence of both species are considered as $z_i = \pm 1$ for $i = 1, 2$. Though our model, i.e. Poisson-Nernst-Plank model has the flexibility to handle multivalent ions, for the sake of simplicity, we have considered here $1 : 1$ electrolyte like, $Nacl + H_2O$ solution.

The modified Navier-Stokes equation (7) along with the continuity equation (8) (in non-dimensional form) become,

$$Re\,(\mathbf{q}.\nabla\mathbf{q}) + \nabla p - \nabla^2 \mathbf{q} + \frac{\kappa h_i^{\,2}}{2\Lambda} \rho_e \nabla(\phi + \psi) = 0 \tag{7}$$

$$\nabla.\mathbf{q} = 0 \tag{8}$$

Here, $\mathbf{q} = (u, v)$ is the dimensionless velocity vector. Both u and v are scaled by the Helmholtz-smolouchowski velocity $U_{HS} = \frac{\varepsilon_e \phi_0 E_0}{\mu}$. Pressure p is scaled by $\frac{\mu U_{HS}}{h_i}$ and the external potential(ψ) as well as the induced potential(ϕ) are scaled by the same parameter $\phi_0 = \frac{k_B T}{e}$. ρ_e is the scaled net charge density where $\rho_e = n_1 - n_2$, n_i represents the concentration of the ith ionic species which is scaled by n_0 i.e., the bulk number concentration. Here, E_0, μ, k_B, T, e stand for the externally applied electric field, fluid viscosity, Boltzman's constant, absolute temperature and elementary charge respectively. Re and Λ are two non-dimensional numbers described as $Re = \frac{\rho U_{HS} h_i}{\mu}$ and $\Lambda = \frac{E_0 h_i}{\phi_0}$ and $\kappa = \sqrt{\frac{2en_0}{\epsilon_e \phi_0}}$ is the inverse of EDL thickness (λ). Moreover, ∇ and ∇^2 stand for,

$$\nabla \equiv \left(\frac{\partial}{\partial X} - Y \frac{h'}{h} \frac{\partial}{\partial Y}, \frac{1}{h} \frac{\partial}{\partial Y} \right), \tag{9}$$

$$\nabla^2 \equiv \frac{\partial^2}{\partial X^2} - 2Y \frac{h'}{h} \frac{\partial^2}{\partial X \partial Y} + \left[Y^2 \left(\frac{h'}{h} \right)^2 + \frac{1}{h^2} \right] \frac{\partial^2}{\partial Y^2} + Y \left[2 \left(\frac{h'}{h} \right)^2 - \frac{h''}{h} \right] \frac{\partial}{\partial Y}. \tag{10}$$

2.1 Boundary Conditions

We are assuming no-slip condition at channel wall, i.e. $u = 0$, $v = 0$. At the inlet and outlet of the problem domain, $\frac{\partial u}{\partial x} = 0$ and $v = 0$ has been taken.

In $\xi - \eta$ domain, the above conditions appear as $\frac{\partial u}{\partial \xi} = 0$ and $v = 0$. The fluid is considered to have no ion penetration condition along the boundary of the problem domain, i.e. $(\nabla n_i + z_i n_i \nabla \phi).n = 0$. Whereas at the inlet and outlet of the problem domain, the boundary condition is assumed as, $n_i = 1$ and $\frac{\partial n_i}{\partial x} = 0$ or, $\frac{\partial n_i}{\partial \xi} = 0$ respectively. n represents the unit normal vector indicating towards the channel.

The Poisson equation is subject to the boundary condition, $\phi = \zeta$ at the channel wall and $\frac{\partial \psi}{\partial x} = 0$ ($\frac{\partial \psi}{\partial \xi} = 0\ in\ \xi - \eta\ domain$) at the channel inlet and outlet, where ζ is the zeta potential which is prescribed at the channel wall. Moreover we have considered both the reservoirs are at same pressure, i.e. there is no pressure difference between the reservoirs.

2.2 Transportation of Uncharged Species

To obtain the transport of uncharged eluted species, we solve convection diffusion equation. The non-dimensional form of the species transport equation is,

$$(\mathbf{q}.\nabla)\, C - \frac{1}{Pe_s} \nabla^2 C = 0 \tag{11}$$

$C(c/c_{ref})$ is the non-dimensional species, Pe_s ($= h_i U_{HS}/D_s$) is the Peclet number for uncharged mixing species. D_s represents the diffusion coefficient of eluted species. The above equation can be solved by imposing the boundary conditions as : at inlet of the problem domain : $C = 1$ in the lower half and $C = 0$ in the upper half.

At outlet of the problem domain : $\frac{\partial C}{\partial x} = 0$ or, ($\frac{\partial C}{\partial \xi} = 0\ in\ \xi - \eta\ domain$). At the channel wall and reservoir wall, no mass flux condition is assumed, i.e. $(\nabla C).n = 0$.

The mixing efficiency at different positions of the channel is defined by a parameter σ, as,

$$\sigma(x) = \left[1 - \frac{\int_{lower\ surface}^{upper\ surface} |C - C_\infty| dy}{\int_{lower\ surface}^{upper\ surface} |C_0 - C_\infty| dy} \right] \times 100. \tag{12}$$

where C is the concentration of the eluted species., C_∞ is the concentration at fully mixed state and C_0 is the concentration of unmixed species. Here, $C_\infty = 0.5$ and $C_0 = 0$ or 1. When species are fully mixed $C = C_\infty = 0.5$ that implies $\sigma(x) = 100$ and $\sigma(x) = 0$ implies the state of fully unmixed, i.e. $C = C_0$.

3 Numerical Method and Model Validation

The staggered grid-based finite volume approach is used to solve the non-dimensional governing equations and the boundary conditions numerically. By integrating across each control volume, the linked set of non-linear equations is discretized. Different control volumes are utilised in a staggered grid to integrate the governing equations. The main control volume's center is where pressure and other scalar values are sorted, while the main control volume's normal faces are where velocities are sorted. First order implicit scheme is utilized to discretize time derivatives at each computation step in the control volumes, whereas QUICK scheme and central difference scheme are employed, respectively, for the convective terms and diffusive terms of the governing equations. The discretized set of governing equations is resolved using the SIMPLE approach. The SIMPLE approach relies on a cycle of prediction and correction until we arrive at a velocity field without divergence. By using this approach, we were able to convert the continuity equation to the Pressure-Poisson equation since the velocity field of the Navier-Stokes equation is linked with the pressure components. The Pressure-Poisson equation is solved using the Successive Over Relaxation technique to the level of precision we seek. The convergence criteria here is

$$max_{i,j} \left| e_{i,j}^{k+1} - e_{i,j}^{k} \right| \leq 10^{-6}.$$

where, i, j signify the cell index, k stand for the iteration index and $e\ (= \phi, \psi, u, v, n_i(i = 1, 2))$.

We have verified our simulated results and cross-checked against some of the most reliable existing results. Figure 2a represents the axial velocity profile and compared with the analytical solution of Zhao et al. [20] for different Debye-Hückel parameters ($\kappa h_i = 3,\ 5,\ 10,\ 40$) for $L/h_i = 20$, with the externally applied electric field as 10^4 V/m for fully developed Electro-osmotic flow. For various Debye-Hückel values,

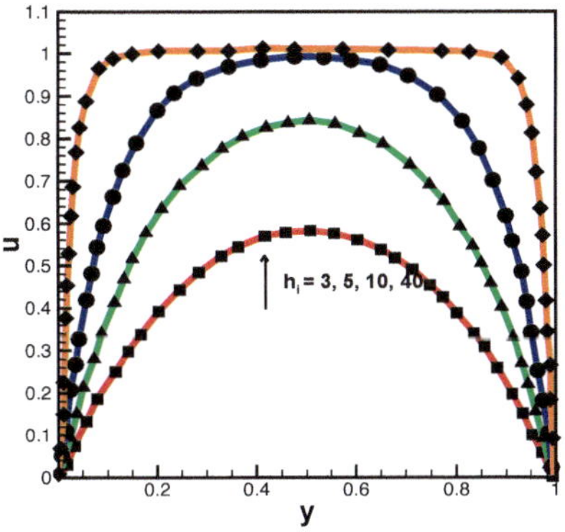

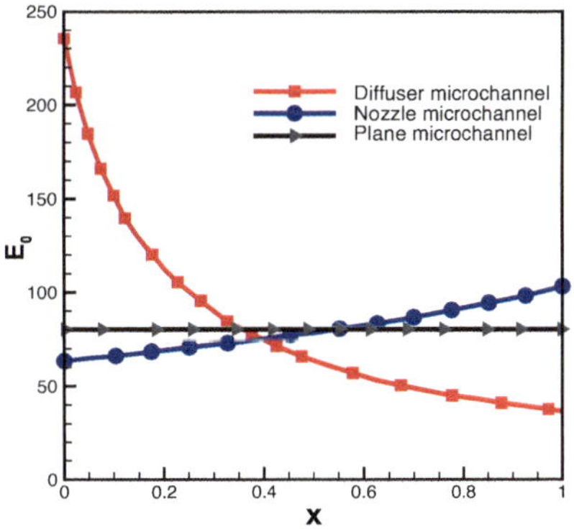

Fig. 2 **a** Comparison of the present axial velocity (lines) with the analytical solution (symbol) obtained by Zhao et al. [20] for different κh_i, **b** Comparison of present (lines) dimensionless imposed electric field (E_i, x) for a nozzle and diffuser with Chen and Conlisk [21]

the velocity profile for u shows a plug-like characteristic and it is shown that the deviation is less than 2%.

We have also verified our computed results for non-dimensional electric potential in nozzle channel, diffuser channel and plane channel with Chen and Conlisk [21] Fig. 2b. To compare the results we have assumed the length of the nozzle is 65 nm; the inlet height is 13 nm and the outlet height of the channel is 8 nm. The length of the diffuser is taken 650 μm; the inlet height is 20 μm and the outlet height is 130 μm. The results found good agreement with the results due to Chen and Conlisk and deviation is less than 3%.

4 Results and Discussions

In this section, the results for four different types of channel, i.e. Nozzle, Diffuser including every channel is connected with reservoirs are presented. The majority of earlier studies on electro-osmotic flow and mixing focused only on micro/nanochannels, which are separated from reservoirs and do not have genuine entrances and exits. In contrast, the channel should be thought of as having a limited length when it is linked to reservoirs. The fluid flows through abrupt converging and diverging portions where the fluid flow in the channel is greatly influenced by reservoirs and in many situations the flow pattern varies as a result of reservoirs being present at the channel's upstream and downstream ends. The parameters that we have considered for simulation are as follows, average height of the channel is considered as 10μm, channel length is considered to be vary from $10 \times h_i$ to $60 \times h_i$, h_i is the channel inlet height. κh_i is ranging from 2 to 50. The wall zeta

Table 1 Physical parameters used for computation [22, 23]

Parameter	Value	Unit
h	10	μm
R	8.314	$J\text{mol}^{-1}K^{-1}$
T	300	K
F	96485	$C\text{mol}^{-1}$
K_B	1.38×10^{-23}	$\text{m}^2\text{kgs}^{-2}K^{-1}$
ϵ_e	695.4×10^{-12}	$CV^{-1}m^{-1}$
e	1.602×10^{-19}	C
N_A	6.023×1023	mol^{-1}
ρ	10^3	$\text{kg}m^{-3}$
m	10^{-3}	Pas
Di	1×10^{-9}	m^2s^{-1}
E_0	104	Vm^{-1}
D_s	1×10^{-11}	m^2s^{-1}

potential has been taken −25.6mV. Dielectric constant of electrolyte is $\epsilon_e = 80\epsilon_0$, where $\epsilon_0 = 8.85 \times 10^{-12}\text{F/m}$ is dielectric constant for vacuum. The electrolyte solution is considered as symmetric and binary and the diffusion coefficient is considered equal i.e., $D_1 = D_2 = 2 \times 10^{-9}\ \text{m}^2/\text{s}$. All other parameters related to this study are presented below in Table 1.

4.1 End Effect Due to Reservoirs in Microchannel

In this section, the results for end effect due to sudden expansion and compression effects are clearly observed when the consequences happen in micro-channel due to reservoirs connecting at both end of the channel. The EOF representations in reservoirs connected to microchannel in terms of the contour plots of velocity vectors, cations, anions, potential have significant changes due to the reservoirs located at both end of the channel are represented in Fig. 3. Here we have considered reservoirs connected Nozzle microchannel with the channel length is 20 times of the channel inlet height and there is no pressure difference between two reservoirs. Also the Debye-Hückel parameter is considered as 30. Figure 3a represents the contour plot of externally applied electric field which has a tremendous effect on the flow behavior. A denser arrangement of external potential has been seen in the regions where the channel meets the upstream and downstream reservoirs. That implies the electric field strength increases at the junctions of reservoirs and channel although it is clear from the Fig. 3a that the electric potential distribution in the channel is not affected by the reservoirs.

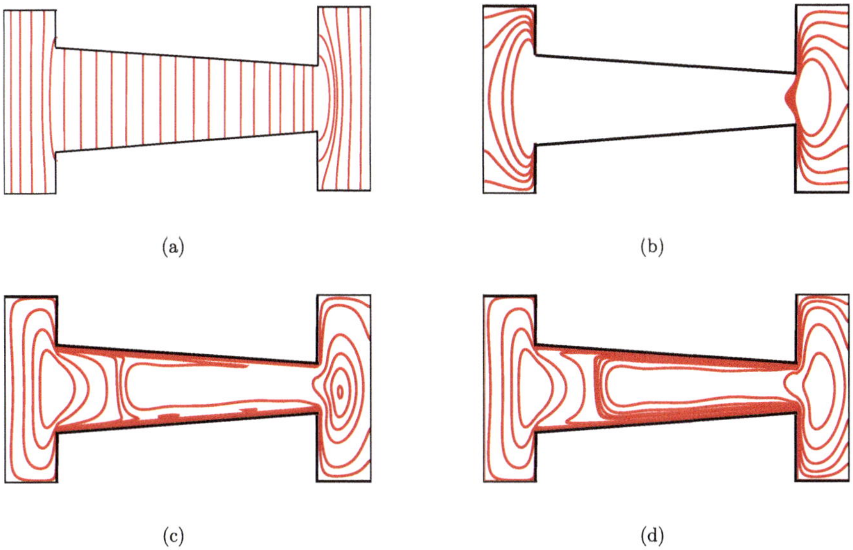

Fig. 3 EOF profile of contour plots for **a** externally applied electric potential, **b** induced potential, **c** cation, **d** anion in reservoir connected microchannel when $L/h_i = 20$ and $\kappa h_i = 30$

The contour plotting of induced potential has been shown in Fig. 3b. The distribution of induced potential at the reservoirs are different that of the the channel and it affects the induced potential distribution of the channel. At the middle of the channel, the induced potential is parallel to the flow but at inlet and outlet of the channel, there is a small effect due to the reservoirs within a distance $h_i/2$. This effect is due to the non-axial velocity which has the influence on co-ions and counter-ions in the advection term of the Nernst-Plank equation. The contour plottings of co-ions and counter-ions have been shown in Fig. 3c and Fig. 3d respectively. Unlike the potential distribution, the cations and anions in the reservoirs affect the ions distribution of the channel. This will also affect the flow through the channel as the ions are coupled with the velocity vectors. Hence it can be deduced that if we do not consider the reservoirs, the flow pattern may be differ a lot. The velocity vector plot has been shown in Fig. 4. The velocity in the microchannel has some great impact of reservoirs within a distance of h_i whereas the reservoir flow affected the non-axial velocity in distance less than $h_i/2$. At the upstream of the channel, it shows ideal EOF profile (plug like profile). But the flow become pressure dominated while pointing towards the channel end and even in the fully developed regions, the ideal EOF profile is not seen. The streamline distribution throughout the channel and reservoirs are represented in Fig. 5 for different micro-channel (nozzle, diffuser). There are some flow recirculation at the corners of reservoirs due to the sudden contraction of the channel whereas the streamlines along the microchannel are parallel to both the boundaries.

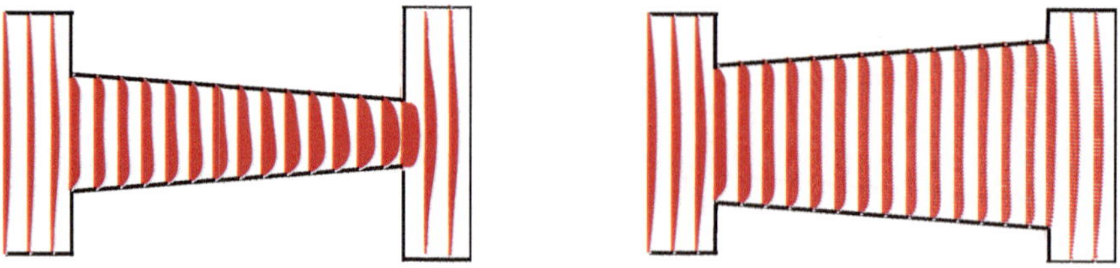

(a) (b)

Fig. 4 Velocity vector for reservoir connected **a** nozzle channel, **b** diffuser channel channel of electro-osmotic flow when $L/h_i = 20$ and $\kappa h_i = 30$

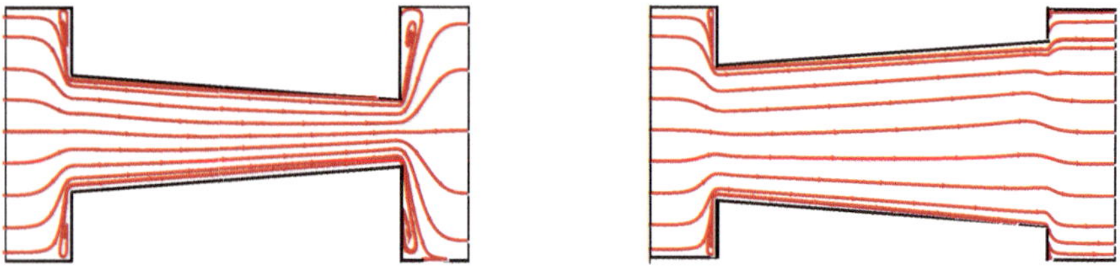

(a) (b)

Fig. 5 Streamline patterns for reservoir connected **a** nozzle channel, **b** diffuser channel channel of electro-osmotic flow when $L/h_i = 20$ and $\kappa h_i = 30$

4.2 Mixing Due to Different Geometry

Due to the low Reynolds number and high Peclet number in the micro/nano channel, EOF mixing is a difficult task. As a consequence, only the diffusion process may cause mixing, which takes longer time since the channel height is low. We require either a very long channel or surface patterned heterogeneity even for shorter length channels to get a high value of mixing efficiency in the downstream reservoir. In this section, the mixing of uncharged eluted species are described for various Debye-Hückel parameter (κh_i) and various channel slope/ conical angle. Figure 6 describes mixing efficiency for different Debye-Hückel parameters ($\kappa h_i = 2, 5, 10, 30, 50$) and we have compared the mixing efficiency among the two types of channel (reservoir connected microchannel) mentioned above. The figures depict that the mixing efficiency has a high jump when the flow reaches the channel inlet from reservoir. At the sharp corners of upstream reservoir, the flow can not pass through the sudden contraction of the channel and creates back flow that will create vortex at the reservoir corner. The same thing happens at channel outlet. As the flow area increases when the flow reaches downstream reservoir from channel, the mixing efficiency slows down a bit due to the sharp corners of reservoir, this also creates back flow and the mixing efficiency again enhanced. For thicker EDL case ($\kappa h_i = 2$), the mixing efficiency is very high and also the jump in mixing efficiency at channel inlet is high enough.

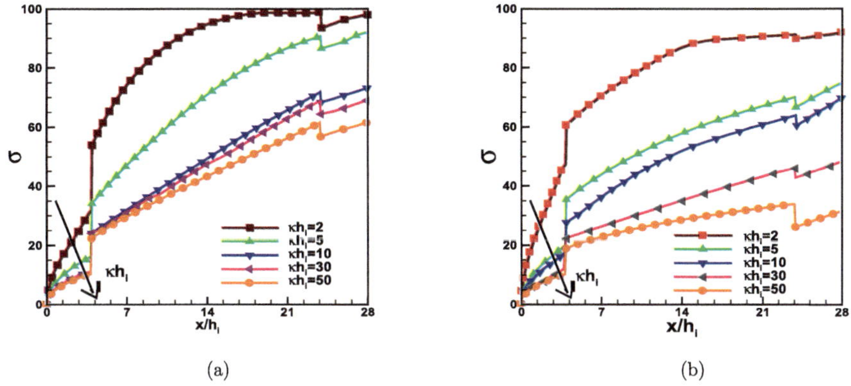

Fig. 6 Mixing efficiency with $\kappa h_i = 2,\ 5,\ 10,\ 30,\ 50$ for **a** nozzle, **b** diffuser microchannel connected with reservoirs when $\kappa h_i = 10$, $L/h_i = 10$, $m = 0.02$, $H_R/h_i = 4$, $E_0 = 10^4$ V/m

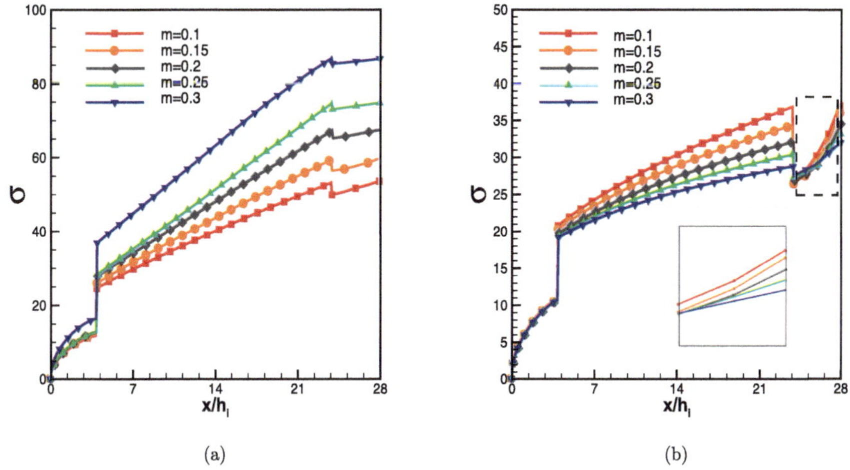

Fig. 7 Mixing efficiency with different conical angles (m=0.01, 0.015, 0.02, 0.025, 0.03) for **a** nozzle, **b** diffuser micro-channel connected with reservoirs when $\kappa h_i = 10$, $L/h_i = 10$, $m = 0.02$, $H_R/h_i = 4$, $E_0 = 10^4$ V/m

For overlapped EDL, the fluid experience very low momentum to drive the flow, so the average flow rate increases with increment of κh_i. The flow gets less retention time in the channel as flow rate increases. So the mixing efficiency decreases with the increment of Debye-Hückel parameter. Mixing efficiency is highest for nozzle channel for a fixed κh_i due to the converging nature of the channel at downstream region.

Figure 7 shows mixing efficiency for different conical angle of nozzle and diffuser channel when $\kappa h_i = 10$, $L/h_i = 10$ and $E_0 = 10^4$ V/m. For nozzle channel, the mixing efficiency increases as the conical angle increases. Because the fluid in the

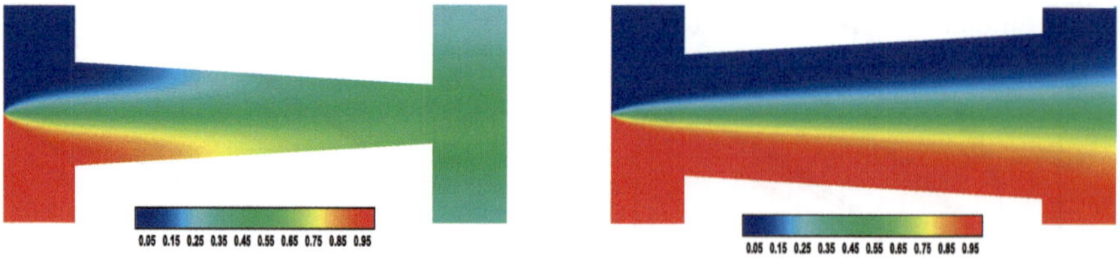

(a) (b)

Fig. 8 Contour plot of species transport mixing in **a** nozzle, **b** diffuser microchannel connected with reservoirs when $\kappa h_i = 10$, $L/h_i = 10$, $m = 0.02$, $H_R/h_i = 4$, $E_0 = 10^4$ V/m

microchannel needs to pass through very constricted region at inlet and outlet of the micro-channel. This phenomenon induces more retention time while the flow passes through the microchannel, thus improves mixing efficiency. But for diffuser channel, mixing efficiency gets worse for higher conical angle, as the fluid passes more rapidly through channel in this case. The mixing efficiency in nozzle channel for $m = 0.3$ is approximately 85% without adding any hurdles or surface heterogeneity at microchannel wall, which is sufficiently high. The corresponding concentration contour plots for different reservoir connected microchannel (nozzle, diffuser) are presented for $\kappa h_i = 10$, $L/h_i = 10$, $m = 0.02$ and $E_0 = 10^4$ V/m in Fig. 8.

5 Conclusion

In this paper, a Poisson-Nernst-Plank-based Navier-Stokes model is used to build a mathematical model for the electroosmotic flow of Newtonian fluid in a reservoir linked to variously shaped nozzles and diffusers positioned in a microdomain. Due to the existence of reservoirs, particular attention is paid to the end wall impact of the channel. When reservoirs are included in the computational domain, the contour plottings of potential, co-ions, counter-ions, and velocities are displayed to provide an overview of the changes in flow structures. According to our investigation, there is a negative pressure gradient impact at the channel's entry and exit that gradually alters the electrokinetic flow behaviour over the whole microchannel. When the channel length to height ratio is less, the result of the induced pressure is greater and is found to be at its highest when the reservoir is linked to the nozzle microchannel. The fluid streams went via different folded and expanded zones at the junction of the reservoirs and channel, according to the simulated findings, and the flow rate is also determined to be maximum for diffuser microchannel. Mixing efficiency decreases with increment of Debye-Hückel parameter (κh_i) for all the considered channel geometries. For thicker EDL, the fluid encounters less momentum, resulting lower flow rate and higher mixing efficiency. The mixing efficiency is maximum for nozzle microchannel with a fixed κh_i and electric field strength. Also the mixing efficiency vary differently for different conical angles or slope (m) of the microchannel. The mixing efficiency is proportionate with slope of the channel for a specific κh_i and channel length. It can be concluded that, mixing efficiency for nozzle channel with appropriate conical angle and debye-Hückel parameter, is sufficiently high without adding any hurdles or potential patch along the channel surface which can be used to gain a deeper understanding of the underlying framework of mixing in microfluidic systems.

Data Availability

The data that support the findings of this study are available within the article and its supplementary material.

References

1. Lim, Y.C., Kouzani, A.Z., Duan, W.: Lab-on-a-Chip : A Component View Microsystem Technologies, 16, pp. 1995–2015 (2010)
2. Hunten R.J.: Zeta potential in colloid science: principles and applications. Academic London, pp. 4–7 (1981)
3. Probstien, R.F.: Physiochemical Hydrodynamics. Butterworth Publishers, Boston (1999)
4. Lehnert, T., Gijs, M.A.M., Netzer, R., Bischoff, U.: Realization of hollow SiO_2 micronozzles for electrical measurments on living cells. Appl. Phys. Lett. **81** (2002)
5. Chen, Y.T., Kong, S.W., Wu, L.C., Lee, S.H.: Fabrication and investigation of PDMS micro-diffuser/nozzle. J. Mater. Process. Technol. **198**(1–3), 478–484 (2007)
6. Derakhshan, S., Beigzadeh, B., Rashidi, M., Pourrahmani, H.: Performance improvement and two phase flow study of piezoelecric micropump with tesla nozzle-diffuser microvalve. J. Appl. Fluid Mech. **12**(2), 341–350 (2019)
7. Yang, K.S., Chao, T.F., Chen, I.Y., Wang, C.C., Shyu, J.C.: A comparative study of nozzle-diffuser micropumps with novel valves. Molecules **17**, 2178–2187 (2012)
8. Ahmed, F., Kim, K.Y.: Analysis of a two layer nozzle and diffuser electroosmotic micromixer. Chem. Eng. Technol. **2164** (2018)
9. Yeh, H.C., Chang, C.C., Yang, R.J.: Electroosmotic pumping and ion concentration polarization based on conical nanopores. Phys. Rev. E **91**, 062302 (2015)
10. Basati, Y., Mohammadipour, O.R., Niazmand, H.: Numerical investigation and simultaneous optimization of geometry and zeta potential in electroosmotic mixing flows. Int. J. Heat Mass Trans. **133**, 786–799 (2019)
11. Mirbozorgi, S.A., Niazmand, H., Renksizbulut, M.: Electro osmotic flow in reservoi-connected flat microchannels with nonuniform zeta potential. J. Fluid Eng. **128**, 1133 (2006)
12. Yan, D.G., Yang, C., Huang, X.Y.: Effect of finite reservoir size on electroosmotic flow in microchannels. Microfluid Nanofluid **3**, 333–340
13. Mirbozorgi, S.A., Niazmand, H., Renksizbulut, M.: Streaming electric potential in pressure driven flows through reservoir connected microchannels. J. Fluid Eng. **129**, 1346 (2007)
14. Yang, R.J., Tseng, T.I., Chang, C.C.: End effects on electroosmotic flows in micro-channels. J. Micromech. Microeng. **15**, 254–262 (2005)
15. Eribol, P., Uguz, A.K.: Experimental investigation of electrohydrodynamic instabilities in micro channels. Eur. Phys. J. Spec. Top. **224**, 425–434 (2015)

16. Nayak, A.K., Haque, A., Banerjee, A., Weigand, B.: Flow mixing and electric potential effect of binary fluids in micro/nano channels. Colloids Surf. AâŁ¯: Physiochem. Eng. Aspects **512**, 145–157 (2017)
17. Chang, C.C., Yang, R.J.: Computation analysis of electrokinetically driver flow mixing in microchannels with patterned blocks. J. Micromach. Microeng. **14**, 550–558 (2004)
18. Chen, C.K., Cho, C.C.: Electrokinetically driven flow mixing in microchannels with wavy surface. J. Colloid Interf. Sci. **312**, 470–480 (2007)
19. Patankar, S.: Numerical Heat Transfer and Fluid Flow. Hemisphere Publishing Corporation (1980)
20. Zhao, C., Zholkovskij, E., Masliyah, J.H., Yang, C.: Analysis of electroosmotic flow of power-law fluids in a slit microchannel. J. Colloid Interface Sci. **326**(2), 503–510 (2008)
21. Lei, C., Conlisk, A.T.: Electroosmotic flow and particle transport in micro/nano nozzles and diffuser. Biomed Microdevices. **10**, 289–298 (2008)
22. Majhi, M., Nayak, A.K., Banerjee, A.: Enhanced electro-osmotic flow of power-law fluids in hydrophilic patterned nanochannel. J. Fluids Eng. **142**, 101201–1 (2020)
23. Haque, A., Nayak, A.K., Bhattacharyya, S.: Numerical study on ion transport and electro-convective mixing of power-law fluid in a heterogeneous micro-constrained channel. Phys. Fluids **33**, 122014 (2021)

Analytical Solution of Multi-species Pollutant Transport Problem Coupled with Linear Reactions and Additional Source/Sink Term

Manish Chaudhary and Hari Shankar Mahato

Abstract This article provides an analytical solution for the multi-species pollutant transport model in the presence of additional source/sink term in one-dimensional semi-infinite medium. The contaminant concentration of reactive species is assumed in liquid and solid phases, both. Model is formulated for constant dispersion coefficient and velocity profile. The effect of average porosity and bulk density of the porous formation is included with the retardation factor and distinct value of retardation factor is considered with each reactive species. Analytical solution is derived for three-species reactive system with first order decay rate by using Laplace transform technique along with different transformation processes and graphs are plotted for concentration variation by using MATLAB software.

Keywords Pollutant · Reactive species · Analytical solution

1 Introduction

Nowadays, the soil and groundwater contamination, which is arises due to various human activities like extensive use of fertilizers and pesticides in the irrigation processes, toxic substances from the industrial and chemical sites, venomous wastes from the septic tanks, has become the universal concern. Also, it is deep rooted threat for the human health and ecological environment. For the remediation measures of the groundwater pollution, it is quite vital to understand the pollutant transport behaviour in the groundwater and its physical properties. Mathematical modelling based on solution of advection-dispersion equation (ADE) under various types of analytical and numerical approaches plays an important role in the investigation of

M. Chaudhary (✉)
Department of Pure and Applied Mathematics, Alliance University, Bengaluru, Karnataka, India
e-mail: mani5014@gmail.com

H. S. Mahato
Department of Mathematics, Indian Institute of Technology, Kharagpur, West Bengal, India
e-mail: hsmahato@maths.iitkgp.ac.in

S. Bhattacharyya and H. S. Mahato (eds.), *Modeling, Analysis and Simulations of Multiscale Transport Phenomena*, Springer Proceedings in Mathematics & Statistics 491, https://doi.org/10.1007/978-981-96-3098-1_7

pollutant migration behaviour in the geological structures. An adequate number of analytical and numerical solutions have been presented in last few decades under the influence of various physical parameters and model conditions [1–11].

In the groundwater system, there are various types of reactive pollutants like chlorine solvents and radio-nuclides which produces their successive components under certain reactions or degradation processes. The mathematical models based on single species migration system are not capable to formulate and define the mass transformation from the initial (or parent) reactive species to its produced (or daughter) components of any type of degradable pollutants. This necessity leads to evolve the multi-species pollutant transport models for more accurate evaluation of transport phenomenon of reactive species. At very first, Cho [12] modelled the multi-species reactive system to simulate the nitrification process. Later, Clement [13] presented the computer codes to analyse the multi-species migration system. Various types of decomposition techniques and transformation methods were employed to get the solutions for such systems [14–19].

Most of the literature for analytical and numerical solutions of multi-species models are corresponding to liquid phase concentration, only. In this work, we have included the source/sink term with the multi-species model and expressed it in terms of solid phase concentration in the geological formations. Concept of linear isotherm is adopted with the model. The analytical solution is presented for the three-species reactive system. Model is formulated for distinct retardation factor value, depending on bulk density and porosity of the geological formation, for each reactive species. The solution procedure is adopted from the study Quezada et al. [20] in which the coupled set of equations are first transform into uncoupled form and then solution is formulated by using Laplace transform technique and various linear transformations. Proposed analytical solution of the considered model can describe the pollutant transport behaviour of multispecies reactive system in more realistic and generalized form.

2 Governing Equation of Multi-species Reactive System

The generalized form of 1-D reactive migration system, predicting the fate and movement of degradable pollutants with arbitrary number of pollutant species, which undergoes with first order degradation reaction network can be written as a system of partial differential equations as follows:

$$\frac{\partial C_i}{\partial t} = D\frac{\partial^2 C_i}{\partial x^2} - V\frac{\partial C_i}{\partial x} + k_{i-1}C_{i-1} - k_i C_i - G_i \ ; \quad i = 1, 2, 3, \ldots.., n \tag{1}$$

where, $C_i\left[ML^{-3}\right]$ represent the pollutant concentration of ith species in liquid phase; $D\left[L^2T^{-1}\right]$ and $V\left[LT^{-1}\right]$ are the dispersion and velocity components, respectively; $k_i\left[T^{-1}\right]$ represents the first order decay rate of the ith species; n is the total number of species in the reactive system; G_i represents the source/sink term which describe the rate at which the dissolved pollutant species is removed from the groundwater system and it is represented as the mass of solute per unit mass per unit volume.

Under sorption conditions in the porous medium, the term $G_i\left[ML^{-3}T^{-1}\right]$ is expressed as [6, 21]:

$$G_i = \frac{\rho}{\varphi}\frac{\partial S_i}{\partial t} \tag{2}$$

where, $\rho\left[ML^{-3}\right]$ and $\varphi\,[-]$ are the bulk density and average porosity of considered porous structure, respectively and $S_i\left[MM^{-1}\right]$ represents the pollutant concentration strength in solid phase.

Under the condition of isotherm, we have [6]:

$$\frac{\partial S_i}{\partial t} = \frac{\partial S_i}{\partial C_i}\frac{\partial C_i}{\partial t} \tag{3}$$

By using the value of G_i from relation given by Eq. (2)-(3), governing equation of considered model is transformed as follows:

$$\left(1 + \frac{\rho}{\varphi}\frac{\partial S_i}{\partial C_i}\right)\frac{\partial C_i}{\partial t} = D\frac{\partial^2 C_i}{\partial x^2} - V\frac{\partial C_i}{\partial x} + k_{i-1}C_{i-1} - k_i C_i\ ;\quad i = 1, 2, 3, ..., n \tag{4}$$

3 Problem of Three-Species Reactive System

In this study, three-species reactive migration system is considered with the concept of linear isotherm for all three species i.e., $S_i = K_{d_i}C_i\ ;\ i = 1, 2, 3$, where, $K_{d_i}\left[M^{-1}L^3\right]$ is the proportional parameter, also known as distribution coefficient for ith reactive pollutant species. Also, the dispersion and velocity components are assumed to be constant throughout the medium. Under these assumption, governing equation presenting the three species reactive system in semi-infinite porous structure is defined as:

$$R_1\frac{\partial C_1}{\partial t} = D\frac{\partial^2 C_1}{\partial x^2} - V\frac{\partial C_1}{\partial x} - k_1 C_1 \tag{5}$$

$$R_2\frac{\partial C_2}{\partial t} = D\frac{\partial^2 C_2}{\partial x^2} - V\frac{\partial C_2}{\partial x} - k_2 C_2 + k_1 C_1 \tag{6}$$

$$R_3\frac{\partial C_3}{\partial t} = D\frac{\partial^2 C_3}{\partial x^2} - V\frac{\partial C_3}{\partial x} - k_3 C_3 + k_2 C_2 \tag{7}$$

where, $R_i = 1 + \frac{\rho}{\varphi}K_{d_i}\ ;\ i = 1, 2, 3$ is the retardation factor of ith species.

The initial and boundary conditions for the considered three-species model are considered as follows:

$$C_i(x,t) = c_i^0; \quad x > 0, t = 0; \quad i = 1, 2, 3 \tag{8}$$

$$C_i(x,t) = c_{i0}; \quad x = 0, t > 0; \quad i = 1, 2, 3 \tag{9}$$

$$\lim_{x\to\infty} \frac{\partial C_i(x,t)}{\partial x} = 0; \quad t \geq 0; \quad i = 1, 2, 3 \tag{10}$$

4 Analytical Solution of the Model

The constructed model, given by Eqs. (5)–(7) can be written in matrix form as follows:

$$\begin{bmatrix} R_1 & 0 & 0 \\ 0 & R_2 & 0 \\ 0 & 0 & R_3 \end{bmatrix} \frac{\partial}{\partial t} \begin{Bmatrix} C_1 \\ C_2 \\ C_3 \end{Bmatrix} = D \frac{\partial^2}{\partial x^2} \begin{Bmatrix} C_1 \\ C_2 \\ C_3 \end{Bmatrix} - V \frac{\partial}{\partial x} \begin{Bmatrix} C_1 \\ C_2 \\ C_3 \end{Bmatrix} + \begin{bmatrix} -k_1 & 0 & 0 \\ k_1 & -k_2 & 0 \\ 0 & k_2 & -k_3 \end{bmatrix} \begin{Bmatrix} C_1 \\ C_2 \\ C_3 \end{Bmatrix} \tag{11}$$

Now taking the Laplace transformation on both sides of above Eq. (11) and using the considered initial condition given by Eq. (8) and thereafter by simplifying, Eq. (11) yields,

$$V \frac{\partial}{\partial x} \left\{\overline{C_i}\right\} - D \frac{\partial^2}{\partial x^2} \left\{\overline{C_i}\right\} = [B_1] \left\{\overline{C_i}\right\} + [B_2] \tag{12}$$

where, $\left\{\overline{C_i}\right\}$ is the column matrix representing the concentration strength of all three species, respectively, in the Laplacian domain, defined by

$$\overline{C_i}(x,p) = \int_0^\infty e^{-pt} C_i(x,t)\, dt \; ; \; i = 1, 2, 3 \tag{13}$$

and

$$[B_1] = \begin{bmatrix} -k_1 - pR_1 & 0 & 0 \\ k_1 & -k_2 - pR_2 & 0 \\ 0 & k_2 & -k_3 - pR_3 \end{bmatrix}; \; [B_2] = \begin{bmatrix} R_1 c_1^0 \\ R_2 c_2^0 \\ R_3 c_3^0 \end{bmatrix} \tag{14}$$

The corresponding boundary conditions are also transformed in Laplacian domain (say, $\overline{C}$—domain) as follows:

$$\overline{C_i}(x,p) = c_{i0} \frac{1}{p} = \overline{C_{i,0}}\,(say) \; ; \; x = 0 \tag{15}$$

$$\frac{\partial \overline{C_i}(x,p)}{\partial x} = 0; \quad x \to \infty, t \geq 0 \tag{16}$$

This Eq. (12) for the three-species model is in coupled form. To construct the analytical solution of the system, a linear transformation is applied to transform the model in uncoupled form which will be much convenient to solve.

For this, linear transformation is considered as follows [20]:

$$\{W_i\} = [A]^{-1}\left\{\overline{C_i}\right\} \quad \text{or} \quad \left\{\overline{\mathrm{C_i}}\right\} = [\mathrm{A}]\{\mathrm{W_i}\} \tag{17}$$

where, the columns of matrix $[A]$ are constructed by using the eigen vectors of the matrix $[B_1]$. Under this relationship, the Eq. (12) can be transformed in new domain, say W—domain, as follows:

$$D\frac{\partial^2}{\partial x^2}\begin{Bmatrix} W_1 \\ W_2 \\ W_3 \end{Bmatrix} - V\frac{\partial}{\partial x}\begin{Bmatrix} W_1 \\ W_2 \\ W_3 \end{Bmatrix} + \begin{bmatrix} -k_1 - pR_1 & 0 & 0 \\ 0 & -k_2 - pR_2 & 0 \\ 0 & 0 & -k_3 - pR_3 \end{bmatrix}\begin{Bmatrix} W_1 \\ W_2 \\ W_3 \end{Bmatrix} = \begin{bmatrix} -R_1c_1^0 \\ \alpha_1R_1c_1^0 - R_2c_2^0 \\ -\alpha_4R_1c_1^0 + \alpha_3R_2c_2^0 - R_3c_3^0 \end{bmatrix} \tag{18}$$

where,

$$\begin{aligned} \alpha_1 &= \frac{k_1}{p(R_2-R_1)+k_2-k_1}\ ; \quad \alpha_2 = \frac{k_1k_2}{\{p(R_3-R_1)+k_3-k_1\}\{p(R_2-R_1)+k_2-k_1\}} \\ \alpha_3 &= \frac{k_2}{p(R_3-R_2)+k_3-k_2}\ ; \quad \alpha_4 = \alpha_1\alpha_3 - \alpha_2 \end{aligned} \tag{19}$$

This Eq. (18) presents the system of equations in uncoupled form similar to the standard form of advection-dispersion equations in W—domain.

Also, the corresponding boundary conditions are constructed in W—domain by using the algorithm [20]:

$$W_{i,0} = \overline{C_{i,0}} - \sum_{j=1}^{n-1} A_{ij}W_{j,0} \tag{20}$$

where, A_{ij} are the terms of matrix $[A]$.

Now, by solving these uncoupled equations in W—domain, thereafter transforming the results into $\overline{C}$—domain and then by using inverse Laplace transform, the solution for all three species is constructed as follows:

Solution for first species:

$$C_1(x,t) = \left[c_{10}P_{11} - c_1^0P_{21}\right] + c_1^0P_{81} \tag{21}$$

Solution for second species:

$$\begin{aligned} C_2(x,t) &= \tfrac{c_{10}k_1}{(k_2-k_1)}[P_{11}-P_{31}] - \tfrac{c_1^0 k_1}{(R_2-R_1)}\tfrac{1}{(\beta_1^2+\psi_1^2)}[P_{21}-P_{31}] \\ &+ \tfrac{c_1^0 k_1}{(R_2-R_1)}\tfrac{1}{(\beta_1^2+\psi_1^2)}[P_{81}-P_{91}] + c_{20}[P_{12}] - \tfrac{c_{10}k_1}{(k_2-k_1)}[P_{12}-P_{32}] - c_2^0[P_{22}] \\ &+ \tfrac{R_1 c_1^0 k_1}{R_2(R_2-R_1)}\tfrac{1}{(\beta_1^2+\psi_2^2)}[P_{22}-P_{32}] + c_2^0 P_{82} - \tfrac{R_1 c_1^0 k_1}{R_2(R_2-R_1)}\tfrac{1}{(\beta_1^2+\psi_2^2)}[P_{82}-P_{91}] \end{aligned} \quad (22)$$

Solution for third species:

$$\begin{aligned} C_3(x,t) &= \tfrac{k_1k_2}{(R_2-R_1)(R_3-R_1)}\begin{bmatrix} c_{10}\{M_1P_{11}+M_2P_{31}-M_3P_4\} \\ -c_1^0\{M_4(P_{21}-P_{81})+M_5(P_{31}-P_{91})-M_6(P_4-P_{92})\}\end{bmatrix} \\ &+ \tfrac{k_1k_2}{(R_2-R_1)(R_3-R_2)}\begin{bmatrix} -c_{10}\{M_7P_{12}+M_8P_{32}-M_9P_6\} \\ +\tfrac{R_1}{R_2}c_1^0\{M_{10}(P_{22}-P_{82})+M_{11}(P_{32}-P_{91})-M_{12}(P_6-P_{93})\}\end{bmatrix} \\ &- \tfrac{c_2^0k_2}{(R_3-R_2)}\tfrac{1}{(\beta_3^2+\psi_2^2)}[\{P_{22}-P_6\}-\{P_{82}-P_{93}\}] + \tfrac{c_{20}k_2}{(k_3-k_2)}[\{P_{12}-P_6\}-\{P_{13}-P_7\}] \\ &+ \tfrac{R_2c_2^0k_2}{R_3(R_3-R_2)}\tfrac{1}{(\beta_3^2+\psi_3^2)}[\{P_{23}-P_7\}-\{P_{83}-P_{93}\}] + c_{30}[P_{13}] - c_3^0[P_{23}] + c_3^0[P_{83}] \\ &- \tfrac{R_1c_1^0k_1k_2}{R_3(R_3-R_2)(R_3-R_1)}[M_{13}\{P_{23}-P_{83}\}+M_{14}\{P_5-P_{92}\}-M_{15}\{P_7-P_{93}\}] \\ &+ \tfrac{c_{10}k_1k_2}{(R_3-R_2)(R_3-R_1)}[P_{13}+P_5-P_7] \end{aligned} \quad (23)$$

where, values of all unknowns are as follows:

$$P_{1i} = \frac{1}{2}\exp\left(\frac{Vx}{2D}\right)\begin{bmatrix} \exp(-\lambda_i\mu_i x)\,erfc\left(\frac{\lambda_i x}{2\sqrt{t}}-\mu_i\sqrt{t}\right) \\ +\exp(\lambda_i\mu_i x)\,erfc\left(\frac{\lambda_i x}{2\sqrt{t}}+\mu_i\sqrt{t}\right)\end{bmatrix};\ i=1,2,3 \quad (24)$$

$$P_{2i} = \frac{1}{2}\exp\left(\frac{Vx}{2D}\right)\begin{bmatrix} \exp(\psi_i^2-\zeta_i\lambda_i x)\,erfc\left(\frac{\lambda_i x}{2\sqrt{t}}-\zeta_i\sqrt{t}\right) \\ +\exp(\psi_i^2+\zeta_i\lambda_i x)\,erfc\left(\frac{\lambda_i x}{2\sqrt{t}}+\zeta_i\sqrt{t}\right)\end{bmatrix};\ i=1,2,3 \quad (25)$$

$$P_{3i} = \frac{1}{2}\exp\left(\frac{Vx}{2D}\right)\begin{bmatrix} \exp(-\beta_1^2-\delta_i\lambda_i x)\,erfc\left(\frac{\lambda_i x}{2\sqrt{t}}-\delta_i\sqrt{t}\right) \\ +\exp(-\beta_1^2+\delta_i\lambda_i x)\,erfc\left(\frac{\lambda_i x}{2\sqrt{t}}+\delta_i\sqrt{t}\right)\end{bmatrix};\ i=1,2 \quad (26)$$

$$P_4 = \frac{1}{2}\exp\left(\frac{Vx}{2D}\right)\begin{bmatrix} \exp(-\beta_2^2-\delta_3\lambda_1 x)\,erfc\left(\frac{\lambda_1 x}{2\sqrt{t}}-\delta_3\sqrt{t}\right) \\ +\exp(-\beta_2^2+\delta_3\lambda_1 x)\,erfc\left(\frac{\lambda_1 x}{2\sqrt{t}}+\delta_3\sqrt{t}\right)\end{bmatrix} \quad (27)$$

$$P_5 = \frac{1}{2}\exp\left(\frac{Vx}{2D}\right)\begin{bmatrix} \exp(-\beta_2^2-\delta_6\lambda_3 x)\,erfc\left(\frac{\lambda_3 x}{2\sqrt{t}}-\delta_6\sqrt{t}\right) \\ +\exp(-\beta_2^2+\delta_6\lambda_3 x)\,erfc\left(\frac{\lambda_3 x}{2\sqrt{t}}+\delta_6\sqrt{t}\right)\end{bmatrix} \quad (28)$$

$$P_6 = \frac{1}{2}\exp\left(\frac{Vx}{2D}\right)\begin{bmatrix} \exp\left(-\beta_3^2 - \delta_4\lambda_2 x\right) erfc\left(\frac{\lambda_2 x}{2\sqrt{t}} - \delta_4\sqrt{t}\right) \\ +\exp\left(-\beta_3^2 + \delta_4\lambda_2 x\right) erfc\left(\frac{\lambda_2 x}{2\sqrt{t}} + \delta_4\sqrt{t}\right) \end{bmatrix} \tag{29}$$

$$P_7 = \frac{1}{2}\exp\left(\frac{Vx}{2D}\right)\begin{bmatrix} \exp\left(-\beta_3^2 - \delta_5\lambda_3 x\right) erfc\left(\frac{\lambda_3 x}{2\sqrt{t}} - \delta_5\sqrt{t}\right) \\ +\exp\left(-\beta_3^2 + \delta_5\lambda_3 x\right) erfc\left(\frac{\lambda_3 x}{2\sqrt{t}} + \delta_5\sqrt{t}\right) \end{bmatrix} \tag{30}$$

$$P_{8i} = \exp\left(\psi_i^2 t\right)\ ;\ P_{9i} = \exp\left(-\beta_i^2 t\right)\ ;\ i = 1, 2, 3 \tag{31}$$

$$\begin{aligned} M_1 &= \tfrac{1}{\beta_1^2\beta_2^2};\ M_2 = \tfrac{1}{\beta_1^2(\beta_1^2-\beta_2^2)};\ M_3 = \tfrac{1}{\beta_2^2(\beta_1^2-\beta_2^2)};\ M_4 = \tfrac{1}{(\beta_1^2+\psi_1^2)(\beta_2^2+\psi_1^2)}; \\ M_5 &= \tfrac{1}{(\beta_1^2+\psi_1^2)(\beta_1^2-\beta_2^2)};\ M_6 = \tfrac{1}{(\beta_2^2+\psi_1^2)(\beta_1^2-\beta_2^2)};\ M_7 = \tfrac{1}{\beta_1^2\beta_3^2}; \\ M_8 &= \tfrac{1}{\beta_1^2(\beta_1^2-\beta_3^2)};\ M_9 = \tfrac{1}{\beta_3^2(\beta_1^2-\beta_3^2)};\ M_{10} = \tfrac{1}{(\beta_1^2+\psi_2^2)(\beta_3^2+\psi_2^2)}; \\ M_{11} &= \tfrac{1}{(\beta_1^2+\psi_2^2)(\beta_1^2-\beta_3^2)};\ M_{12} = \tfrac{1}{(\beta_3^2+\psi_2^2)(\beta_1^2-\beta_3^2)};\ M_{13} = \tfrac{1}{(\beta_2^2+\psi_3^2)(\beta_3^2+\psi_3^2)}; \\ M_{14} &= \tfrac{1}{(\beta_2^2+\psi_3^2)(\beta_2^2-\beta_3^2)};\ M_{15} = \tfrac{1}{(\beta_3^2+\psi_3^2)(\beta_2^2-\beta_3^2)} \end{aligned} \tag{32}$$

$$\begin{aligned} \lambda_i &= \sqrt{\tfrac{R_i}{D}};\ \mu_i = \sqrt{\tfrac{V^2}{4R_i D} + \tfrac{k_i}{R_i}};\ \psi_i = \sqrt{-\tfrac{k_i}{R_i}};\ \zeta_i = \sqrt{\mu_i^2 + \psi_i^2};\ \ i = 1, 2, 3 \\ \beta_1 &= \sqrt{\tfrac{k_2-k_1}{R_2-R_1}};\ \beta_2 = \sqrt{\tfrac{k_3-k_1}{R_3-R_1}};\ \beta_3 = \sqrt{\tfrac{k_3-k_2}{R_3-R_2}}; \\ \delta_1 &= \sqrt{\mu_1^2 - \beta_1^2};\ \delta_2 = \sqrt{\mu_2^2 - \beta_1^2};\ \delta_3 = \sqrt{\mu_1^2 - \beta_2^2}; \\ \delta_4 &= \sqrt{\mu_2^2 - \beta_3^2};\ \delta_5 = \sqrt{\mu_3^2 - \beta_3^2};\ \delta_1 = \sqrt{\mu_3^2 - \beta_2^2} \end{aligned} \tag{33}$$

By using above developed analytical solutions for all three reactive species, concentration attenuation is graphically observed for the finite spatial domain, $0 \le x\ (cm) \le 150$ by using corresponding MATLAB codes and interpretation is made for following set of input data [1, 20]:

$V = 1\,\mathrm{cm/hr}$; $\mathrm{D} = 0.18\,\mathrm{cm^2/hr}$; $c_1^0 = c_2^0 = c_3^0 = 0$; $c_{10} = 1\,\mathrm{mg/l}$; $c_{20} = 1\,\mathrm{mg/l}$; $c_{30} = 0$; $k_1 = 0.01\,/\mathrm{hr}$; $k_2 = 0.1\,/\mathrm{hr}$; $k_3 = 0.2\,/\mathrm{hr}$; $K_{d_1} = 0.01$; $K_{d_2} = 0.1$; $K_{d_3} = 0.3$.

Also, bulk density and average porosity of the medium is considered to be: $\rho = 2.49$; $\phi = 0.3$(Sandstone medium) [22].

Figure 1 presents the concentration profile for all three reactive species for fixed time period $t = 100\,\mathrm{hr}$ corresponding to the analytical solution formulated as Eq. (21), (22) and (23), respectively. A prominent variation in the concentration attenuation is graphically visualized for the reactive species. The concentration strength of the parent reactive species (species-1) is observed to be higher at each point of the domain as compare to its daughter components (species-2) and same phenomenon is observed for species-3 due to the sorption of the parent reactive component. Figure 2, 3 and 4 illustrate the surface plot for the concentration strength of all three reactive species, respectively. It can be observed that the concentration strength for species-1

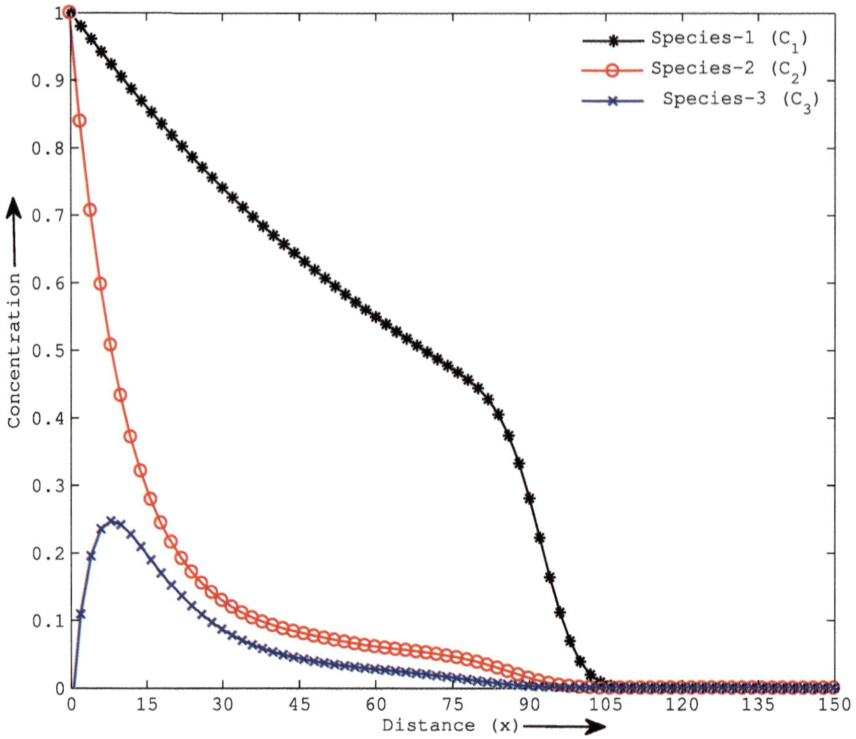

Fig. 1 Plot of Concentration vs Distance for all three reactive species for fixed time period $t = 100\,\text{hr}$

and species-2 decreases with positional value throughout the medium but for reactive species-3, the concentration strength increases initially with distance and after attaining its peak concentration strength, it decreases asymptotically with position. For all three species, concentration value attains its minimum strength at the end of the domain. Also, the concentration strength for all three reactive species is observed to be higher for the high temporal value as depicted from the figures.

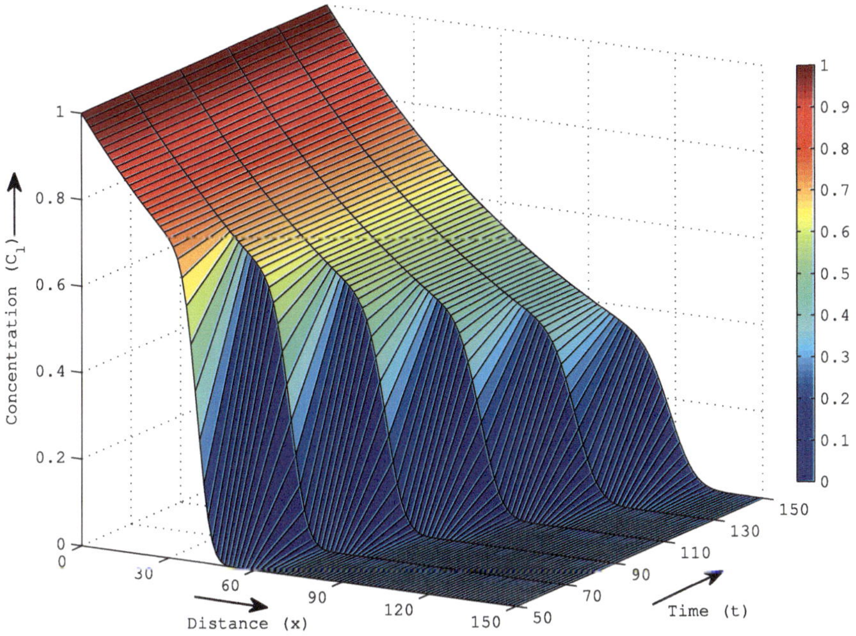

Fig. 2 Surface evolution species-1 for sandstone medium

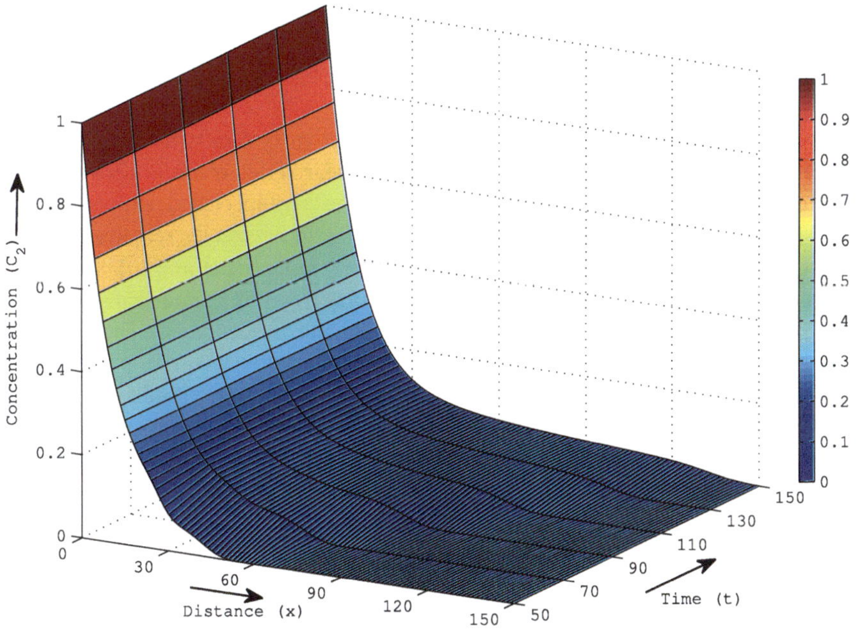

Fig. 3 Surface evolution species-2 for sandstone medium

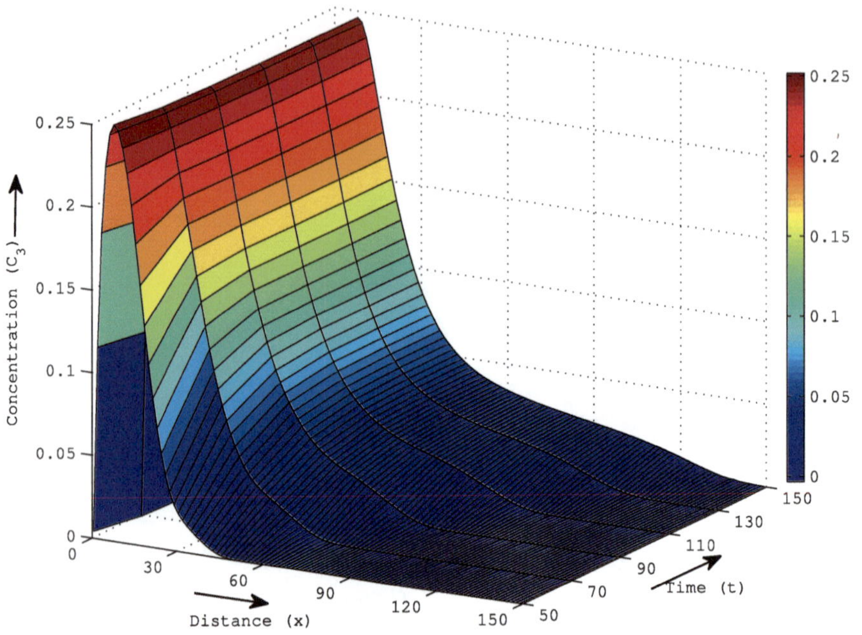

Fig. 4 Surface evolution species-3 for sandstone medium

5 Summary and Conclusions

One-dimensional model for the multi-species reactive system is formulated in the presence of additional source/sink term under linear series reaction network. The concept of linear isotherm is incorporated with all reactive species to remove the solid phase concentration. The analytical solution for the three-species reactive model is constructed by using the methodology proposed by Quezada et al. [20], where, the coupled system is transformed into uncoupled form by using certain transformation and solution is constructed by using Laplace transform technique. The proposed model for multi-species system is helpful to analyse the concentration attenuation of the reactive species in any particular geological formations with their respective bulk density and average porosity. Also, the model can further enhanced for the variable dependent transport parameter and also for the generalized solution for arbitrary number of reactive species. One may also use such type of modelling for various remedial measures in terms of removing the reactive pollutant species in the geological formations.

References

1. Van Genuchten, M.T.: Analytical solutions of the one-dimensional convective-dispersive solute transport equation (No. 1661). US Department of Agriculture, Agricultural Research Service
2. Barry, D.A., Sposito, G.: Analytical solution of a convection-dispersion model with time-dependent transport coefficients. Water Resour. Res. **25**(12), 2407–2416 (1989)
3. Yeh, G.T., Tripathi, V.S.: A model for simulating transport of reactive multispecies components: model development and demonstration. Water Resour. Res. **27**(12), 3075–3094 (1991)
4. Wexler, E.J.: Analytical solutions for one-, two-, and three-dimensional solute transport in ground-water systems with uniform flow. US Government Printing Office (1992)
5. Lunn, M., Lunn, R.J., Mackay, R.: Determining analytic solutions of multiple species contaminant transport, with sorption and decay. J. Hydrol. **180**(1–4), 195–210 (1996)
6. Batu, V.: Applied Flow and Solute Transport Modeling in Aquifers: Fundamental Principles and Analytical and Numerical Methods. CRC Press (2005)
7. Singh, M.K., Singh, P., Singh, V.P.: Analytical solution for two-dimensional solute transport in finite aquifer with time-dependent source concentration. J. Eng. Mech. **136**(10), 1309–1315 (2010)
8. Chen, J.S., Ni, C.F., Liang, C.P., Chiang, C.C.: Analytical power series solution for contaminant transport with hyperbolic asymptotic distance-dependent dispersivity. J. Hydrol. **362**(1–2), 142–149 (2008)
9. Suk, H.: Developing semi-analytical solutions for multispecies transport coupled with a sequential first-order reaction network under variable flow velocities and dispersion coefficients. Water Resour. Res. **49**(5), 3044–3048 (2013)
10. Thakur, C.K., Chaudhary, M., Van Der Zee, S.E.A.T.M., Singh, M.K.: Two-dimensional solute transport with exponential initial concentration distribution and varying flow velocity. Pollution **5**(4), 721–737 (2019)
11. Chaudhary, M., Thakur, C.K., Singh, M.K.: Analysis of 1-D pollutant transport in semi-infinite groundwater reservoir. Environ. Earth Sci. **79**(1), 1–23 (2020)
12. Cho, C.M.: Convective transport of ammonium with nitrification in soil. Canadian J. Soil Sci. **51**(3), 339–350 (1971)
13. Clement, T.P.: A modular computer code for simulating reactive multi-species transport in 3-dimensional groundwater systems (No. PNNL-11720; EW4010). Pacific Northwest National Lab., Richland, WA (US) (1999)
14. Sun, Y., Clement, T.P.: A decomposition method for solving coupled multi-species reactive transport problems. Trans. Porous Media **37**(3), 327–346 (1999)
15. Sun, Y., Petersen, J.N., Clement, T.P., Skeen, R.S.: Development of analytical solutions for multispecies transport with serial and parallel reactions. Water Resour. Res. **35**(1), 185–190 (1999)
16. Slodicka, M., Balazova, A.: Singular value decomposition method for multi-species first-order reactive transport with identical decay rates. Trans. Porous media **73**(2), 161–172 (2008)
17. Slodicka, M., Balazova, A.: Decomposition method for solving multi-species reactive transport problems coupled with first-order kinetics applicable to a chain with identical reaction rates. J. Comput. Appl. Math. **234**(4), 1069–1077 (2010)
18. Srinivasan, V., Clement, T.P.: Analytical solutions for sequentially coupled one-dimensional reactive transport problems-Part I: mathematical derivations. Adv. Water Resour. **31**(2), 203–218 (2008)
19. Chaudhary, M., Singh, M.K.: Study of multispecies convection-dispersion transport equation with variable parameters. J. Hydrol. **591**, 125562 (2020)
20. Quezada, C.R., Clement, T.P., Lee, K.K.: Generalized solution to multi-dimensional multi-species transport equations coupled with a first-order reaction network involving distinct retardation factors. Adv. Water Resour. **27**(5), 507–520 (2004)
21. van Genuchten, M.T.: Mass Transfer Studies in Sorbing Porous Media, Ph.D. dissertation, New Mexico State University, Las Cruces, New Mexico, December, 148 pp (1974)
22. Manger, G.E.: Porosity and bulk density of sedimentary rocks (1963)

A Basic Theoretical Framework with a Incline to the Fermi Energy for Low-Dimensional (L-D) Nano-materials

Subhamoy Singha Roy and Nibhash Kumar

Abstract In the modern era, low-dimensional elements with quantum incarceration in multi-dimensions, such as narrow films, displaced layers, quantum wires, and dots, have been studied by Metal Organic Chemical Vapor Deposition, Molecular-Beam Epitaxy, Fine-Line Lithography, and other analyses. These elements not only help identify new phenomena in polymeric electronics but also prompt the submission of devices in heterostructure-based count-less nano materials, which are currently being assessed due to the enhancement of carrier mobility and ex-digital networks found by such quantum captive systems, visual regulators, and other devices. In line with the recently assessed electron dispersion law, this paper aims to comprehend the existence of the Fermi–Dirac distribution function in degenerate semiconductors that form band tails $[f_2]$. The well-defined Fermi–Dirac function $[f_1]$, is primarily attracted to carrier degeneracy conditions. Since I support degenerate semiconductors and electrical devices, the Fermi–Dirac distribution function inquiry is looking at whether the transport coefficient or electron dynamics will alter.

Keywords Fermi-Dirac distribution function · Degenerate semiconductors · Low-dimensional elements

1 Introduction

In light of this, I would like to show and explain that band tail appearance in semiconductors that have degenerated is a reality that requires further research and is frequently explained as the impurity band projecting onto the conduction and valance bands [1–3]. The semi-classical hypothesis in semiconductors due to band tailing through unbroken bands of parabolic energy has been separately expanded by Kane and Bonch Bruevich [4–6]. Kane's semi-classical model offered an explanation for

S. S. Roy (✉) · N. Kumar
Department of Physics, JIS College of Engineering (Autonomous), West Bengal University of Technology, Kalyani, Nadia, India
e-mail: subhomay.singharoy@jisacolege.ac.in

S. Bhattacharyya and H. S. Mahato (eds.), *Modeling, Analysis and Simulations of Multiscale Transport Phenomena*, Springer Proceedings in Mathematics & Statistics 491, https://doi.org/10.1007/978-981-96-3098-1_8

the investigational conclusion of tunneling and visible immersion boundaries within strongly doped semiconductors by taking into account the parabolic density-of-states (DOS) [7]. In contrast to Kane, I have employed the naturalistic representation of the electron's kinetic energy disparity with the local point in space coordinates in this study [8]. Along with the research findings for the Fermi energy, the mean of the kinetic energy is computed over the whole variation zone, assuming a Gaussian type potential energy [9, 10].

2 Theoretical Background

The Fermi–Dirac theory states that the density function provides the likelihood of a fermion in thermal equilibrium with a large reservoir catching an energy level. Fermion classification is frequently applied to particles with half-integer spin $(1/2, 3/2, 5/2, \, etc.)$. According to [11], the function of the predictable Fermi–Dirac distribution is given.

$$f_1(E) = \frac{1}{1 + \exp\left[\{(E) - (E_f)\} / (k_B T)\right]} \tag{1}$$

where, E_f is the equivalent Fermi-energy, the Boltmann constant is represented by k_B, this is the temperature, T and E is the entire electron energy because deliberate from the conduction band border in a perpendicularly upward trend. By integrating the product of the density of states with the probability density function across all conceivable states, one may derive the generic expression of the carrier density in a semiconductor. The integral for electrons in the transmission group is calculated from the bottom, designated E_c, to the top, and may be expressed as

$$n = \int_{E_C}^{\text{top of the conduction band}} n(E)dE = \int_{E_C}^{\text{top of the conduction band}} g_c(E)f(E)dE \tag{2}$$

Fermi function is denoted by $f(E)$.

$$n = \int_{E_C}^{\infty} g_c(E)f(E)dE \tag{3}$$

It is now possible to write the 3-dimensional semiconductor as

$$n = \int_{E_C}^{\infty} \frac{8\pi\sqrt{2}}{h^3} m_e^{*3/2} \sqrt{E - E_C} \frac{1}{1 + e^{\frac{E - E_F}{kT}}} dE \tag{4}$$

$$p = \int_{-\infty}^{E_V} g_v(E)[1 - f(E)]dE \quad (5)$$

and

$$p = \int_{-\infty}^{E_V} \frac{8\pi\sqrt{2}}{h^3} m_e^{*3/2} \sqrt{E_V - E} \frac{1}{1 + e^{\frac{E_F - E}{kT}}} dE \quad (6)$$

Right now the electron energy at a certain location $(\vec{r})$ in the degenerate semiconductors that produce band tails is given as

$$E = \left\{\hbar^2 k^2 / \left(2m_c^*\right)\right\} + V(\overline{r}) \quad (7)$$

where $\hbar = \left(h/2\pi\right)$, The Planck constant is represented by h, the electron's wave vector by k, the successful electron mass at the conduction band's boundary by m_c^*, and the impurity potential at a specific local site $(\vec{r})$ by $V(r)$.

One way to express the potential energy is as [5–7].

$$F(V) = \left\{\frac{1}{\sqrt{\left(\pi\eta_e^2\right)}}\right\} \exp\left(-\frac{V^2}{\eta_e^2}\right) \quad (8)$$

The screening potential is represented by η_e. I would like to point out that other writers have developed the Gaussian function for the impurity potential distribution [4–6].

Averaging the local kinetic energy fluctuation, as shown by, yields the average kinetic energy of the entire system.

$$\int_{-\infty}^{E} [\{(E - V) \times F(V)\} \times dV] = \left\{\frac{\left(\hbar^2 k^2\right)}{2m_c}\right\} \times \int_{-\infty}^{E} F(V) \times dV \quad (9)$$

Given that the function $F(V)$ represents the Gaussian allocation and that the restrictions of V extend as of –inf *inity* to +inf *inity*, the accurate plane of Eq. (9).

I raise the higher bound of V beginning $V \to E$ to $V \to$ inf *inity*.

Therefore as of Eq. (9) I can inscribe

$$\left\{\frac{\left(\hbar^2 \times k^2\right)}{\left(2 \times m_c^*\right)}\right\} = \{\gamma(E, \eta_e)\} \quad (10)$$

where

$$\gamma(E,\eta_e)=\left\{\left(\frac{\eta_e}{2\sqrt{\pi}}\right)\exp\left(-\frac{E^2}{\eta_e^2}\right)\right\}+\left(\frac{1}{2E}\right)\left[1+Erf\left(\frac{E}{\eta_e}\right)\right]$$

where the error function is represented by $Erf\left(E/\eta_e\right)$.

The usual way that $\{V(\overline{r})\}$ affects $E\left(\langle E'\rangle\right)$ is like follows

$$\left\langle E'\right\rangle=\frac{\int_{-\infty}^{E}\left[\{(E')\times F(V)\}\times dV\right]}{\int_{-\infty}^{\infty}\left[F(V)\times dV\right]} \tag{11}$$

Equations (10) and (11) can be expressed as

$$\left\langle (E')\right\rangle=\{\gamma(E,\eta_e)\} \tag{12}$$

Since $\{\gamma(E,\eta_e)\}$ exhibits band tailing effects, as I have demonstrated in Eq. (10), $\left[\langle (E')\rangle\right]$ is present for unconstructive standards of E.

The possibility for contamination screening budding $[\eta_e]$, is provided like

$$\eta_e=\left(\frac{e^2}{\varepsilon_d}\right)\left\{(4.\pi.N_i)/K_D\right\}^{1/2} \tag{13}$$

where

$$N_i=\left(\frac{1}{3\pi^2}\right)\left(\frac{2m_c^*}{\hbar^2}\right)^{3/2}\left\{\gamma^{3/2}(\overline{E}_f,\eta_e)\right\} \tag{14}$$

where $\left[\left\langle E'\right\rangle\right]$ is the averaging energy in semiconductors that are degenerate and form band tails and $\overline{E}_f$ is the Fermi energy that corresponds to that energy

$$\text{and}\quad K_D^2=\left\{\left(\frac{e^2}{\varepsilon_d}\right)\left(\frac{1}{4\pi^2}\right)\left(\frac{2m_c^*}{\hbar^2}\right)^{3/2}\right\}\left[1+Erf\left(\overline{E}_f/\eta_e\right)\right] \tag{15}$$

Band tails are absent [11], thus I have,

$$N_i=\left(\frac{1}{3\pi^2}\right)\left(\frac{2m_c^*}{\hbar^2}\right)^{3/2}E_f^{3/2} \tag{16}$$

Comparing Eqs. (13) and (16), find as

$$\eta_e\rightarrow zero,\quad \{\gamma(AverageE_F,\eta_e)\}=\left(E_f\right) \tag{17}$$

Therefore, the Statistics of Fermi–Dirac for degenerate semiconductors and have Gaussian band tails may be expressed as

$$f_2\left(\left\langle\left(E'\right)\right\rangle, \left(\overline{E}_f\right), Temp.\right) = \left[1 + e^{\left(\left\langle\left(E'\right)\right\rangle - \overline{E}_f\right)/\left(k_B T\right)}\right]^{-1} \quad (18)$$

where

$$\left\langle\left(E'\right)\right\rangle = \left\{\frac{1}{(2\pi)^{1/2}}\right\}\eta_e \exp\left(-E^2/\eta_e^2\right) + \left(\frac{1}{2}\right)\left\{E\left[1 + Erf\left(E/\eta_e\right)\right]\right\} \quad (19)$$

Equation (19) suggests that when band tails are absent, $\eta_e \rightarrow zero$, $\left\langle\left(E'\right)\right\rangle = Energy$ and $\left(\overline{E}_f\right) = Fermi.Energy$, correspondingly. In this scenario, when band tails are missing, the Fermi–Dirac distribution function can be easy from Eqs. (18) to (1). Equation (19) makes clear that $\langle E'\rangle$ and E are not the same. Furthermore, the typical energy of an electron $\langle(E')\rangle$ of the conduction band electron differs significantly from E because of the screening of the impurity potential η_e. Consequently, we discover that $f_1\left(E, E_f, T\right)$ and $f_2\left(\left\langle\left(E'\right)\right\rangle, \overline{E}_f, T\right)$ are not the same using Eqs. (1) and (18) [12, 13]. Thus, in addition to the standard E, E_f and T dependences, I can deduce that the $f(E)$ for extrinsic semiconductors generating band tails has an extra concentration dependence due to screening potential.

3 Discussion

For the sake of algebraic calculation, Fig. 1 displays the Fermi function, the parabolic DOS function within $E_c = 0$, the creation of the two, which measures the electron density relative to energy and volume $n(E)$.

The defaced area under the curve corresponds to the integral.

Enchanting n-GaAs for instance, in addition to the specifications $m^* = 0.0670m_o$, $\varepsilon_d = 99.0610\varepsilon_o$, $E = 050.0\,\text{meV}$ $Er\left[f(x)\right] = \left\{\left[1 - \left(a \times t + b \times t^2 + c \times t^3\right)\right]\left[\exp\left(-x^2\right)\right]\right\}$ [14, 15] and $T = 04.20K$, where $a = 0.350$, $b = (-0.095870)$, $c = 0.747850$, $t = \left[\{1 \div (1 + px)\}\right]$ and

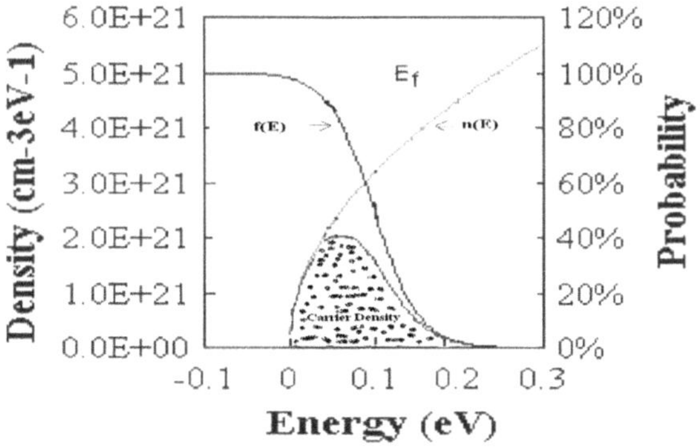

Fig. 1 The carrier density is equal to the cross-hatched area under the curve, and the probability of tenancy $f(E)$ and density per unit energy $n(E)$ are the open to the elements quantities

$p = +0.470470$, f_1 and f_2 as functions of carrier concentration are displayed in Fig. 2.

Through stepladder happening at a inferior N_i for f_1 and next to greater levels of carrier concentration in support of f_2, Fig. 2 illustrates how f_1 and f_2 varies virtually similar to a rung function connecting the standards zero and one. This implies that because of the belongings of carrier concentration, the allocation functions f_1 and f_2 are extra relevant in the container of f_2 than that of f_1 [16, 17].

Thus, when the semiconductor becomes degenerate due to substantial doping and develops band tails, I may deduce so as to the Fermi–Dirac distribution function f_2 is additional efficient in contrast to f_1 at advanced principles of carrier concentration. I would want to note, therefore, that the Fermi–Dirac distribution function f_2, as given by Eq. (18), rather than f_1, as given by Eq. (1), ought to be used to degenerate semiconductors when band tails are present. My conclusion is that cautious discharge of the experiment is necessary to obtain the distribution function values that are close

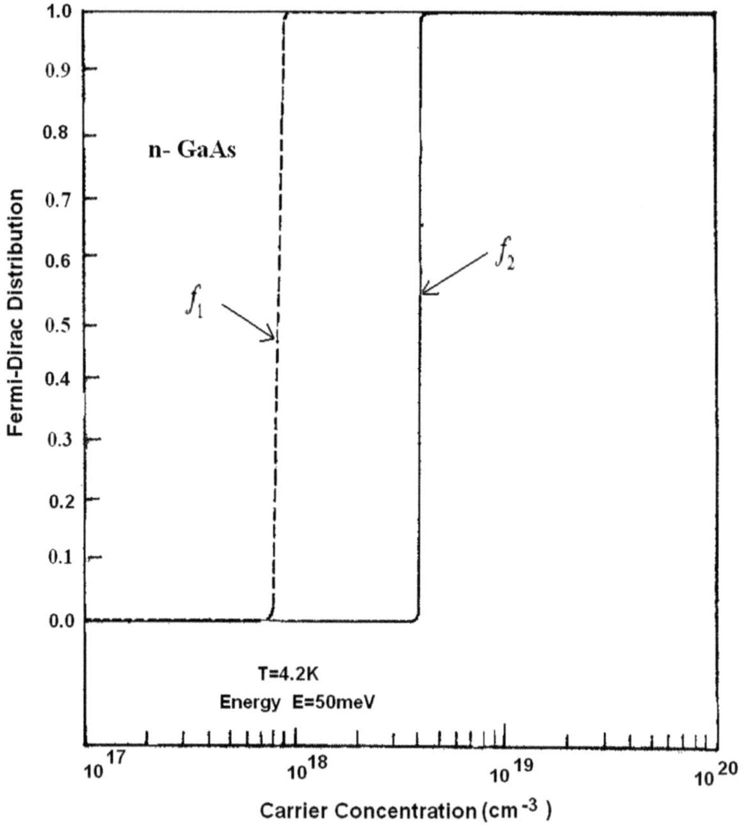

Fig. 2 The dependency of f_2 taking place the carrier-concentration in favor of n-GaAs at 4.20 K is shown in the solid figure. For f_1, the dotted plot shows the similar reliance

to the present theoretical values in degenerate semiconductors that are extensively doped and create band tails. Noteworthy is the fact that the E-k dispersion relation, as described in this work [18, 19], is one such relationship. Even though the theoretical consequences presented at this time would be helpful within examining the experimental information when it appears and can further be applied as a method to investigate the band arrangement in highly doped non-parabolic semiconductors, I am unable to compare our study with experiments because, to the best of our knowledge, there is no information on the experimental outcomes in the literature [20].

References

1. Logan, R.A., Chynoweth, A.G.: Effect of degenerate semiconductor band structure on current-voltage characteristics of silicon tunnel diodes. Phys. Rev. **131**, 89 (1963)
2. Hwang, C.J.: J. Appl. Phys. **40**, 3731 (1969)
3. Kildal, H.: Band structure of CdGeAs near k = 0. Phys. Rev. **10**, 5082–5087 (1974)
4. Willardson, R.K., Beer, A.C.: Semiconductors and Semimetals, vol. 1. Academic Press, New York, p. 102 (1966)
5. Kane, E.O.: Thomas-Fermi approach to impure semiconductor band structure. Phys. Rev. **131**, 79 (1963)
6. Bonch Bruevich, V.L.: Sov. Phys.-Sov. Phys. Solid State 4 (1953)
7. Kane, E.O.: Band tails in semiconductors. Solid State Electron. **28**, 3 (1985)
8. Singha Roy S.: Nanotechnology: Science and Computation, Part-1, ISBN: 978-620-3-84042-1, LAP Lambert Academic Publishing (2021)
9. Hwang, C.J.: Optical properties of n-type GaAs. I. Determination of hole diffusion length from optical absorption and photoluminescence measurements. J. Appl. Phys. **40**, 3731 (1969)
10. Pankove, J.I.: Absorption edge of impure gallium arsenide. Phys. Rev. A **140**, 2059 (1965)
11. Nag, B.R.: Electron Transport in Compound Semiconductors. Springer, Berlin (1980)
12. Roy, S.S.: The simple theoretical analysis of quantum well wires superlattice (QWSL) of communication technology. In: SPIE 8773, Photon Counting Applications IV; and Quantum Optics and Quantum Information Transfer and Processing, vol. 877314 (2013)
13. Tsidilkovski, I.M., Brant, L.M., Kakanov, L.I.: J. Theo. Exp. Phys. **102**, 572 (2000)
14. Roy, S.S.: Ph.D. Thesis, On some Electronic and Optical Properties of Non-Linear Optical and Optoelectronic Materials. Jadavpur University, India (2005)
15. Abramowitz, M., Stegun, I.A.: Handbook of Mathematical Functions with Formulas, Graphs and Mathematical Tables. Wiley, New York (1964)
16. Hook, J.R., Hall, H.F.: Solid State Phys. (Jhon Wiley Inc.) (1990)
17. Koch, F.: Springer Series in Solid State Sciences **53** (1984)
18. Roy, S.S.: J. Electr. Eng. **87**, 1–9. Springer, Germany & USA (2005)
19. Roy, S.S.: Nanosensors, Biosensors, and Info-Tech Sensors and Systems, vol. 7646. SPIE, 76461Q (2010)
20. Roy, S.S.: Proceedings of the SPIE 10111, Quantum Sensing and Nano Electronics and Photonics XIV, p. 1011133 (2017)

A Mixed Convection 2D Flow and Heat Transfer of Powerlaw Fluid Past Through Porous Microchannel

Palash Mondal, Dilip K. Maiti, and Subrata Maiti

Abstract This research investigates the Couette-Poiseuille flow and heat transfer of Newtonian and Non-Newtonian fluids through a porous parallel microchannel under the combined effects of bottom wall injection, magnetic field and buoyancy force. Some typical base fluid temperatures are utilized to determine the physical properties of two base fluids: water and ethanol-glycol -water mixture. The full two-dimensional Navier-Stokes along with the energy equations for incompressible power-law fluid flow are solved with efficient numerical methodologies: non-uniform grid, staggered grid, finite volume method (and then QUICK scheme for convective terms) and SIMPLE algorithm. Effects of various regulatory flow parameters (top wall movement based Reynolds number, inlet pressure gradient, injection rate, power-law index, Richardson number, Prandtl number and Hartman number) on velocity components, vorticity, streamline, temperature distribution, heat transfer and entropy generation rate are investigated here. An effort is made to make the classical one-dimensional channel Couette-Poiseuille flow to a two-dimensional flow under the influence of aforesaid parameters. Under the different flow situations considered here, our intention is to explore (i) active zone of entropy generation inside the channel and (ii) individual contributions of irreversibilities due to fluid friction, Joule heating and thermal gradient to the total entropy generation for both Newtonian and non-Newtonian fluids.

Keywords Couette-poiseuille · Injection · Power-law fluid · Nusselt number · Entropy generation

P. Mondal
Department of Mathematics, Subarnarekha Mahavidyalaya, Gopiballavpur, Jhargram, India

D. K. Maiti (✉)
Department of Applied Mathematics, Vidyasagar University, Midnapur, WB, India
e-mail: d_iitkgp@yahoo.com

S. Maiti
Department of Mathematics College of Engineering and Technology, SRM Institute of Science and Technology, Kattankulathur, Tamil Nadu, India

S. Bhattacharyya and H. S. Mahato (eds.), *Modeling, Analysis and Simulations of Multiscale Transport Phenomena*, Springer Proceedings in Mathematics & Statistics 491, https://doi.org/10.1007/978-981-96-3098-1_9

1 Introduction

One of the core topics of study in science and engineering is the flow and the heat transfer of a fluid with electrical conductivity through a porous micro channels in presence of magnetic field and various boundary conditions [1, 2]. This can be used in thermal engineering scenarios such as water movement in geothermal reservoirs, cooling electronics, thermal insulation engineering, MHD stirring of molten metal, heat pipes, MHD maritime propulsion, and magnetic lavation casting. In several material-processing applications, continuous casting, including extrusion, metal forming, glass and wire fiber drawing, and others, a moving plate continuously exchanges heat with the surrounding fluid, hence the Couette-Poiseuille flows between parallel plates are necessary [3]. Many scientists and engineers have employed injection and suction via a porous plate to modify the properties of boundary layers [1]. Non-Newtonian behavior can be seen in various crucial industrial fluids, including molten plastics, polymers, pulps, and food [4].

Makinde and Chinyoka [5] explored the effects of an incompressible electrically conducting viscous fluid, uniform suction/injection through the walls, a pressure gradient, and a magnetic field on the flow through a vertical channel. The natural convection flow and heat transfer of a cavity with a magnetic field and buoyancy force are examined by Dimitrienko and Li [6].

Zhao [7] investigated the mixed convection boundary layer flow and temperature distribution of a flat plate using finite volume method. Kefayati [8] numerically solved the two-dimensional mixed convection power-law fluid problem in an enclosure with the sinusoidal boundary conditions and discuss the effects on power law fluid flow and heat transfer. Kefayatyi and Sidik [9] investigated the entropy generation of a cavity filled with pseudoplastic fluid with the effect of the magnetic field applied at various angles. Hajji et al. [10] investigated the entropy generation within a rectangular enclosure due to both natural convection and thermal radiation. Ullah et al. [11] investigated numerically with finite difference method the flow, heat transfer and entropy generation of a laminar free convection flow of a power law fluid.

Mokarizadeh et al. [12] considered the effect of viscous dissipation of a fully developed Couette-Poiseuille flow and solved the momentum equation analytically whereas the energy equation was solved by the finite volume method. Coelho and Poole [13] provided profiles for velocity and temperature analytically in a fully-developed laminar C-P power-law fluid flow between parallel plates. They considered favourable as well as adverse pressure gradient. The behavior of the non-Newtonian boundary layer flow with the derivatives up to the second order are investigated by Al-Ashhab [14]. In our last two studies [15, 16] we have investigated the MHD Couette-Poiseuille flow through a porous microchannel with slip, suction-injection and radiation. We kept the injection plate at rest while the suction plate moving in upward/downward direction under the action of both buoyancy force and transverse magnetic field [15]. We applied variable pressure gradient (favourable/adverse) due to C-P flow. We implemented the slip conditions at plates for both moving-plate directions and pressure gradients based on the classical C-P flow. Singularities in the variation of Nusselt number (at both plates) while varying the Reynolds number were reported there [15]. In [16], we reported that heat transfer irreversibility is leading the global entropy generation rate for water, while for EG:W base fluid frictional irreversibility is the major (around 70%) contributing part of the global entropy. Rahamati and Derikvand [17] have examined the non-Newtonian fluid flow, heat transfer and entropy generation of a long slippy channel and they discussed the fluid friction and heat transfer irreversibility of the system properly. Zehtabiyan-Rezaie et al. [18] modelled the effect of a magnetic field on various angles on the heat transfer and entropy generation at every point of a long channel. They reported that the magnetic field enhances heat transfer.

It is clear from the literature review that the physical problem of Couette-Poiseuille (C-P) flow and heat transfer of Newtonian and Non-Newtonian fluids through a porous microchannel under the combined influence of bottom wall injection, magnetic field and buoyancy force has not been paid much attention. Most of the previous studies on the microchannel flow are based on either similarity transformation or approximation of the governing equations. Primitive variable formulation technique on the full two-dimensional Navier-Stokes along with the energy equations in channel flow has not been applied for Non-Newtonian fluids, which is essential for a detail analysis on velocities, vorticity, streamline and other flow characteristics throughout the entire channel. One of the main objective of this study is to explore the possibilities of converting the classical one-dimensional channel C-P flow to a two-dimensional flow under the influence of governing parameters considered here. For the scientists/ engineers , it becomes challenging and essential task how to limit the wastage or loss of useful power for non-Newtonian materials' flow and heat-transfer characteristics inside a microchannel. We put an effort to use the second laws of thermodynamics (local and global entropy) due to inherent irreversibilities in fluid friction, heat transfer and Joule heating. Thereby, we explore the active zone of entropy generation inside the channel as well as individual contributions of three irreversibilities to the total entropy generation for both Newtonian and non-Newtonian fluids.

2 Formulation of the Problem and Its Solution

2.1 *Problem Formulation*

Let us consider a mixed convection flow of a horizontal microchannel, whose top wall is moving at a constant velocity U_0 . The Cartesian coordinate system is chosen, with the x'-axis along the channel walls and the y'-axis perpendicular to the channel walls (ref. Fig. 1). The bottom wall is fixed at $y' = 0$, and the top wall is placed at $y' = a$. At the inlet, u'-velocity boundary condition is considered from the classical Couette-Poiseuille flow condition, while the v'-velocity and temperature are considered as 0 and T_0, respectively. At the outlet, the zero gradient boundary condition for all the unknown variables are applied. The top wall is considered as solid whereas a fixed

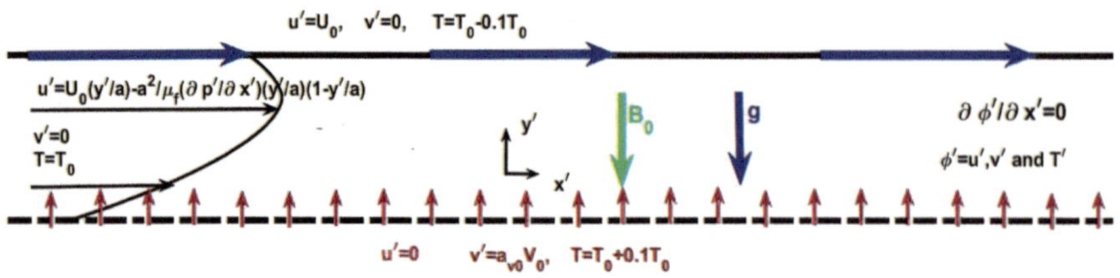

Fig. 1 The physical sketch of the problem

injection is applied at the bottom wall. The base fluids are considered as water and EG:W (water and ethylene glycol-water mixture, 60:40 by weight). The top wall temperature is considered less than the bottom wall temperature.

The electrically conducting fluid is considered here, so the Lorentz force $\mathbf{F} = \mathbf{J} \times \mathbf{B}$, where $\mathbf{J}$ is the electric current and $\mathbf{B}$ is the magnetic field, is to be incorporated in the momentum equation [19]. Here a $\mathbf{B} = (0, B_0)$, where B_0 is the initial intensity of the magnetic field imposed along y direction, is applied to the flow. The Lorentz force (electromagnetic force) can be simplified as $-\sigma_f B_0^2 u$ where σ_f is the electrical conductivity of the fluid.

2.2 *Governing Equations with Boundary Conditions*

In reference to Kefayati [8] and based on the above assumption, the dimensional governing continuity, x-momentum, y-momentum and energy equations are respectively as follows:

$$\frac{\partial u'}{\partial x'} + \frac{\partial v'}{\partial y'} = 0 \tag{1}$$

$$\rho_f(\frac{\partial u'}{\partial t'} + u'\frac{\partial u'}{\partial x'} + v'\frac{\partial u'}{\partial y'}) = -\frac{\partial p'}{\partial x'} + \frac{\partial}{\partial x'}(2\mu'_{af}\frac{\partial u'}{\partial x'}) + \frac{\partial}{\partial y'}[\mu'_{af}(\frac{\partial v'}{\partial x'} + \frac{\partial u'}{\partial y'})] - \sigma_f B_0^2 u' \tag{2}$$

$$\rho_f(\frac{\partial v'}{\partial t'} + u'\frac{\partial v'}{\partial x'} + v'\frac{\partial v'}{\partial y'}) = -\frac{\partial p'}{\partial y'} + \frac{\partial}{\partial y'}(2\mu'_{af}\frac{\partial v'}{\partial y'}) + \frac{\partial}{\partial x'}[\mu'_{af}(\frac{\partial v'}{\partial x'} + \frac{\partial u'}{\partial y'})] + g(\beta\rho)_f(T - T_0) \tag{3}$$

$$(\rho C)_f(\frac{\partial T'}{\partial t'} + u'\frac{\partial T'}{\partial x'} + v'\frac{\partial T'}{\partial y'}) = \kappa_f(\frac{\partial^2 T}{\partial x'^2} + \frac{\partial^2 T}{\partial y'^2}) + \sigma_f B_0^2 u'^2, \tag{4}$$

For a purely viscous non-Newonian power law fluid, the shear stress tensor τ is $\tau_{xy} = \mu_{eff}(\frac{\partial u'}{\partial y'} + \frac{\partial v'}{\partial x'})$, where $\frac{1}{2}(\frac{\partial u'}{\partial y'} + \frac{\partial v'}{\partial x'})$ is the deformation tensor rate, $\mu_{eff} = \mu_f \mu'_{af}$ is the effective viscosity of the power law fluid, at which

$$\mu'_{af} = L'[2\{(\frac{\partial u'}{\partial x'})^2 + (\frac{\partial v'}{\partial y'})^2\} + \{\frac{\partial u'}{\partial y'} + \frac{\partial v'}{\partial x'}\}^2]^{(m-1)/2} \tag{5}$$

with L' (=$(\frac{a}{u_0})^{(m-1)}$) [8, 9] is the consistency coefficient, μ_f is the dynamic viscosity of the newtonian fluid, and m is the power-law index.

Boundary conditions:

- *At inlet*: $u' = U_0(y'/a) - (a^2/\mu_f)\frac{\partial p'}{\partial x'}(y'/a)(1 - y'/a),\ v = 0,\ T = T_0$
- *At outlet*: $\frac{\partial u'}{\partial x'} = 0,\ \frac{\partial v'}{\partial x'} = 0,\ \frac{\partial T'}{\partial x'} = 0$
- *Top wall*: $u' = U_0,\ v' = 0,\ T = T_0 - 0.1T_0$
- *Bottom wall*: $u' = 0,\ v' = a_{v_0}V_0,\ T = T_0 + 0.1T_0$

Here, ρ_f is the density, β_f is the thermal expansion coefficient, C is the heat capacitance and κ_f is the thermal conductivity of the fluid. g is the gravitational acceleration, a_{v_0} is the non-dimensional injection rate, a is the width of the channel, U_0 is the top wall velocity, V_0 is the bottom wall injection velocity and T_0 is the reference temperature of the fluid. The following non-dimensional variables are introduced before moving on to the system's numerical solution:

$u = u'/U_0,\ v = v'/V_0,\ x = x'/a,\ y = y'/a,\ t = U_0 t'/a,\ \theta = (T - T_0)/T_0,\ p = p'/(\rho_f U_0^2)$. After applying these non-dimensional variables the above continuity, momentum equations, energy equation along with the boundary condition reduce to:

$$Re_u\frac{\partial u}{\partial x} + Re_v\frac{\partial v}{\partial y} = 0 \tag{6}$$

$$(\frac{\partial u}{\partial t} + u\frac{\partial u}{\partial x} + \frac{Re_v}{Re_u}v\frac{\partial u}{\partial y}) = -\frac{\partial p}{\partial x} + \frac{1}{Re_u}[\frac{\partial}{\partial x}(2\frac{\mu_{af}}{\mu_f}\frac{\partial u}{\partial x}) + \frac{\partial}{\partial y}\{\frac{\mu_{af}}{\mu_f}(\frac{\partial v}{\partial x} + \frac{\partial u}{\partial y})\}] - \frac{Ha^2}{Re_u}u \tag{7}$$

$$(\frac{\partial v}{\partial t} + \frac{Re_v}{Re_u}u\frac{\partial v}{\partial x} + \frac{Re_v^2}{Re_u^2}v\frac{\partial v}{\partial y}) = -\frac{\partial p}{\partial y} + \frac{1}{Re_u}[\frac{\partial}{\partial y}(2\frac{\mu_{af}}{\mu_f}\frac{\partial v}{\partial y}) + \frac{\partial}{\partial x}\{\frac{\mu_{af}}{\mu_f}(\frac{\partial v}{\partial x} + \frac{\partial u}{\partial y})\}] + Ri\theta \tag{8}$$

$$(\frac{\partial \theta}{\partial t} + u\frac{\partial \theta}{\partial x} + \frac{Re_v}{Re_u}v\frac{\partial \theta}{\partial y}) = \frac{1}{Re_u Pr}(\frac{\partial^2\theta}{\partial x^2} + \frac{\partial^2\theta}{\partial y^2}) + \frac{Ha^2 Ec}{Re_u}u^2, \tag{9}$$

where,

$$\mu_{af} = \mu_f[2\{(\frac{\partial u}{\partial x})^2 + (\frac{\partial v}{\partial y})^2\} + \{\frac{\partial u}{\partial y} + \frac{\partial v}{\partial x}\}^2]^{(m-1)/2} \tag{10}$$

Boundary conditions:

- *At inlet*: $u = y + Re_u Py(1 - y), v = 0, \theta = 0$
- *At outlet*: $\frac{\partial u}{\partial x} = 0, \frac{\partial v}{\partial x} = 0, \frac{\partial \theta}{\partial x} = 0$
- *Top wall*: $u = 1.0, v = 0, \theta = -0.1$
- *Bottom wall*: $u = 0, v = a_{v_0}, \theta = 0.1$

2.3 Parameters Values

Here the operational dimensionless parameters are defined as (i) $Re_u = U_0 a\rho_f/\mu_f$: the Reynolds number due to plate movement ($\in$ [0.62,4]) [15], (ii) $Re_v = V_0 a\rho_f/\mu_f$: the Reynolds number due to injection and suction (= 1.24) [15], (iii) a_{v_0}: bottom wall injection rate ($\in$[0,0.1]), (iv) $P = -dp'/dx'$: the pressure gradient at the inlet due to Couette-Poiseuille flow (=1.2, 0.8, 0.4 & 0.2), (v) $Pr = \mu_f C/\kappa_f$: the Prandtl number(=7.97 for water & 57.92 for EG:W at $T_0 = 290$ K), (vi) $Gr = g(\rho_f\beta_f)a^3 T_0/(\rho_f v_f^2)$: the Grashof number (=63.70 for water & 5.59 for EG:W at $T_0 = 290$ K), (vii) $Ha = B_0 a\sqrt{(}\sigma/\mu_f)$: the Hartman number ($\in$[0.2,1]), $(viii)$ m: the power law fluid index ($\in$ [0.2,1.4]), (ix) $Ec = {U_0}^2/(T_0 C_f)$: the Eckart number(=0.5), (x) $Br = \frac{\mu U_0^2}{\kappa T_0}$: the Brinkmann Number (=1.0) and (xi) Ri: Gr/Re_u^2: the Richardsons number. The non-dimensional channel length is considered as 25. The reference temperature of the fluid is considered as 290 K, and the Pr and Gr are calculated at this temperature for water and EG:W.

2.4 Numerical Methodology

The iterative pressure correction-based third order Semi-Implicit Method for Pressure Linked Equations (SIMPLE) technique is used numerically to solve the governing equations. Small control volumes are created inside the computational area. The variables are kept in a staggered grid configuration. The finite volume approach is then used to discretize the equations across each control volume (FVM). The convective terms using a third-order accurate Quadratic Upstream Interpolation for Convective Kinematics (QUICK) and central difference for diffusion terms are discretized. The time derivatives are discretized using a completely implicit third-order technique.

The pressure correction between continuity and momentum is made by changing the continuity equation into a Poisson equation for pressure. A divergent velocity field's pressure is corrected using the Poisson equation. We have thoroughly analyzed the numerical methods (staggered grid, FVM, QUICK and SIMPLE algorithms) applied here in our previous studies [20, 21].

2.4.1 Grid Distribution

An uniform grid is considered along the y-direction, while a non uniform grid is distributed along the x-direction starting with grid size 0.01 from the channel entrance. Then gradually the size of grid along the x-direction increases. The numerical code used in our previous study [21] is adopted here. We have conducted the grid independence study there [21]. For non-dimensional channel length as 100, we reported there that the change in the value of velocity for different grid size occurs at third decimal place. We have also conducted the same with grids: 120×60, 350 × 20, and 650×120 for the case of $Re_u = 2.46$, $Re_v = 1.24$, Ha = 0.02, Ec = 0.5, $a_{v0} = 0.1$, Ri = 10.57, Pr −7.97 and m = 1.0 with channel length 25 and presented the u-velocity at x = 10 in Fig. 5b. Here 120x60 grids indicates the total number of grid at which 120 grids along the x-direction and 60 grids along the y direction. The difference between the results is found minor when it varying between 120×60, 350 × 120 and 650 × 120. In other words, the difference between the results is tending to zero when the number of grids increase. It can also be noted that finer grid takes much more time than medium grids. So the medium grid 350×120 is considered as a reasonable finer grid, and applied for all the computations here.

2.5 *Validation of Numerical Code*

The numerical code used in the previous studies (chapter#5) has been used here to investigate the flow, heat transfer and entropy generation of a Couette-Poiseuille flow of a Newtonian/non-Newtonian fluid in a long porous channel with injection/suction, FHD and MHD. This code was validated for the case of: (i) v-velocity, stream line, local and total Nusselt number for non-Newtonian fluid with Kefayatyi [8] in Mondal and Maiti [22] (ref. Fig. 2 and Table-1) and (ii) irreversibilities due to the fluid friction and heat transfer for non-Newtonian fluid with Kefayatyi and Sidik [9] in Mondal and Maiti [22] (ref. Fig. 2).

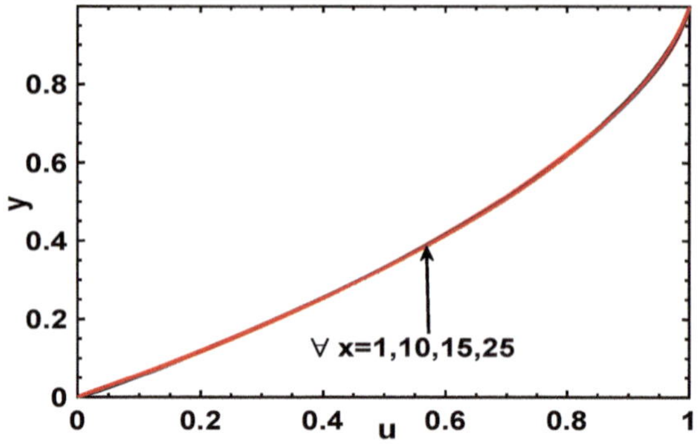

Fig. 2 u-velocity profile with no injection at bottom wall, m = 1.0, $Re_u = 0.62$, $Re_v = 1.24$, Ri = 165.71, Pr = 7.97, P = 1.2 for water (Pr = 7.97, Ri = 165.71)

3 Results and Discussion

3.1 Velocity Distribution

Figure 2 displayes the u-velocity profile at different positions of the channel with no injection-suction on bottom wall with m = 1.0, $Re_u = 0.62$, $Re_v = 1.24$, Ri = 10.57, Pr = 7.97, P = 0.4. The u-velocity is found fully developed after a minimal distance from the entrance of the channel, i.e., the u-velocity of the fluid is only the function of y after a minimal distance from the entrance of the channel.

Figure 3a depicts the u-velocity profile at different $x = 0.5$, 10 & 24, for different values of $Re_u = 0.62$, 1.24 & 2.48 with $a_{v0} = 0.1$ and m = 1.0. In contrast to Fig. 2, here in Fig. 3a the u-velocity increases as we move towards the end of the channel at any Re_u. This is not surprising, because additional fluid inside the channel is injected from the bottom wall and the continuity Eq. (6) has to be satisfied. This implies that the u-velocity inside the channel becomes dependent on both x and y here, that is,

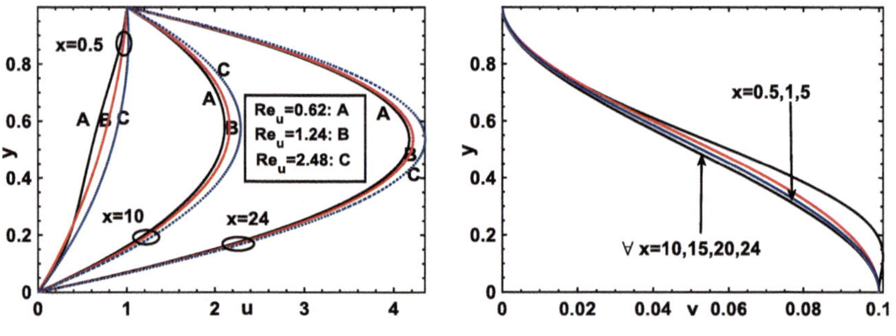

(a) u velocity profile at different values of x for different Re_u.

(b) v velocity profile at different values of x at Re_u=0.62.

Fig. 3 Velocity (u,v) profiles at different values of x with m = 1.0, $Re_v = 1.24$, $a_{v0} = 0.1$, P = 0.4, Ha = 0.02, Ec = 0.5 for water base fluid (Pr = 7.97) u = u(x,y) for the present flow situation. Also, as the value of Re_u increases from 0.62 to 2.48, the u-velocity at every point of the channel also increases. Physically, the fluid gets accelerated throughout the channel at higher plate velocity.

The v-velocity profile at the different positions of the channel for m = 1.0, Re_u = 0.62 and Ri = 165.71 is shown in Fig. 3b. The v-velocity changes up to some distance ($x < 10$) from the entrance of the channel, and then it is almost independent on the channel length (x). It is remarkable to note here that in contrast to the classical C-P flow where v = 0 throughout the channel, here in the present flow configuration the v-velocity is found noticeably high. Similar results are also seen for Re_u = 1.24, 2.48 and 4.0 (not presented here).

Figure 4a presents the u-velocity profile for different power law index m = 1.4,1.0,0.6,0.4 at the different positions of the channel for Re_u = 0.62. At the entrance of the channel (x = 0.5), the u-velocity for all m are almost same, but as the distance increases, the difference is clearly visible, and at the exit of the channel, the difference is found maximum. The influence of the power law index on the u-velocity is maximally visible at the injection (bottom) side and the center of the channel. From the zoom view in Fig. 4a, it is clear that the u-velocity of any value of y, close to the wall, is more for the curve-D(m = 0.4) in comparison to the curve-A(m = 1.4). This is due to the fact that at higher m, the fluid becomes thicker and viscous effect close to the wall becomes more in presence of no-slip condition. An opposite situation in the u-velocity is seen at the channel center. This could be due to the fact that a thicker fluid have a higher momentum, which leads to higher u-velocity at the channel center where wall's no-slip effect is negligible.

The effect of the power law index on the v-velocity profile at x = 0.5 is also depicted in Fig. 4b for the same set of parameter values as in Fig. 4a. The opposite result is seen for the v-velocity profile compared to the u-velocity profile. That is, the maximum effect of power-law index on v-velocity is seen at the entrance of the channel, but beyond some distance from the entrance, the effect is getting significantly

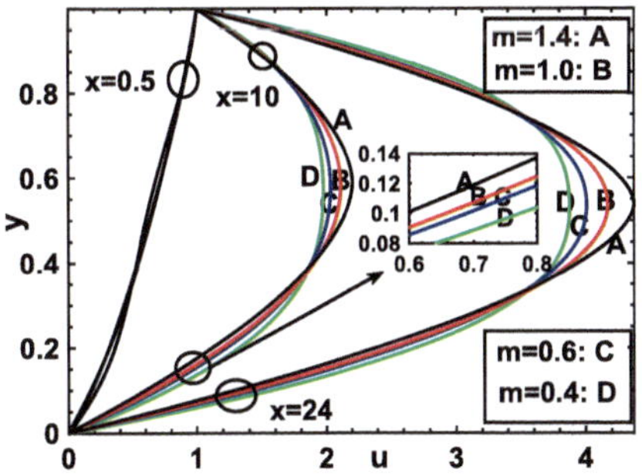

(a) u-velocity profile for different values of x and different powrlaw index.

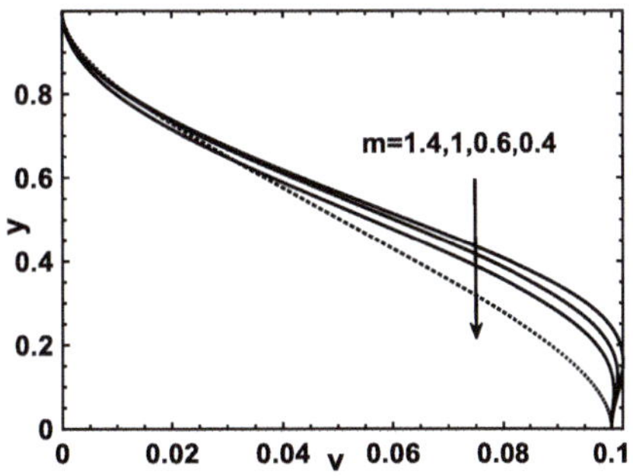

(b) v-velocity profile for different m at $x = 0.5$.

Fig. 4 u and v velocity profiles at different values of x and m, with $Re_u = 0.62$, $Re_v = 1.24$, P = 0.4, Ha = 0.02, Ec = 0.5, $a_{v0} = 0.1$ for water base fluid (Pr = 7.97, Ri = 165.71) less (not presented here). Here also, when the value of m decreases, the velocity of the fluid also decreases. The effects of m at $Re_u = 1.24$, 2.48 and 4.0 on the v-velocity profile are similar, so not presented here.

In Fig. 5a–d, the effect of inlet pressure gradient (P=1.2, 0.8, 0.6 and 0.4), on the u-velocity profile is presented for different values of m = 1.4, 1.0, 0.6 and 0.4 at Re_u=2.48. As can be seen in Fig. 5, the pressure gradients have similar effects on the u-velocity at all values of x and m considered here, i.e., the fluid velocity throughout the channel increases due to the increase of P. Here, the maximum effect is found for m=1.4. It is interesting to note that the u-velocity profiles for different P resembles to those of classical C-P flow only at the channel entrance. However, as we move towards the end of the channel, the u-velocity profile is no longer remain to that of classical C-P flow, rather they looks like a Poiseuille flow. Here due to continuous injection, the u-velocity is overshoot.

Figure 6 presents the vorticity contours (Fig. 6a, b) and streamline contours (Fig. 6c, d) for no injection (first column) as well as injection cases (second column) for the water base fluid with Re_u =0.62, $Re_v = 1.24$, P=0.4, Ha=0.02, Ec=0.5. As seen in Fig. 6a, the value of vorticity remains negative throughout the channel for the no injection case (here a_{v0}=0). In this case, the u-velocity rises with y at this lower value of $P.Re_u$(ref. Fig. 2), inspite of considered C-P flow. Moreover, in

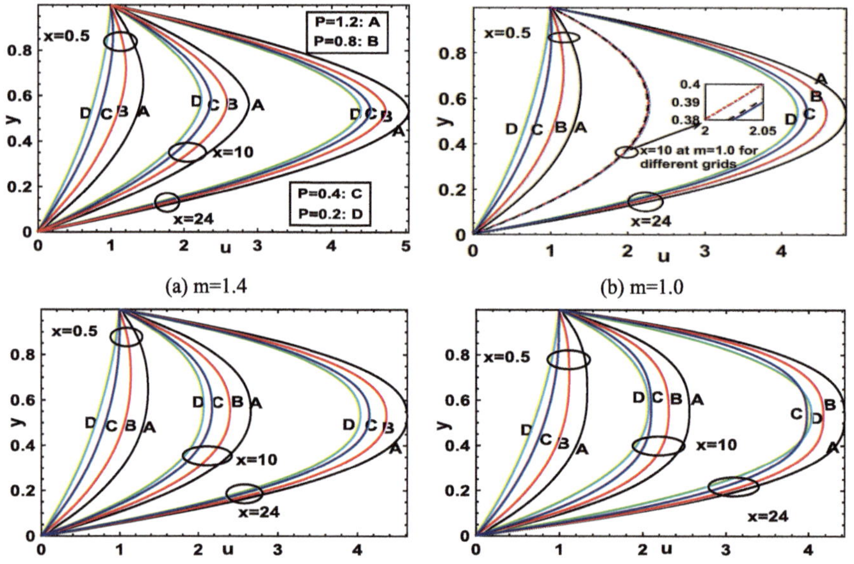

(a) m=1.4 (b) m=1.0

(c) m=0.6 (d) m=0.4

Fig. 5 u-velocity profiles at different x and for different P and m with $Re_u = 2.48$, $Re_v = 1.24$, Ha = 0.02, Ec = 0.5, $a_{v0} = 0.1$ for the water base fluid (Pr = 7.97, Ri = 10.357). In Fig. 5b, u-velocity profiles at x = 10 are for P = 0.4 only and three different grids(120X60: Dotted lines, 350X120: Solid lines, 650x120: Dashed lines)

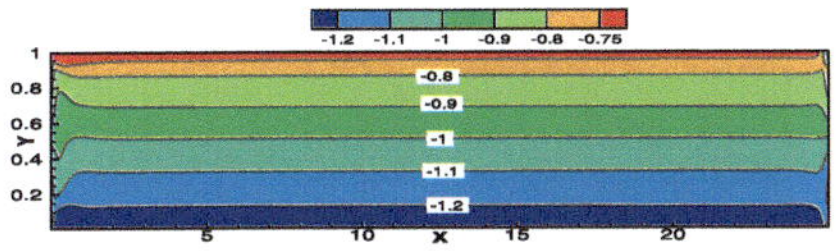

(a) Vorticity for a_{v0}=0

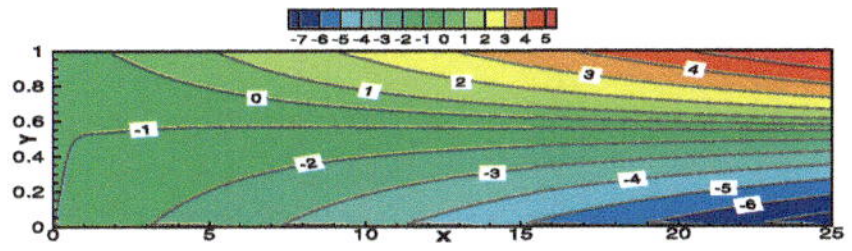

(b) Vorticity for a_{v0}=0.05

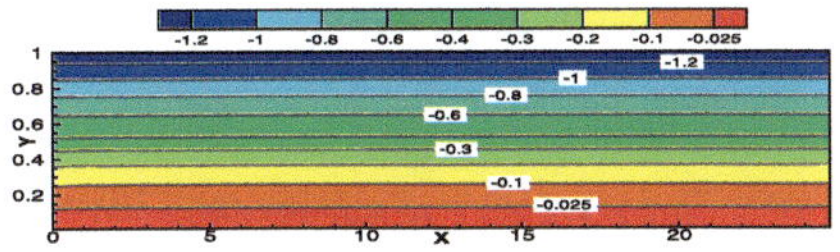

(c) Streamline for a_{v0}=0

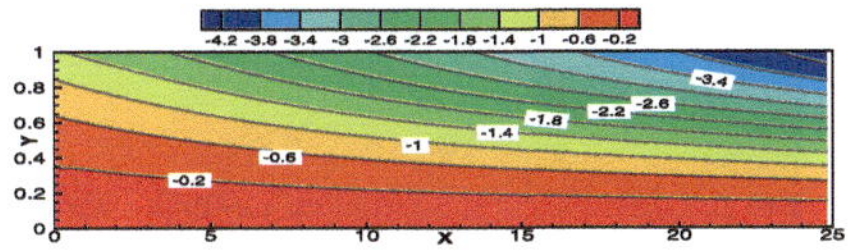

(d) Streamline for a_{v0}=0.05

Fig. 6 Vorticity and Streamline contour for different injection rate $a_{v0} = 0$ and 0.05 with m = 1.4, Re_u=0.62, $Re_v = 1.24$, P=0.4, Ha = 0.02, Ec = 0.5 for the water base fluid (Pr = 7.97, Ri = 165.71)

this case the v-velocity is almost zero throughout the channel. Therefore, the value $J(= \frac{\partial v}{\partial x} - \frac{\partial u}{\partial y} = -\frac{\partial u}{\partial y})$ is expected to be negative throughout the channel. Also, it was shown in Fig. 2 that the u-velocity is only dependent on y in this case. On the other hand in the presence of injection, J becomes positive in the upper half of the channel in Fig. 6b. Here though v is non zero and vary almost linearly with y (ref. Fig. 3b) but its variation with x is almost negligible. Therefore, J value is mainly dependent on $\frac{\partial u}{\partial y}$. As discussed in Figs. 3a, 4a and 5, with the increase of x, the velocity looks like C-P flow at higher pressure gradient, where the u-velocity increases in the lower half while decreases in the upper half with the increase of y. This results to the positive values of J in the upper half of the channel. Also one can understand the cause of magnitude of vorticity is more in the premises of both the walls from the velocity gradients in Figs. 3a, 4a and 5. Only at the top wall, there is a change of sign in J, implying that there is a boundary layer separation on the top wall. It is to note that the forward flow of the channel is mainly affected by the bottom wall injection, and this effect on the flow starts after a certain distance from the entrance of the channel. Pattern of streamlines presented in Fig. 6c, d suggests us that the flow is parallel for the no injection case, while it becomes non parallel for the case of injection.

The effect of speed of top wall movement on the vorticity contour pattern is discussed based on Fig. 6b (at $Re_u = 0.62$) and Fig. 7a–c (Re_u =1.24, 2.48, and 4.0). As seen in these figures, the area of positive vorticity zone increases with Re_u and there is no boundary layer separation on the top wall at higher Re_u As shown in Fig. 3a, when Re_u increases the velocity profile becomes more parabolic, even at starting of channel since we have applied developed velocity profile at the inlet. A similar flow phenomena can be made by increasing the value of P since the roll of P and Re_u mechanically similar [15].

The impact of power law index m on the vorticity contour is analysed using Fig. 6b (m = 1.4) and Fig. 7d, f (m = 1.0, 0.8 and 0.4). It is clearly visible that when m decrease: the point of boundary layer separation shifts slightly from the entrance, the area of positive vorticity decreases and the intensity of vorticity also

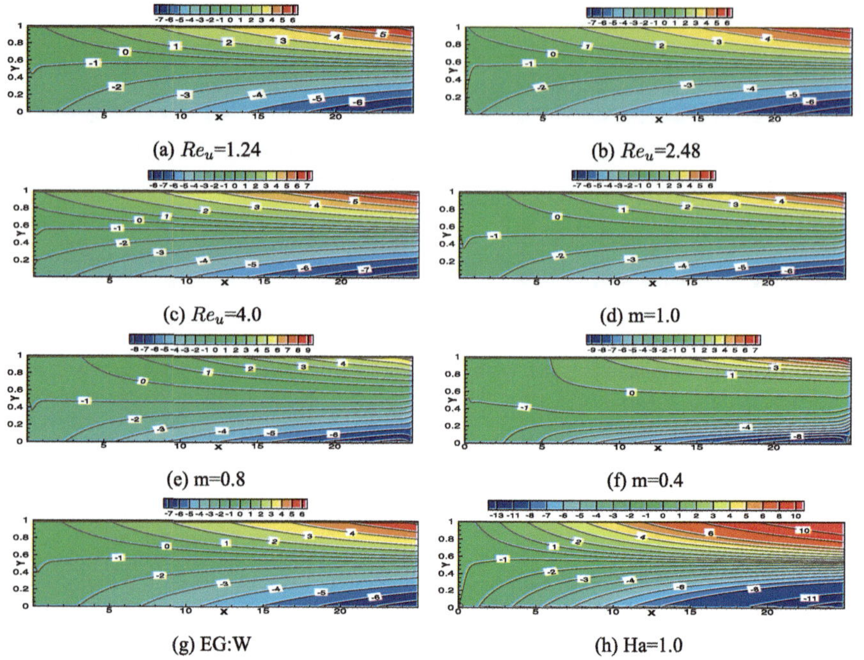

Fig. 7 Vorticity contour for Ec = 0.5, P = 0.4, $Re_v = 1.24$ and a_{v0}=0.05 , **a–c**: Effect of Re_u for water base fluid when Ha = 0.02, m = 1.4; **d**, **f**: Effect of power law index on water base fluid when Ha = 0.02, $Re_u = 0.62$; and **g** for EG:W base fluid when when Ha=0.02, m = 1.4, $Re_u = 0.62$; and **h** for water base fluid at higher Ha = 1.0 when m = 1.4, $Re_u = 0.62$

decreases(mainly in the middle part of the channel). It is seen from Fig. 4a that, the u-velocity profile at the middle becomes flat at lower m. The vorticity pattern presented in Fig. 7g for the EG:W base fluid is almost similar to that in Fig. 6b for the water base fluid. This is not surprising here, because coefficients (i.e., parameters) in the momentum equations are same for both the cases of Figs. 6b and 7g, only differ by the source term $Ri.\theta$. This source term difference here do not have much influence on the vorticity values throughout the channel.

The Hartman number effect on the vorticity contour is discussed based on Fig. 6b (for Ha = 0.02) and Fig. 7h (for Ha = 1.0). Here also not so much difference is found between these two figures, because of the similar reason as mentioned above. Here the source term is $\frac{Ha^2}{Re_u}u$, which do not make much difference in the vorticity values between these two cases. The effect of Re_u on the streamline for m = 1.4 is analyzed by presenting Fig. 8a for $Re_u = 0.62$, Fig. 8b for $Re_u = 1.24$ and Fig. 8c for $Re_u = 4.0$. It is seen that, as Re_u increases from 0.62 to 4.0 the flow becomes more stronger.

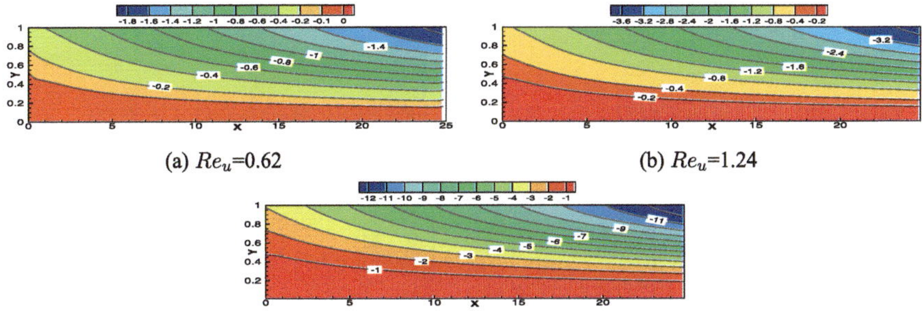

(a) Re_u=0.62 (b) Re_u=1.24 (c) Re_u=4.0

Fig. 8 Streamline contour of m = 1.4 for different Re_u values with $Re_v = 1.24$, P = 0.4, Ha = 0.02, Ec = 0.5, $a_{v0} = 0.1$ for water base fluid (Pr = 7.97)

3.2 *Temperature Distribution*

The effect of Re_u on the isothermal contour for m = 1.4 is analysed based on Fig. 10a for $Re_u = 0.62$, Fig. 9a for $Re_u = 1.24$, Fig. 9b for $Re_u = 2.48$ and Fig. 9c for $Re_u = 4.0$. It is seen that the isothermal lines are affected by Re_u. At the lower Re_u (Fig. 10a for $Re_u = 0.62$), the temperature of the fluid is mainly affected by the hot fluid injected from the bottom wall, but as the top wall velocity increases, the temperature of the fluid starts affecting by the top wall temperature due to the convective activities (i.e., the red zone decreases). Here at any fixed values of x, the temperature of fluid increases from bottom to top wall, since hot fluid injected from the bottom wall. Nu = Nu(x) at the bottom wall is also anticipated to be a decreasing function of x.

The influene of the base fluids, i.e., different Pr (7.97 for water and 57.92 for EG:W at 290 K), Ri (165.71 for water and 14.54 for EG: W at 290 K with $Re_u = 0.62$) and m (= 1.4 and 0.4) on the isothermal contour are elucidated in Fig. 10. The isothermal

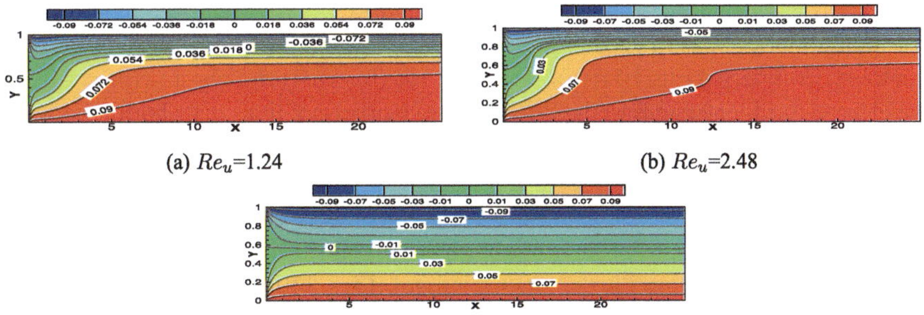

(a) Re_u=1.24 (b) Re_u=2.48 (c) Re_u=4.0

Fig. 9 Isothrmal contour of m = 1.4 for different Re_u values with $Re_v = 1.24$, P = 0.4, Ha = 0.02, Ec = 0.5, $a_{v0} = 0.1$ for water base fluid (Pr = 7.97)

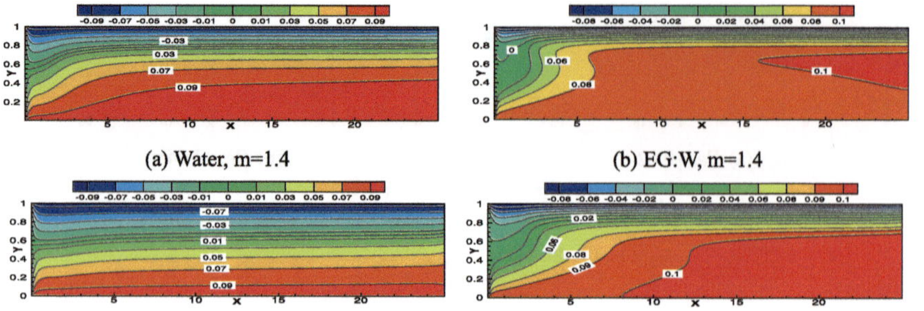

(a) Water, m=1.4 (b) EG:W, m=1.4

(c) Water, m=0.4 (d) EG:W, m=0.4

Fig. 10 Isothrmal contour of m = 1.4 (top row) and 0.6 (bottom row) for water (Pr = 7.97, Ri = 165.71) (for first column) and EG:W (Pr = 57.92, Ri = 14.54) (for second column). Here $Re_u = 0.62$, $Re_v = 1.24$, P = 0.4, Ha = 0.02, Ec = 0.5, $a_{v0} = 0.1$

contours are affected throughout the channel. For a particular m, there are massive differences in the isothermal pattern between water and EG:W. These difference are mainly due to the Prandtl number differences, which largely affect the solution of the energy equation. The isothermal lines for m = 0.4 are less affected than that for m = 1.4 for both base fluids. The base fluid has more impact on the isothermal lines, in comparison to the power law index.

Figure 11a displays the isothermal contour for Ha = 0.02 and Fig. 11b for Ha = 1.0. From Fig. 11a, it is seen that the temperature of the fluid decreases from the bottom to the top wall, and the temperature of the fluid is marginally affected close to the channel walls. Also, it is seen that the temperature is almost same for every x. However, from Fig. 11b, it is clear that the the fluid's temperature is largely affected by Ha. The temperature of the fluid becomes more than the walls temperature, and the fluid becomes more and more hotter as we move towards the exit of the channel. In fact, the temperature of the fluid at the exit of channel becomes more than five times of the temperature at the entry. Here, the Hartman number (Ha) and Eckart number (Ec) have similar role in the energy equation (9) and both are contributing as in the source term. This source term significantly enhance the internal fluid temperature. Hartman number effect is more on isothermal contours, in comparison to vorticity contours. Comparing Fig. 10a (m = 1.4), Fig. 10c (m = 0.4) and Fig. 11a (m = 1.0) for water base fluid, one can find the difference in isothermal lines distribution between Newtonian and non-Newtonian fluids.

The influence of injection rate on isothermal contours is analysed based on Figs. 11c, d and 10a for $a_{v_0} = 0$, 0.05 and 0.1 respectively. It is seen that for no injection case ($a_{v_0} = 0$, ref. Fig. 11c) the isothermal lines are almost parallel to each other, but as the injection rate a_{v_0} increases (ref. Figs. 11d, and 10a), i.e., the more hot fluid is injected from the bottom wall, so the lower half fluids temperature will also increases (red zone increases). In our considered boundary condition, injection flow is controlled by two parameters: a_{v_0} and v_0. So the temperature of the fluid is also dependent on a_{v_0} and Re_v.

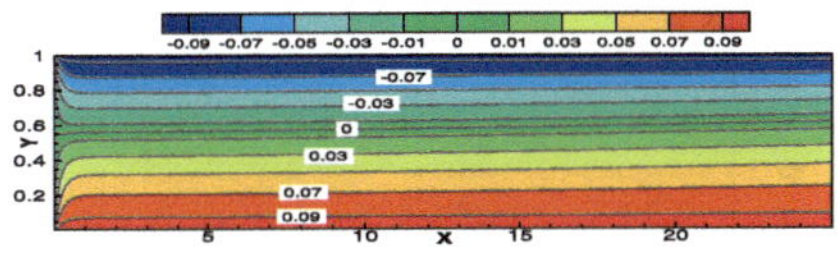

(a) Isothermal, Ha=0.02

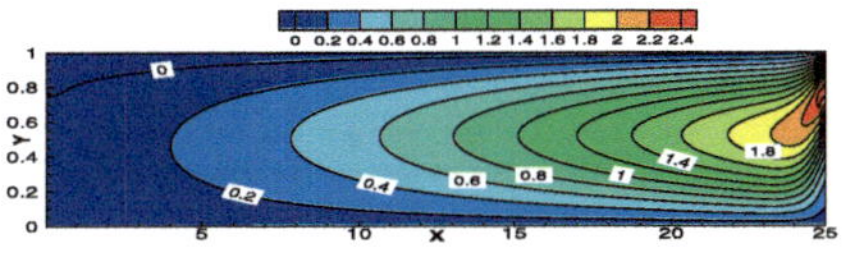

(b) Isothermal, Ha=1.0

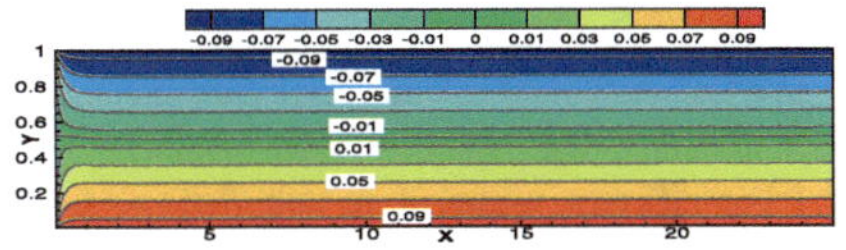

(c) Isothermal lines for a_{v0}=0

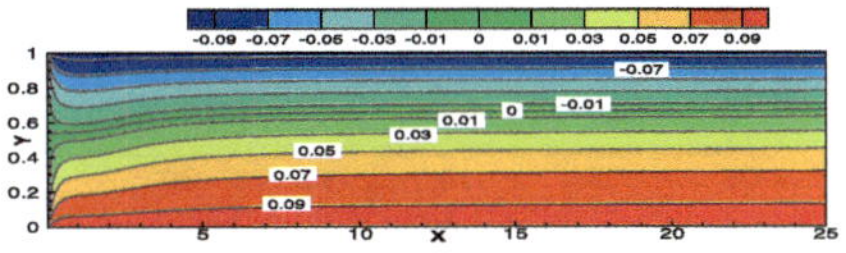

(d) Isothermal lines for a_{v0}=0.05

Fig. 11 Isothermal contour with $Re_u = 0.62$, $Re_v = 1.24$, P = 0.4, Ec = 0.5 for water base fluid (Pr = 7.97, Ri = 165.71): **a**, **b**: For different Hartman number with $a_{v_0} = 0.1$ and m = 1.0.; **c**, **d**: for different injection velocities with Ha = 0.02 and m = 1.4

3.3 *Heat Transfer Analysis*

The non-dimensional local Nusslt number at the both bottom and top walls are defined respectively as [8]: $Nu|_{y=0} = -\frac{\partial\theta}{\partial y}$ and $Nu|_{y=1} = \frac{\partial\theta}{\partial y}$.

Figure 12a is portrayed to present the impact of plate movement (Re_u) on the local heat transfer (Nusselt number) profile on both plates. The fluid with temperature $\theta = 0$ is injected from the left side and the fluid with temperature $\theta = 0.1$ is also injected from the bottom wall, while the top wall temperature is kept fixed with $\theta = -0.1$. Naturally, a positive heat transfer from the heated bottom wall to the fluid is expected and seen in the Fig. 12a. Again, the temperature of the top wall is always less than the fluid's temperature. Therefore, a negative heat transfer from the fluid to the top wall of the channel is noticed here. The variation of the heat transfer due to change in Re_u is more at the top wall than that at the bottom wall. This is not surprising since the convective activities are found to be more close to the moving wall. One can see from the isothermal contour (ref. Fig. 9) that isothermal lines are more cluster near the top moving wall. At lower Re_u, the variation of Nu (at both the walls) with x is almost negligible, while this variation is noticeable at higher Re_u. As can be seen from the Fig. 9 that isothermal lines are almost uniformly distributed at lower Re_u, but at higher Re_u this isothermal lines distributions is scattered throughout the channel. If one examine the isothermal lines, especially at lower wall, in Fig. 9b–c, it will be clear why Nu=Nu(x) behave non-linearly for $x < 15$. The maximum heat transfer is observed at the entry of the channel since there is sudden difference in the temperature between the incoming fluid and walls. As seen in Fig. 12a that the magnitude of space-averaged Nusselt number (total heat transfer from walls) is found to be more at the top wall, in comparison to that of the bottom wall at any Re_u. Eventually the magnitude of the total heat transfer for both the walls increases with Re_u due to enhance convective activities inside the channel at higher Re_u.

Figure 12b describes the heat transfer quantity for different power-law fluids at both walls. For $m \leq 1$, the magnitude of heat transfer increases with the decrease

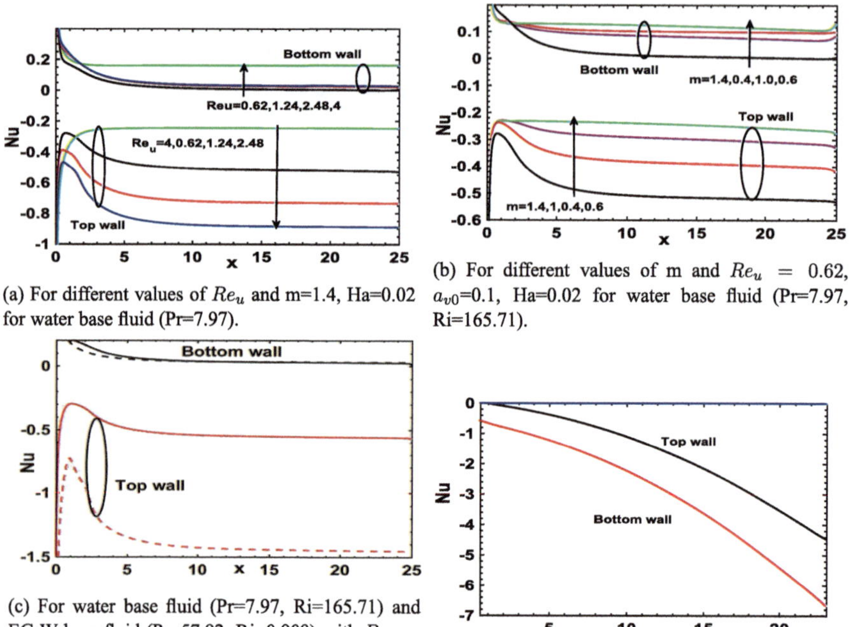

(a) For different values of Re_u and m=1.4, Ha=0.02 for water base fluid (Pr=7.97).

(b) For different values of m and Re_u = 0.62, a_{v0}=0.1, Ha=0.02 for water base fluid (Pr=7.97, Ri=165.71).

(c) For water base fluid (Pr=7.97, Ri=165.71) and EG:W base fluid (Pr=57.92, Ri=0.909) with Re_u = 0.62, a_{v0}=0.1, m=1.4. Solid lines for water and dashed lines for EG:W

(d) For m=1.0, Re_u =0.62, Ha=1.0, Pr=7.97 for water base fluid (Pr=7.97, Ri=165.71).

Fig. 12 Nusselt number distribution along both walls for $Re_v = 1.24$, P = 0.4, Ec = 0.5 and $a_{v0} = 0.1$

of m at the top wall, while the trend is almost independent to m at the bottom wall. The maximum heat transfer (magnitude) is seen for m = 1.4 at the top moving wall whereas it is seen lowest at the fixed bottom wall. Except at the entrance of the channel, the Nusselt number variation is almost independent to x for Newtonian fluid, indicating that the temperature of fluid almost remain same throughout the channel, while this variation is very much noticeable up to middle of the channel for both types (Dilatant and Pseudoplastic) non-Newtonian fluids. For non-Newtonian fluids, the magnitude of heat transfer increases towards the channel exit at the top wall, while the trend is reverse at the bottom wall. One can get justifications for these results from the isothermal lines presented in Fig. 10a, c. As an example (ref. Fig. 10c), due to uniform injection throughout the bottom wall, the thermal boundary layer becomes more thicker towards the channel exit. At the same time due to this additional fluid inside the channel as well as movement of the top wall, the thermal boundary layer at this wall becomes more thinner in the direction of the channel's exit

The influence of base fluid on the heat transfer is illustrated in Fig. 12c. Here solid line for water and dash line for EG:W. Here also the effect at the top is much more than that at the bottom wall. Apparently, the heat transfer is more for the case of

EG:W than the case of water. Here Nu = Nu(x) is strongly dependent on x (up to middle of the channel) for EG:W base fluid, in comparison to water. It is shown in Fig. 10 that the thermal boundary layer for the case of water is thicker than that for EG:W. As reported in our previous study [15] the inlet pressure gradient has significantly less impact on the Nusselt number. Here also, a similar result is seen (not presented here), i.e., a very minimal effect of the pressure gradient is found on the heat transfer.

Figure 12d presents the Nu distribution at both the walls for m = 1.0, $Re_u = 0.62$, $Re_v = 1.24$, P = 0.4 at higher Ha (= 1.0) for the water base fluid. As reported in Fig. 11b that the fluids' temperature becomes very high compared to the walls, so the heat transfer takes place from the fluid to both the walls in Fig. 12d, i.e., the negative heat transfer is seen here. While looking into the effect of Ha at a fixed m (= 1.0) by comparing Fig. 12b, d, it is observed that the Hartman number have an important role on the heat transfer at both the walls. Reason behind these facts can be found from the isothermal lines presented in Fig. 11a, b. From Fig. 11b it is also clear that isothermals lines are clustering in the bottom wall's direction, indicating more heat transfer at the exit of the channel. Comparing all the Nusselt number distribution in Fig. 12, it is remarkable to note here that the magnitude of the heat transfer from the bottom wall is continuously increasing in the direction of the channel's exit.

3.4 Local and Global Entropy Generation

Because of immense importance in the process of industrial and engineering, we are also interested to explore the role of entropy generated in the microchannel. In the present study the rate of local entropy generation (S_S) is calculated by the fluid friction (S_F), Joule heating (S_J) and heat transfer irreversibilities (S_T). The non-dimensional form of the local entropy generation as used by [9, 23] is given by (with normalized factor $\frac{\kappa_f}{u^2}$):

$$S_S = S_F + S_J + S_T \tag{11}$$

where, $S_F = Br[2\{(\frac{\partial u}{\partial x})^2 + (\frac{Re_v}{Re_u}\frac{\partial v}{\partial y})^2\} + \{\frac{\partial u}{\partial y} + \frac{Re_v}{Re_u}\frac{\partial v}{\partial x}\}^2]^{(m+1)/2}$, $S_J = Br Ha^2 u^2$ and $S_T = \{(\frac{\partial \theta}{\partial x})^2 + (\frac{\partial \theta}{\partial y})^2\}$. Now, integrating S_S, S_F, S_J and S_T over the whole region, the global entropy generation $< S_S >$, $< S_F >$, $< S_J >$ and $< S_T >$ are found. Then, normalize the global entropy generations by its maximum value to get the normalized global entropy generation.

Figure 13 illustrates the effect of different injection rate a_{v_0} (from the bottom wall) on the local entropy generation rate (S_S). The case of no injection is also considered here (ref. Fig. 13a). The lowest entropy is found at the top, then it gradually increases, and the highest S_S is seen at the bottom of the channel. Contour lines in Fig. 13a are almost parallel to each other, indicating that S_S is almost independent to x and only dependent on y. But, in Fig. 13b–d, as the injection rate at the bottom wall increases

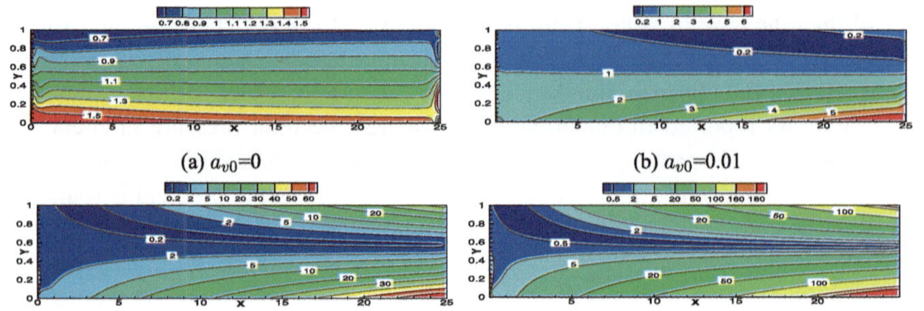

(a) a_{v0}=0 (b) a_{v0}=0.01

(c) a_{v0}=0.05 (d) a_{v0}=0.1

Fig. 13 Local entropy generation rate (S_s) for different bottom wall injection rates a_{v_0} with m = 1, $Re_u = 0.62$, $Re_v = 1.24$, $a_{v0} = 0.1$, P = 0.4, Ha = 0.02, Ec = 0.5 for water base fluid (Pr = 7.97, Ri = 165.71)

from 0.01 to 0.1, the S_S also increases at the both walls. Here S_S is a function of y as well as x. At any y, S_s increases in the direction of the exit of the channel(x). The minimum entropy generation zone identified at the top wall for the no injection case is shifting towards the center of the channel while increasing the injection rate, and at the same time, S_S is getting higher towards the exit of the both walls. Also the active zone of entropy generation found throughout the bottom wall at $a_{v_0} = 0$ is shifting towards the exit along that wall with the increase of injection rate. At a higher injection rate, the channel becomes almost symmetry about the centreline in terms of entropy generation rate.

Figures 14, 15 and 16 present the individual contribution of the fluid friction(S_F), Joule heating (S_J) and heat transfer irreversibility (S_H), respectively for different injection rates a_{v_0}. By looking into the numerical values of S_S, S_F, S_J and S_T from Figs. 13, 15 and 16, it is very much clear that the frictional irreversibility (S_F) is the major contributing part of the total local entropy generation rate (S_S). When no injection at the bottom wall is considered, a fully developed flow is found after a minimal distance from the channel's entrance, consequently the S_F is found uniform throughout the channel in Fig. 14a. But, as the bottom wall injection increases, the velocity of the fluid throughout the channel also increases and the velocity of the fluid becomes different at every point of the channel. As a result, the fluid friction also increases. Figure 14a–d are analogous to their respective contour of total entropy generation rate presented in Fig. 13a–d quality and quantity, that is, the position of active zone of S_F falls at the same position as of S_S. Therefore, a sufficient amount of energy is needed to overcome the fluid's friction at those zones. Eventually, S_F increases throughout the channel when the injection rate increases (ref. Fig. 14b–d). The top wall S_F is less than the bottom wall S_F. This is physically justified, since the top wall is moving towards the direction of fluid while the bottom wall is fixed. It is also seen in Fig. 14b–d that, towards the channel's exit S_F increases. Due to injection from the bottom wall, more fluid is getting into the channel, leading to increase the velocity gradient towards the exit of the channel. From the entropy generation

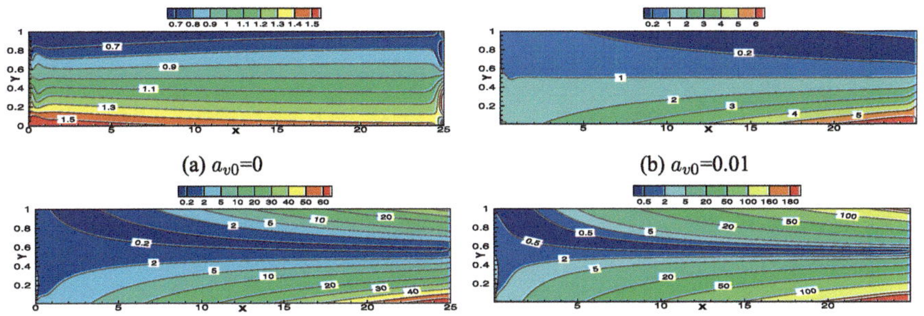

(a) a_{v0}=0 (b) a_{v0}=0.01

(c) a_{v0}=0.05 (d) a_{v0}=0.1

Fig. 14 Individual contribution of fluid friction irreversibility(S_F) to the local entropy generation rate for different bottom wall injection rates a_{v_0} with m = 1, $Re_u = 0.62$, $Re_v = 1.24$, P = 0.4, Ha = 0.02, Ec = 0.5 for water base fluid (Pr = 7.97, Ri = 165.71)

formulae, it is seen that the fluid friction is dependent on Br. The value of Br is considered here as 1.0, but the original value of Br is very less [15]. So, as less as the Br is considered, the fluid friction part and hence, the local entropy generation will be less accordingly.

The contribution of S_J on S_S for different injection rates a_{v_0} is plotted in Fig. 15a–d. The variation of S_J throughout the channel is significantly less and it is directly proportional to the injection factor a_{v_0}. For the case of no injection, S_J contribution at the top wall is more than that of the bottom wall. The position of active zone for $S_J$inside the channel is opposite to that for S_S or S_F when $a_{v_0} = 0$, but for injection cases this active zone is seen at the channel outlet part. Due to continuous injection from the bottom wall, the velocity is found to be maximum at the channel center exit position, which eventually enhances S_J (re. Eq. 11).

Figure 16 presents the individual contribution due to thermal gradients S_T for different injection rates. The maximum heat transfer irreversibility is found at the channel's top and bottom entrance, and the minimum is seen at the middle of the channel entrance. This is due to the temperature jump at the boundary condition. Also, for $a_{v_0} > 0.01$, the maximum and minimum irreversibilities are found at the channel's top and bottom exit, respectively. Active zone for S_J lies at the exit of the top wall where the heat transfer is reported more. Overall, by looking into the color bars of all entropy generation rate figures (Figs. 14, 15 and 16), one can infer that $S_F >> S_T > S_J$ for injection velocity rate up to $a_{v0} = 0.01$, but $S_F >> S_J > S_T$ for higher injection velocity $a_{v0} \geq 0.05$.

Contour of local entropy generation rate for different values of Re_u= 0.62 and $Re_u = 1.24 - 4.0$ is presented in Fig. 13(d) and Figs. 17(a)–(c), respectively. Iso-entropy generation rate contours presented in Fig. 17a–c are almost similar to that in Fig. 13d at $Re_u = 0.62$, that is, active zone for the entropy generation rate inside the channel is not affected by the enhanced speed of the top wall. As shown in Fig. 9, isothermal lines are congregated on both the wall with the increase in Re_u and a successive thinning of the thermal boundary layers at both walls leads to increase

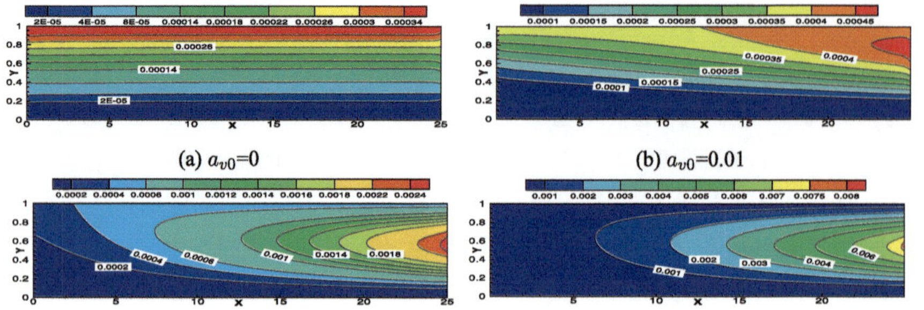

(a) a_{v0}=0 (b) a_{v0}=0.01

(c) a_{v0}=0.05 (d) a_{v0}=0.1

Fig. 15 Individual contribution of Joule heating irreversibility(S_J) to the local entropy generation rate for different bottom wall injection rates a_{v_0} with m = 1, Re_u = 0.62, Re_v = 1.24, P = 0.4, Ha = 0.02, Ec = 0.5 for water base fluid (Pr = 7.97, Ri = 165.71)

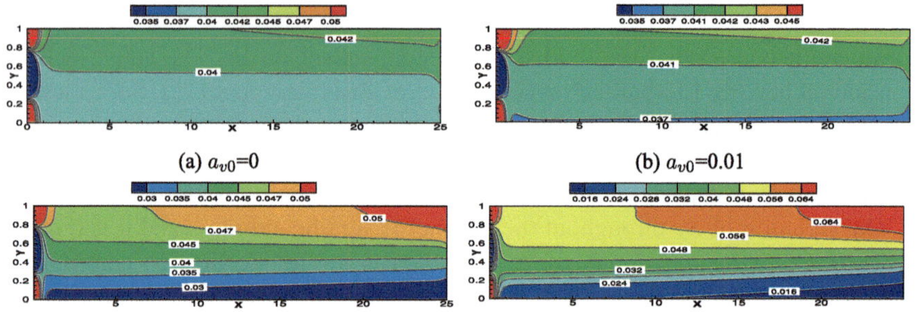

(a) a_{v0}=0 (b) a_{v0}=0.01

(c) a_{v0}=0.05 (d) a_{v0}=0.1

Fig. 16 Individual contribution of the heat transfer irreversibility(S_T) to the local entropy generation rate for different bottom wall injection rates a_{v_0} with m = 1, Re_u = 0.62, Re_v = 1.24, P = 0.4, Ha = 0.02, Ec = 0.5 for water base fluid (Pr = 7.97, Ri = 165.71)

the entropy generation rate at higher Re_u. Here also, the main contributing part is S_F, which is very high at both walls of the channel. However, S_S at the middle of the channel also depends on the S_J and this is significantly less in this situation. Consequently S_S in the middle of the channel is less. The impact of inlet pressure gradient on S_S for different P(not presented here) is similar to different Re_u. Joule heating and heat transfer irreversibility contributing parts are much less than the fluid friction part.

The impact of power-law index m = 1.4, 1.0, 0.8, 0.6 and 0.4 on the local entropy generation rate S_S is discussed based on Fig. 18(a), Fig. 13(d), Fig. 18(b), Fig. 18(c) and Fig. 18(d), respectively. As we reported in Fig. 4, the fluid's velocity interior to the channel is an increasing function of m; so the fluid friction increases with the increase of m. Here also S_F is the major contributing part to the local entropy generation (not presented here). Moreover, S_F (Eq. (11)) is directly proportional to m. Therefore, the dependency of S_S on the power law index m is very natural, which

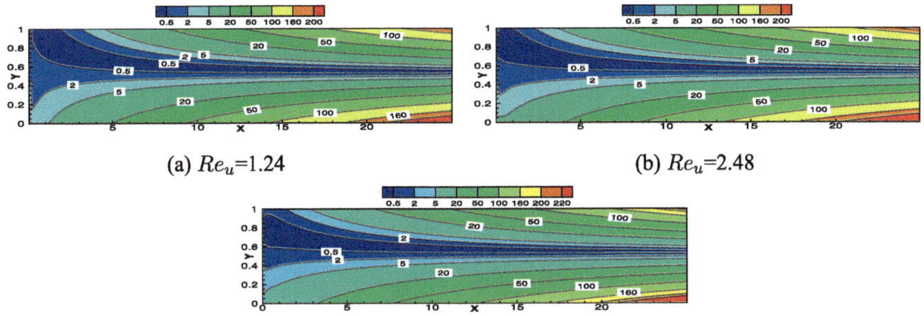

(a) Re_u=1.24 (b) Re_u=2.48 (c) Re_u=4.0

Fig. 17 Local entropy generation rate(S_S) for different values of Re_u with m = 1.0, $Re_v = 1.24$, P = 0.4, Ha = 0.02, Ec = 0.5, $a_{v0} = 0.1$ for water base fluid (Pr = 7.97)

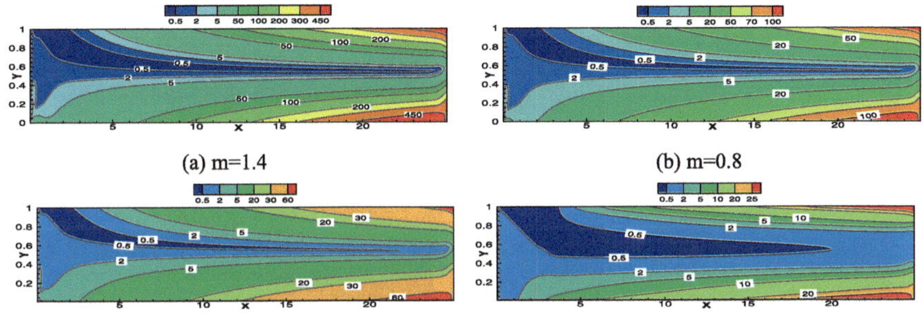

(a) m=1.4 (b) m=0.8 (c) m=0.6 (d) m=0.2

Fig. 18 Local entropy generation rate(S_S) for different power-law index with Re_u=0.62, Re_v=1.24, P = 0.4,Ha = 0.2, Ec = 0.5, $a_{v0} = 0.1$ for water base fluid (Pr = 7.97, Ri = 165.71)

is exhibited in these figures. Here also active zones for entropy generation remains at both walls and towards the exit for all m considered here.

Figure 19a displays the local entropy generation rate at high Prandtl numbers Pr = 57.2 and Ri = 0.9 (i.e., EG:W). Apparently, the entropy generation rate at every point of the channel is found more for EG:W, in comparison to water base fluid since the heat transfer is reported to be more for EG:W than water. While comparing Fig. 14d (water) and Fig. 19b (EG:W) for S_F; Fig. 15d (water) and Fig. 19c (EG:W) for S_J, and Fig. 16d (water) and Fig. 19d (EG:W) for S_T, it is seen that each and every individual components becomes more for EG:W than their respective component of water.

Figure 20 presents the local entropy generation rate (S_S) as well as its all three components (S_F, S_J and S_T) at higher Ha. Eventually, the local entropy generation rate is enhanced significantly at this higher Ha, in comparison to its counter part in Fig. 13d at lower Ha (= 0.02). Though the active zones for the entropy generation rate remain at the end of both walls, but minimum entropy generation zone is represented only at the channel's entry part. Unlike to the case of lower Ha, here in the case of

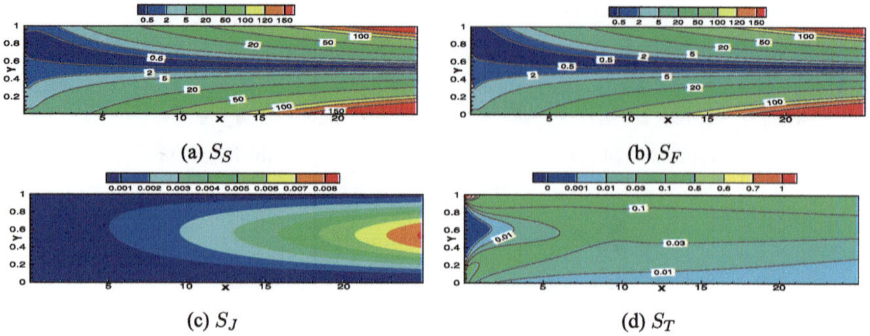

(a) S_S (b) S_F (c) S_J (d) S_T

Fig. 19 Local entropy generation rate(S_S) with its individual contributions(S_F, S_J, S_T) for m = 1.0, $Re_u = 0.62$, Re_v=1.24, P = 0.4, Ec = 0.5, Ha = 0.02, $a_{v0} = 0.1$ and EG:W base fluid (Pr = 57.97, Ri = 0.909)

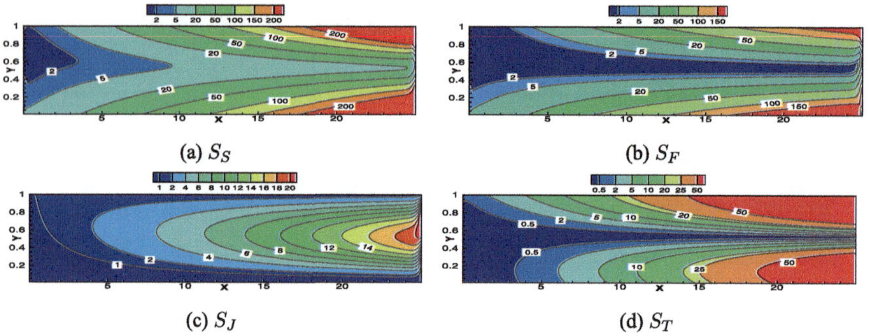

(a) S_S (b) S_F (c) S_J (d) S_T

Fig. 20 Local entropy generation rate(S_S) with its individual contribution(S_F, S_J, S_T) for m = 1.0, $Re_u = 0.62$, $Re_v = 1.24$, P = 0.4, Ha = 1.0, Ec = 0.5, $a_{v0} = 0.1$ for water base fluid (Pr = 7.97, Ri = 10.57)

higher Ha, irreversibilities due to heat and Joule heating are also contributing significantly to the total entropy generation rate at every points of the channel. Precisely, because of active zone of Joule heating irreversibility at the channels center(mainly), the minimum entropy generation zone is restricted to only at the channel entry part. These are not surprising, because isothermal distribution inside the channel is largely affected by Ha (ref. Fig. 11b), owing to increase the thermal gradient. At the same time, S_J formulae (Eq. (11)) is directly proportional to Ha.

4 Conclusion

This research work focused on the flow and heat transfer of a Couette-Poiseuille (C-P) power-law fluid that passes through a horizontal porous microchannel with the combined effects of injection, buoyancy force and magnetic field. The full two-dimensional Navier-Stokes along with the energy equations for incompressible power-law fluid flow are solved numerically in primitive variable formulation technique. The following is a summary of the significant findings from this investigation:

- The u-velocity profile(u = u(y)) of classical C-P flow with no injection can be made more parabolic pattern by increasing the injection rate or/and changing the position x towards the end of the channel. In other words, a one-dimensional classical C-P flow inside a microchannel is made two dimensional under the present flow conditions.
- For the case of no-injection, the vorticity is found negative throughout the channel at lower P and Re_u, while positive vorticity is reported in the upper half of the channel by increasing P or Re_u or injection rate (a_{v_0}). Separation of boundary layer is seen at the upper wall for Dilatant type of fluids.
- At lower Re_u, isothermal lines are almost parallel, while this uniformity is lost at higher Re_u. The base fluid has more impact on the isothermal lines, in comparison to the power law index value. The temperature of the fluid can be made more than the walls' temperature at higher magnetic field (Ha), and the temperature of the fluid becomes more and more hotter as we move towards the exit of the channel.
- Nusselt number distribution along the walls almost remain same throughout the channel for Newtonian fluid, while this variation is noticeable up to middle of the channel for non- Newtonian fluid (both types: Dilatant and Pseudoplastic). The total heat transfer at the top wall is reported more than the bottom wall at any Re_u and these are more for the EG:W than the water base fluid.
- For the no-injection case, contour lines of local entropy generation rate (S_S) are only dependent on y, while with the increase of injection rate, S_S becomes function of y as well as x. Also the active zone of entropy generation found on the bottom wall for the non-injection case is shifting towards the exit along that wall with the increase of a_{v_0}. Each and every individual components of S_S becomes more for EG:W than their respective components for water. S_S at both walls is increasing with x as well as while increasing either a_{v_0} or Re_u or m or Ha. The change of type of fluid (due to m) makes the enormous changes in S_S.
- The contributions of irreversibilities due to Joule heating($< S_J >$) and thermal gradient ($< S_T >$) may be neglected while calculating the global entropy generation rate for the cases of variations of a_{v_0}, m and Re_u. However, at higher Ha, $< S_J >$ and $< S_T >$ are contributing around 8% and 40%, respectively, to the global entropy. Overall, the maximum entropy is generated in the system for the case of higher m, a_{v_0}, Re_u and Ha.

References

1. Ibáñez, G., López, A., Pantoja, J., Moreira, J., Reyes, J.A.: Optimum slip flow based on the minimization of entropy generation in parallel plate microchannels. Energy **50**, 143–149 (2013)
2. Makinde, O.D., Chinyoka, T.: Numerical investigation of buoyancy effects on hydromagnetic unsteady flow through a porous channel with suction/injection. J. Mech. Sci. Technol. **27**(5), 1557–1568 (2013)
3. Davaa, G., Shigechi, T., Momoki, S.: Effect of viscous dissipation on fully developed heat transfer of non-newtonian fluids in plane laminar poiseuille-couette flow. Int. Commun. Heat Mass Transf. **31**(5), 663–672 (2004)
4. Nakayama, A., Koyama, H.: An analysis for friction and heat transfer characteristics of power-law non-newtonian fluid flows past bodies of arbitrary geometrical configuration. Wärme-und Stoffübertragung **22**(1), 29–36 (1988)
5. Makinde, O., Chinyoka, T.: Analysis of unsteady flow of a variable viscosity reactive fluid in a slit with wall suction or injection. J. Pet. Sci. Eng. **94**, 1–11 (2012)
6. Dimitrienko, Y.I., Li, S.: Numerical simulation of MHD natural convection heat transfer in a square cavity filled with carreau fluids under magnetic fields in different directions. Comput. Appl. Math. **39**(4), 252 (2020)
7. Zhao, J.: Finite volume method for mixed convection boundary layer flow of viscoelastic fluid with spatial fractional derivatives over a flat plate. Comput. Appl. Math. **40**(1), 10 (2021)
8. Kefayati, G.R.: Mesoscopic simulation of double-diffusive mixed convection of pseudoplastic fluids in an enclosure with sinusoidal boundary conditions. Comput. Fluids **97**, 94–109 (2014)
9. Kefayati, G., Sidik, N.A.C.: Simulation of natural convection and entropy generation of non-newtonian nanofluid in an inclined cavity using buongiorno's mathematical model (part ii, entropy generation). Powder Technol. **305**, 679–703 (2017)
10. Hajji, F., Mazgar, A., Sakly, A., Nejma, F.B.: Entropy generation due to combined natural convection and thermal radiation within a rectangular enclosure. Heat Trans. Eng. **39**(19), 1698–1714 (2018)
11. Ullah, H., Hayat, T., Ahmad, S., Alhodaly, M.S.: Entropy generation and heat transfer analysis in power-law fluid flow: finite difference method. Int. Commun. Heat Mass Transf. **122**, 105111 (2021)
12. Mokarizadeh, H., Asgharian, M., Raisi, A.: Heat transfer in couette-poiseuille flow between parallel plates of the giesekus viscoelastic fluid. J. Non-Newton. Fluid Mech. **196**, 95–101 (2013)
13. Coelho, P.M., Poole, R.J.: Heat transfer of power-law fluids in plane couette-poiseuille flows with viscous dissipation. Heat Trans. Eng. **41**(13), 1189–1207 (2020)
14. Al-Ashhab, S.: Properties of boundary-layer flow solutions for non-newtonian fluids with non-linear terms of first and second-order derivatives. J. Eng. Math. **123**(1), 29–39 (2020)
15. Mondal, P., Maiti, D.K., Shit, G., Ibáñez, G.: Heat transfer and entropy generation in a MHD couette-poiseuille flow through a microchannel with slip, suction-injection and radiation. J. Therm. Anal. Calorim. **147**(6), 4253–4273 (2022)
16. Mondal, P., Maiti, D.K.: Base fluids, its temperature and heat source on mhd couette-poiseuille nanofluid flow through slippy porous microchannel with convectiveradiative condition: entropy analysis. J. Eng. Thermopys. vol. Accpted, not publised (2022)
17. Rahmati, A.R., Derikvand, M.: Numerical study of non-newtonian nano-fluid in a micro-channel with adding slip velocity and porous blocks. Int. Commun. Heat Mass Transf. **118**, 104843 (2020)
18. Zehtabiyan-Rezaie, N., Mirzaei, M., Saffar-Avval, M.: Numerical investigation of magnetic field effect on heat transfer and entropy generation in channel; new approach for fluid and length scale selections. Heat Trans. Eng. **38**(13), 1222–1232 (2017)
19. Tzirtzilakis, E.: A mathematical model for blood flow in magnetic field. Phys Fluids **17**(7) (2005)
20. Kundu, M., Vamsee, K., Maiti, D.: Simulation and analysis of flow through microchannel. Asia-Pacific J. Chem. Eng. **4**(4), 450–461 (2009)

21. Bhattacharyya, S., Maiti, D., Dhinakaran, S.: Influence of buoyancy on vortex shedding and heat transfer from a square cylinder in proximity to a wall. Numer. Heat Trans. Part A: Appl. **50**(6), 585–606 (2006)
22. Mondal, P., Maiti, D.K.: Variable injection-suction and temperature on couette-poiseuille non-newtonian flow through slippy microchannel: heat transfer and entropy generation. J. Therm. Anal. Calorim. **148**(21), 12225–12246 (2023)
23. Khan, M.I., Hayat, T., Alsaedi, A.: Numerical investigation for entropy generation in hydromagnetic flow of fluid with variable properties and slip. Phys. Fluids **30**(2), 023601 (2018)

Electroosmotic Flow of Generalized Maxwell Fluids in Polyelectrolyte Grafted Nanopore Modulated by Ion Partitioning Effects Under AC Electric Field

Priyanka Koner and Subrata Bera

Abstract The article presents a theoretical study on the electroosmotic flow of generalized Maxwell fluids through a polyelectrolyte layer-coated nanopore, considering ion partitioning under an alternating current electric field. The relative permittivity of the polyelectrolyte layer is assumed to be significantly lower than that of the electrolyte solution, creating ion partitioning effects. The Born formula, related to the change in ion free energy between regions of different permittivities, accounts for this effect. Ionic species distribution follows the modified Boltzmann distribution for low surface charge density. The Born formula, integrated with the modified Poisson-Boltzmann equation, determines the induced potential distribution. Using the Debye-Hückel approximation, an analytic solution of the modified Poisson-Boltzmann equation is derived within and outside the polyelectrolyte layer. The Cauchy momentum equation and the Maxwell constitutive relationship's stress components are considered for fluid flow, presenting an analytic solution for the full domain. The study adopts cylindrical coordinates, assuming axisymmetric, fully developed flow. Solutions use the Modified Bessel function of the first and second kinds. The effects of permittivity differences, PEL thickness, fixed charge density, and the softness parameter are analyzed. Comparisons between generalized Maxwell and Newtonian fluid models show increased permittivity differences reduce axial velocity, more prominently in Maxwell fluids. Increased PEL thickness and fixed charge density decrease average flow, while oscillating Reynolds numbers increase flow oscillation, reducing average flow.

P. Koner
School of Applied Sciences, Kalinga Institute of Industrial Technology (KIIT) Deemed to be University, Bhubaneswar 751024, India
e-mail: priyanka.konerfma@kiit.ac.in

S. Bera (✉)
Department of Mathematics, National Institute of Technology Silchar, Silchar, India
e-mail: subrata@math.nits.ac.in

S. Bhattacharyya and H. S. Mahato (eds.), *Modeling, Analysis and Simulations of Multiscale Transport Phenomena*, Springer Proceedings in Mathematics & Statistics 491, https://doi.org/10.1007/978-981-96-3098-1_10

1 Introduction

Due to the growing interest in developing nanofluidic systems for controlling ionic transport, electrokinetic transport of liquid or ion through nanoscaled pores or channels perform a crucial task. The imposition of an electric field along the micro capillary leads the mobile ions to experience an electrical body force. For the ability to detect and analyze biomolecules that pass through these devices, there are enormous practical benefits of fluid and ions transport phenomena in biophysics and biochemistry [1–3] originated by electric field. Few of them are energy conversion [4], seawater desalination [5], biomolecule concentration and separation [6], and nanofluidic diodes [7].

Whenever an electrolyte solution is brought into touch with charged wall, the redistribution of ions causes the formation of electric double layer (EDL) at the wall surface. Electroosmosis is associated with the fluid motion initiation by introducing an applied electric field on diffuse layer of the EDL. It is wildly used alternative techniquefor manipulating liquids flowing in a microfluidic devices. The amplitude and direction of fluid flow depends many factor such as electric field strength, surface charge density, fluid viscosity, permittivity and many more. Several authors investigated different aspect of EOF under imposed DC and AC electric field [8–10]. Stroock et al. [11] explored the EOF in microchannels in two distinct patterns: (a) When surface charge fluctuates perpendicular to the electric field. These sort of arrangement initiates multidirectional flow parallel to the field direction. (b) When the charge pattern on the surface fluctuates along a line parallel to the field. Bhattacharyya et al. [12] have looked into how electrokinetics and mixing are influenced in charged microchannel by a pressure gradient and electric field. They did detailed computer simulations relying upon the Poisson, Nernst-Planck, and Navier-Stokes equations. Also have addressed the electrokinetic transport and other hydrodynamic phenomena.

Despite its many applications, the use of DC electric flux can result in bubble formation and electrode degradation. AC electric fields are used as an alternative mechanism to overcome these obstacles. Over the last few decades, there have been numerous applications for the utilisation of AC electric fields in sectors such as colloidal science and biotechnology. Loucaides et al. [13] examined the dielectrophoretic entrapment of nanoparticles in parallel electrode arrays using AC electric field. The steric impact of particle collection is taken into consideration in the numerical simulations, resulting in more realistic findings than the non-steric model. Dielectrophoretic DNA trapping has also been considered on this model. Jian et al. [14] have considered linear Poisson-Boltzmann (PB) and Navier-Stokes equation to look at how the EOF through a microannulus changes over time. The EOF velocity is modelled analytically as a function of radial distance, periodic time, and other partinent factors. For thin EDL, they have shown that the EOF amplitude has a square plug like contour when Reynolds number is small. The variations of EOF velocityare mostly limited inside narrow layer around two solid surfaces at high Reynolds numbers. Using the backwards-Euler time stepping numerical method, Luo [15] explored two-dimensional, time-periodic EOF generated by AC electric field along patchwise surface heterogeneities spread across wall of the micro-channel confinements. The creation of localised stream circulations inside bulk flow is demonstrated to be caused by the occurrence of surface heterogeneities on oppositely charged micro-channel confinement.

A rigid nanochannel surrounded by polyelectrolyte membrane is called polyelectrolyte layer (PEL) coated nanochannel or soft nanochannel. The electrokinetic movement across PEL coated nanopores gives comprehensive information for modification of nanopore-based sensing devices and analysis of physiological mechanisms in living beings. Generally, these type of soft nanochannel or pore contains a fixed charge density. Chanda et al. [16] developed a hypothesis to explore the influence of the steric effect on the electrostatic potential distribution of a charged porous surface which is in touch with an electrolyte liquid. This argument is grounded on free energy model which authorizes both the structural characteristics of soft interface and the impact of finite ion sizes. Keh and Ding [17] provided an analysis of the steady flow of electricity through a long, uniform capillary tube. The internal capillary wall is coated with an adsorbate or charge-regulating polymer layer that is in equilibrium with the surrounding electrolyte liquid. It is illustrated in this paper that at uniform densities in the frictional segments and PEL ionogenic functional groups are scattered on this solvent-permeable and ion-penetrable surface. Saha et al. [18] studied EOF in a nanochannel with continuous negative surface charge at the walls, filled with electrolyte liquid and encircled by non-miscible oil layer, and powered by a DC or AC electric field. The nonlinear PB equation can be employed to determine electrostatic potential. Here, they considered the unstable Stokes equation to compute the axial velocity distribution using the known electric potential. Since the PEL's dielectric permittivity is often lower than that of the electrolyte solution, a successive impact will take place. The consequences that arise due to the partitioning effects on flow field and electrostatic interaction are reported theoritically and experimentally by [19–21]. Leunissen el el. [19] have discovered that unequal partitioning of micro-ions between two liquid phases is the operative factor underlying the majority of the observed electrostatic effects. Sin [21] have discussed the dependency of the electrostatic properties on the electrolyte ion size and distance between soft surfaces.

In most physiology and medicine applications, non-Newtonian bio fluids are frequently utilised in microfluidic devices. Significant research on non-Newtonian fluids have been accounted to illustrate the complicated stream patterns in microfluidic devices. Liu et al. [22] have used the separation of variables method to analyse AC magnetohydrodynamic EOF between micro-parallel plates for incompressible Maxwell fluids. Fluids were propelled by the Lorentz force, which was created by the interactions of two horizontal electrical fields and a vertical magnetic field. They employed graphical representation to illustrate the velocity distributions and average flow rates. Under Debye-H*ü*ckel linear approximation, analytic solutions for time-bounded EOF of generalized Maxwell fluids via a rectangular microconfinement are discovered by [23] in two-dimension. The study includes the governing equations of linearized PB and the Cauchy momentum (CM) equation with general Maxwell constitutive relationship. The velocity distributions and average flow rates are numerically reported in this paper in rectangular microchannel. Peralta et al. [24] developed a analytic process for the mass transfer of electroneutral solvent in micro-annulus microdevice motivated by oscillatory EOF (OEOF) of Maxwell fluid behaviour. In this setup, the annular microdevice connects two reservoirs that contain solute with varying concentrations. The mathematical modelling of the OEOF assumes that the micro-wall annulus's zeta potentials can be symmetric or asymmetric. Sun et al. [25] have offered analytical solutions through cylindrical micro-pipe for time-periodic EOF of linear viscoelastic fluids defined by generic Maxwell model. The linearized PB equation, alongside the CM equation in association with generalized Maxwell shear stress relationship is used to solve the problem. The flow region is separated into a depletion layer and a bulk flow exterior of the depletion layer in this case and the depletion effect brought on by the Maxwell fluid's macromolecules interacting with the channel surface is considered.

The aims of this paper is to examine the ion partitioning effect of a transient non-Newtonian Maxwell fluid throughout a PEL surrounded nanopore operated by externally enforced AC electric field. The solutions of the induced electric potential is acquired from the linearized Poisson-Boltzman equation in the inside and outside of the electrolyte layer. Accordingly, the solution for the velocity profile using the known induced potential is derived by solving CM equation along with stress component represented by the Maxwell constitutive relationship in cylindrical coordinates. Here we have presented the significance of the permittivity ratio, softness parameter, PEL thickness and scaled charge density on the OEOF. The viscoelastic properties of the Maxwell fluid are provided through a discussion of the domination of the Reynolds number and normalized relaxation time. The impacts of PEL thickness on the Maxwell and the Newtonian fluids are discussed in this study and also the difference in velocity profile with rigid nanopore are shown. We calculate the average flow rate and give a review of the influence of the various dimensionless parameters.

2 Mathematical Model

We have considered cylindrical nanopore of radius h, inner surface of which is surrounded by a brush like polymer structure of thickness p (as Fig. 1). We have considered that the nanopore walls experience a surface charge density σ^*. The PEL molecules exert a fixed charge $\rho_{fix} = ZeN$ on the liquid that is identically distributed throughout the PEL where N and Z indicate the molar concentration and valence of the PEL ions. Both within and outside of the PEL, the electrolyte ions may be present. We have assumed that the permittivity of the PEL is different from electrolyte. The nanopore is filled with a symmetric aqueous solute of KCl salt with dielectric constant ϵ_f. For such KCl solute the radius of cations and anions are approximately equivalent. The dielectric constant of the PEL is ϵ_p. The viscoelastic nature of the fluid is employed here by applying Maxwell constitutive relation for stress component. We have denoted here the the bulk ionic concentration, density and zero shear rate viscosity of the fully developed and axisymmetric as n^0, ρ and η_0 respectively. The flow is generated by OEOF instigated by the EDL, formed at the wall-electrolyte

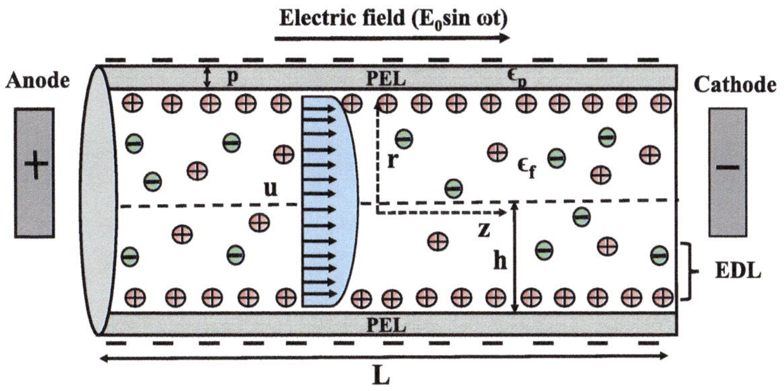

Fig. 1 The schematic diagram of the annular region between two concentric cylinders

interface due to ion redistribution and the imposition of the time-dependent electric field $E_z(t^*) = E_0 sin(\omega t^*)$. Here, E_0, $\omega(= 2\pi \$)$ and t^* are magnitude, angular frequency of electric field with frequency $\$$ and time scale respectively. EDL thickness is denoted as $\lambda_D = \sqrt{\epsilon_0 \epsilon_f k_B T / \Sigma_i (z_i e)^2 n^0}$. Here, e is the electron charge, z_i (i = 1, 2) is valence, T is absolute temperature, k_B is Boltzmann constant and ϵ_0 identifies as permittivity of the free space. The Debye-Huckel parameter κh is denoted as h/λ_D.

2.1 Analytic Expression for Induced Potential

In this study, we assume the convective transport of ions is negligible. The variations of cations, anions and electric potential are occurred in the normal direction to the nanopore wall. These assumptions allow the steady Boltzmann distribution of the ions. Under the above-mentioned assumptions for steady EOF through cylindrical nanopore, the electric potential can be modelled using the PB equation.

$$\nabla^2 \psi^* = \begin{cases} -\frac{\Sigma z_i e n_i^*}{\epsilon_0 \epsilon_f} & 0 \le r^* \le h - p^* \\ -\frac{\Sigma z_i e n_i^*}{\epsilon_0 \epsilon_p} f_i - \frac{ZeN}{\epsilon_0 \epsilon_p} & h - p^* < r^* \le h. \end{cases} \tag{1}$$

The above equation is generalized PB equation including the ion partitioning effect. Here, $f_i = \exp(-\Delta W_i / k_B T)$ [26, 27] is the ion partitioning coefficient of ith ionic species which relates the ionic concentration n_i between two different reference medium and is written as $n_i^* \mid_{r=h-p^{*-}} = f_i n_i^* \mid_{r=h-p^{*+}}$. Here, ΔW_i is the Born energy [28] that measure the electrostatic interaction of charge transfer from electrolyte to PEL phase and expressed as

$$\Delta W_i = \frac{(z_i e)^2}{8\pi r_i \epsilon_0} \left(\frac{1}{\epsilon_p} - \frac{1}{\epsilon_f} \right) \tag{2}$$

Here r_i denotes the hydration radii of the ith ionic species. The boundary conditions corresponding to the above PB equations are

$$\begin{aligned} &\frac{d\psi_f^*}{dr^*}|_{r^*=0} = 0, \qquad \psi_p^*|_{r^*=(h-p^*)^+} = \psi_f^*|_{r^*=(h-p^*)^-} \\ \epsilon_0\epsilon_p \frac{d\psi_p^*}{dr^*}|_{r^*=(h-p^*)^+} &= \epsilon_0\epsilon_f \frac{d\psi_f^*}{dr^*}|_{r^*=(h-p^*)^-}, \qquad \frac{d\psi_p^*}{dr^*}|_{r^*=h} = -\frac{\sigma^*}{\epsilon_0\epsilon_f} \end{aligned} \tag{3}$$

The induced potential at the PEL and electrolyte phase are identified as ψ_p^* and ψ_f^*. For describing the relation between the number concentration and electric potential of ion in equilibrium, the Boltzmann distribution is employed and it is defined as

$$n_i^* = n_i^0 \exp(\frac{-z_i e\psi^*}{k_B T}) \tag{4}$$

The normalization factor for ionic concentration n_i^*, approaches to n^0 (i.e., $n_+^0 = n_-^0 = n^0$). Substituting above relation to the Poisson Eq. 1 and imposing the Debye-Hückel approximation we obtained dimensionless PB equation for symmetric electrolyte as

$$r^2\frac{d^2\psi_f}{dr^2} + r\frac{d\psi_f}{dr} = (\kappa hr)^2\psi_f \qquad 0 \le r \le 1-p \tag{5a}$$

$$r^2\frac{d^2\psi_p}{dr^2} + r\frac{d\psi_p}{dr} = \frac{1}{\epsilon_r}\left[(\kappa hr)^2\exp(-\Delta w)\psi_p - q_{fix}r^2\right] \qquad 1-p < r \le 1 \tag{5b}$$

Here, ϵ_r is the permittivity ratio between the PEL and electrolyte layer. Boundary conditions in non- dimensional form are:

$$\begin{aligned} &\frac{d\psi_f}{dr}|_{r=0} = 0, \qquad \psi_p|_{r=(1-p)^+} = \psi_f|_{r=(1-p)^-} \\ \epsilon_r\frac{d\psi_p}{dr}|_{r=(1-p)^+} &= \frac{d\psi_f}{dr}|_{r=(1-p)^-}, \qquad \frac{d\psi_p}{dr}|_{r=1} = -\sigma \end{aligned} \tag{6}$$

We scale electric potential ψ^* by $\psi_0(= k_BT/e)$ and surface charge density σ^* by and $\epsilon_0\epsilon_f\psi_0/h$, and Born energy ΔW_i by k_BT. The dimensionless Born energy is denoted by m here. We express the PEL fixed charge density as $q_{fix} = Ne^2zZh^2/\epsilon_0\epsilon_f k_BT$. Solution of Eqs. 5a and 5b considering to the above conditions (6) is

$$\psi_f = \mathcal{A}I_0(\kappa hr) + \mathcal{B}K_0(\kappa hr) \qquad 0 \le r \le 1-p \tag{7a}$$

$$\psi_p = \mathcal{C}I_0\left(\kappa h\sqrt{\frac{m}{\epsilon_r}}r\right) + \mathcal{D}K_0\left(\kappa h\sqrt{\frac{m}{\epsilon_r}}r\right) + \frac{q_{fix}}{m(\kappa h)^2} \qquad 1-p < r \le 1 \tag{7b}$$

Arbitrary constnts are stated as

$$\mathcal{A} = \left(\frac{q_{fix}}{(\kappa h)^2 m} + F - \frac{ES}{Q}\right)\left(\frac{Q}{QR - PS}\right)$$

$$\mathcal{B} = 0$$

$$C = \left(\frac{q_{fix}}{(\kappa h)^2 m} + F - \frac{ER}{P}\right)\left(\frac{P}{QR - PS}\right)$$

$$\mathcal{D} = \frac{1}{K_1\left(\kappa h\sqrt{\frac{m}{\epsilon_r}}\right)}\left[\frac{\sigma}{\kappa h\sqrt{\frac{m}{\epsilon_r}}} + CI_1\left(\kappa h\sqrt{\frac{m}{\epsilon_r}}\right)\right]$$

accordingly the known variables are

$$P = I_1(\kappa h(1 - p))$$

$$Q = \sqrt{m\epsilon_r}\left(I_1\left(\kappa h\sqrt{\frac{m}{\epsilon_r}}(1 - p)\right) - \frac{I_1\left(\kappa h\sqrt{\frac{m}{\epsilon_r}}\right)}{K_1\left(\kappa h\sqrt{\frac{m}{\epsilon_r}}\right)}K_1\left(\kappa h\sqrt{\frac{m}{\epsilon_r}}(1 - p)\right)\right)$$

$$R = I_0(\kappa h(1 - p))$$

$$S = I_0\left(\kappa h\sqrt{\frac{m}{\epsilon_r}}(1 - p)\right) + \left(\frac{I_1\left(\kappa h\sqrt{\frac{m}{\epsilon_r}}\right)}{K_1\left(\kappa h\sqrt{\frac{m}{\epsilon_r}}\right)}\right)K_0\left(\kappa h\sqrt{\frac{m}{\epsilon_r}}(1 - p)\right)$$

$$E = -\frac{K_1\left(\kappa h\sqrt{\frac{m}{\epsilon_r}}(1 - p)\right)}{K_1\left(\kappa h\sqrt{\frac{m}{\epsilon_r}}\right)}\frac{\sigma\epsilon_r}{\kappa h}$$

$$F = -\frac{K_0\left(\kappa h\sqrt{\frac{m}{\epsilon_r}}(1 - p)\right)}{K_1\left(\kappa h\sqrt{\frac{m}{\epsilon_r}}\right)}\frac{\sigma}{\kappa h\sqrt{\frac{m}{\epsilon_r}}}$$

2.2 *Analytic Expression for Velocity Field*

We have analyzed the OEOF of an electrolyte in a capillary tube with inner confinement enclosed with polymer brushes under the influence of an AC electric field when an external pressure gradient and gravitational influences are not taken into consideration. The EOF in the electrolyte and PEL can be described by the one-directional CM equation with electric body force term, which is represented as

$$\rho\frac{\partial}{\partial t^*}v_f^*\left(r^*, t^*\right) + \frac{1}{r^*}\frac{\partial}{\partial r^*}\left(r^*\tau_{rz}^*\right) = \rho_e^* E_z(t^*) \qquad 0 \le r^* \le h - p^* \quad (8a)$$

$$\rho\frac{\partial}{\partial t^*}v_p^*\left(r^*,t^*\right)+\frac{1}{r^*}\frac{\partial}{\partial r^*}\left(r^*\tau_{rz}^*\right)=\rho_e^*E_z(t^*)-\gamma v_p^*(r^*,t^*)\qquad h-p^*<r^*\leq h \tag{8b}$$

Here, we identify v_p^* and v_f^* as velocity components in the PEL and electrolyte regions.

To solve the momentum equation we consider the boundary conditions as

$$\begin{aligned}\frac{\partial v_f^*}{\partial r^*}|_{r^*=0}=0,\qquad v_p^*|_{r^*=(h-p^*)^+}=v_f^*|_{r^*=(h-p^*)^-}\\ \frac{\partial v_p^*}{\partial r^*}|_{r^*=(h-p^*)^+}=\frac{\partial v_f^*}{\partial r^*}|_{r^*=(h-p^*)^-},\qquad v_p^*|_{r^*=h}=0\end{aligned} \tag{9}$$

Here, we consider the following constitutive equation to describe the stress tensor which is stated as

$$(1+\lambda\frac{\partial}{\partial t^*})\tau_{r,z}^*=-\eta_0\frac{\partial v^*}{\partial r^*} \tag{10}$$

We identify the oscillatory velocity field of as $v(r^*,t^*)$ and λ as the relaxation time [29]. The term γ in the RHS of the Eq. 8b is the hydrodynamic drag coefficient [30] related to flow resistance incited by the PEL. The normalization factors for time t^* and velocity v^* are $1/\omega$ and $U_{HS}=\epsilon_0\epsilon_f E_0\psi_0/\eta_0$ (Helmholtz-Smoluchowski velocity). The stress component $\tau_{r,z}^*$ is non-dimensionlized by $\eta_0 U_{HS}/h$. the kinematic viscosity is denoted as $\nu=\eta_0/\rho$. We defined the angular Reynolds number by $Re_w=\rho\omega h^2/\eta_0$. for oscillatory electric field the complex form of electric field and velocity is given as

$$E_z(t^*)=Im\left[E_0e^{i\omega t^*}\right]\quad\&\quad v^*(r^*,t^*)=Im\left[v_0^*(r^*)e^{i\omega t^*}\right] \tag{11}$$

Here, the imaginary part of a complex component is specified by Im. The amplitude of the velocity is $v_0^*(r)$ where i is the imaginary number.

In non-dimensional form CM is illustrated as

$$Re_w\frac{\partial v_f}{\partial t}=-\frac{1}{r}\frac{\partial}{\partial r}(r\tau_{rz})-(\kappa h)^2\psi_f Im(e^{it})\qquad 0\leq r\leq 1-p \tag{12a}$$

$$Re_w\frac{\partial v_p}{\partial t}=-\frac{1}{r}\frac{\partial}{\partial r}(r\tau_{rz})-(\kappa h)^2 m\psi_p Im(e^{it})-\lambda_s^2 v_p\qquad 1-p<r\leq 1 \tag{12b}$$

non dimensional constitutive relation for stress component is

$$\left(1+\lambda\omega\frac{\partial}{\partial t}\right)\tau_{rz}=-\frac{\partial v}{\partial r} \tag{13}$$

The term $\lambda\omega$ is the normalized relaxation time, it is also called Deborah or elasticity number. The dimensional frictional coefficient is indicated by λ_s=h/λ_0^{-1} inside the

PEL where we identify as $\lambda_0^{-1} = \sqrt{\eta_0/\gamma}$ degree of softness. Making use of the Eq. 13 in (12a) and (12b) then considering (11) we have the non-dimensional velocity equation in the form

$$\frac{d^2(v_0)_e}{dr^2} + \frac{1}{r}\frac{d(v_0)_e}{dr} - (v_0)_e i Re_w(1+i\lambda\omega) = (\kappa h)^2(1+i\lambda\omega)\mathcal{A}I_0(\kappa hr) \qquad 0 \le r \le 1-p \tag{14a}$$

$$\frac{d^2(v_0)_p}{dr^2} + \frac{1}{r}\frac{d(v_0)_p}{dr} - (v_0)_p\left((\lambda_s^2 + iRe_w)(1+i\lambda\omega)\right) = (\kappa h)^2 m(1+i\lambda\omega) *$$

$$\left[CI_0\left(\kappa h\sqrt{\frac{m}{\epsilon_r}}r\right) + \mathcal{D}K_0\left(\kappa h\sqrt{\frac{m}{\epsilon_r}}r\right) + \frac{q_{fix}}{(\kappa h)^2 m}\right] \qquad 1-p < r \le 1 \tag{14b}$$

Boundary conditions in non-dimensional form are stated as

$$\begin{aligned} &\frac{d(v_0)_e}{dr}\Big|_{r=0} = 0, \qquad (v_0)_p|_{r=(1-p)^+} = (v_0)_e|_{r=(1-p)^-} \\ &\frac{d(v_0)_p}{dr}\Big|_{r=(1-p)^+} = \frac{d(v_0)_e}{dr}\Big|_{r=(1-p)^-}, \qquad (v_0)_p|_{r=1} = 0 \end{aligned} \tag{15}$$

Solving of Eqs. 14a and 14b in the presence of (15) we get

$$(v_0)_e = C_1 I_0\left(\sqrt{iRe_w(1+i\lambda\omega)}r\right) + D_1 K_0\left(\sqrt{iRe_w(1+i\lambda\omega)}r\right) + GI_0(\kappa hr) \qquad 0 \le r \le 1-p \tag{16a}$$

$$(v_0)_p = C_2 I_0\left(\sqrt{(\lambda_s^2 + iRe_w)(1+i\lambda\omega)}r\right) + D_2 K_0\left(\sqrt{(\lambda_s^2 + iRe_w)(1+i\lambda\omega)}r\right)$$

$$+ MI_0\left(\kappa h\sqrt{\frac{m}{\epsilon_r}}r\right) + NK_0\left(\kappa h\sqrt{\frac{m}{\epsilon_r}}r\right) - q_{fix}\frac{1+i\lambda\omega}{(\lambda_s^2 + iRe_w)(1+i\lambda\omega)} \qquad 1-p < r \le 1 \tag{16b}$$

Arbitrary constants introduced in Eqs. 16a and 16b are

$$\begin{aligned} C_1 &= \frac{M_{10}M_7 - M_4M_9}{M_3M_7 - M_4M_6} \\ C_2 &= \frac{M_6M_{10} - M_9M_3}{M_3M_7 - M_4M_6} \\ D_1 &= 0 \\ D_2 &= -C_2\frac{I_0\left(\sqrt{(\lambda_s^2 + iRe_w)(1+i\lambda\omega)}\right)}{K_0\left(\sqrt{(\lambda_s^2 + iRe_w)(1+i\lambda\omega)}\right)} - \frac{M_2}{K_0\left(\sqrt{(\lambda_s^2 + iRe_w)(1+i\lambda\omega)}\right)} \end{aligned}$$

Accordingly the constants are

$$G = \mathcal{A}(\kappa h)^2 \frac{1 + i\lambda\omega}{(\kappa h)^2 - iRe_w(1 + i\lambda\omega)}$$
$$M = C(\kappa h)^2 m\epsilon_r \frac{1 + i\lambda\omega}{(\kappa h)^2 m - \epsilon_r((\lambda_s^2 + iRe_w)(1 + i\lambda\omega))}$$
$$N = \mathcal{D}(\kappa h)^2 m\epsilon_r \frac{1 + i\lambda\omega}{(\kappa h)^2 m - \epsilon_r((\lambda_s^2 + iRe_w)(1 + i\lambda\omega))}$$

$$M_2 = MI_0\left(\kappa h\sqrt{\frac{m}{\epsilon_r}}\right) + NK_0\left(\kappa h\sqrt{\frac{m}{\epsilon_r}}\right) - q_{fix}\frac{1 + i\lambda\omega}{(\lambda_s^2 + iRe_w)(1 + i\lambda\omega)}$$
$$M_3 = I_0\left(\sqrt{iRe_w(1 + i\lambda\omega)}(1 - p)\right)$$
$$M_4 = I_0\left(\sqrt{(\lambda_s^2 + iRe_w)(1 + i\lambda\omega)}(1 - p)\right) - \frac{I_0\left(\sqrt{(\lambda_s^2 + iRe_w)(1 + i\lambda\omega)}\right)}{K_0\left(\sqrt{(\lambda_s^2 + iRe_w)(1 + i\lambda\omega)}\right)}$$
$$* K_0\left(\sqrt{(\lambda_s^2 + iRe_w)(1 + i\lambda\omega)}(1 - p)\right)$$
$$M_5 = MI_0\left(\kappa h\sqrt{\frac{m}{\epsilon_r}}(1 - p)\right) + NK_0\left(\kappa h\sqrt{\frac{m}{\epsilon_r}}(1 - p)\right) - q_{fix}\frac{1 + i\lambda\omega}{(\lambda_s^2 + iRe_w)(1 + i\lambda\omega)}$$
$$M_6 = \sqrt{iRe_w(1 + i\lambda\omega)}I_1\left(\sqrt{iRe_w(1 + i\lambda\omega)}(1 - p)\right)$$
$$M_7 = \sqrt{(\lambda_s^2 + iRe_w)(1 + i\lambda\omega)}I_1\left(\sqrt{(\lambda_s^2 + iRe_w)(1 + i\lambda\omega)}(1 - p)\right)$$
$$+ \frac{I_0\left(\sqrt{(\lambda_s^2 + iRe_w)(1 + i\lambda\omega)}\right)}{K_0\left(\sqrt{(\lambda_s^2 + iRe_w)(1 + i\lambda\omega)}\right)}\sqrt{(\lambda_s^2 + iRe_w)(1 + i\lambda\omega)}$$
$$* K_1\left(\sqrt{(\lambda_s^2 + iRe_w)(1 + i\lambda\omega)}(1 - p)\right)$$
$$M_8 = \kappa h\sqrt{\frac{m}{\epsilon_r}}\left(MI_1\left(\kappa h\sqrt{\frac{m}{\epsilon_r}}(1 - p)\right) - NK_1\left(\kappa h\sqrt{\frac{m}{\epsilon_r}}(1 - p)\right)\right)$$
$$M_9 = \frac{M_2\sqrt{(\lambda_s^2 + iRe_w)(1 + i\lambda\omega)}}{K_0\left(\sqrt{(\lambda_s^2 + iRe_w)(1 + i\lambda\omega)}\right)}K_1\left(\sqrt{(\lambda_s^2 + iRe_w)(1 + i\lambda\omega)}(1 - p)\right)$$
$$+ M_8 - \kappa hGI_1(\kappa h(1 - p))$$
$$M_{10} = M_5 - \frac{M_2}{K_0\left(\sqrt{(\lambda_s^2 + iRe_w)(1 + i\lambda\omega)}\right)}K_0\left(\sqrt{(\lambda_s^2 + iRe_w)(1 + i\lambda\omega)}(1 - p)\right)$$
$$- GI_0(\kappa h(1 - p))$$

We define the average flow rate [31] in the following manner

$$q_v^* = \int_0^h \frac{v^*(r^*, t^*)2\pi r^* dr^*}{\pi h^2} \tag{17}$$

We define q_v^* as $q_v^* = Im\left((q_0^*)_v e^{i\omega t^*}\right)$, Therefore

$$(q_0^*)_v = 2U_{HS}h^2 \int_0^1 v_0(r) r dr \tag{18}$$

Normalizing $q_0^*)_v$ by $U_{HS}h^2$ we define the average flow rate in this manner

$$(q_0)_v = 2\int_0^1 v_0(r) r dr \tag{19}$$

2.3 Results and Discussions

Here, we summarise significant findings in terms of dimensionless factors. While dealing with practical problems, it is necessary to include values for the associated dimensional parameters. The following are the considered parameters: the dimensionless Debye-Huckel parameter, κa; the scaled charge density, q_{fix}; the oscillatory Reynolds number, Re_w; the dimensionless relaxation time, $\lambda_1\omega$; the ratio between the inner and the outer cylindrical radius, h; the zeta potential ratio at the wall between the inner and the outer cylinder, $R_\zeta = \zeta_1/\zeta_0$. The values of the physical parameters that are taken into consideration are as follows: zero shear rate $\eta_0 = 10^{-3}\text{kgm}^{-1}s^{-1}$, permittivity of the electrolyte solution $\epsilon_e = 79.8 \times 8.854 \times 10^{12} Fm^{-1}$, Boltzmann constant $k_B = 1.3806 \times 10^{-23}\text{JK}^{-1}$, density $\rho = 10^3\text{kg/m}^3$, elementary charge $e = 1.602 \times 10^{-19}C$, Peclet number $Pe = 5000$, Faraday's constant $F = 96472\text{Cmol}^{-1}$, absolute temperature T=298K, applied electric field $E_0 = 10^7\text{V/m}$, Helmholtz-Smoluchowski velocity $U_{HS} = 0.18158\text{ms}^{-1}$, molecular diffusion coefficient $D_m = 10^{-9} to 10^{-10} m^2 s^{-1}$, oscillatory frequency in the time periodic electric field $10 \leq \omega \leq 10^3 s^{-1}$ [29], relaxation time is $10^{-4} \leq \lambda_1 \leq 10^3 s$ [32]. The restriction can be expressed as $\lambda_1\omega \leq 2\pi$ [24, 32] to ensure the fundamental assumption in steady state.

The influence of ion partition depending on the permittivity ratio ϵ_r was examined in this work. The effects of the permittivity ratio ϵ_r (= 0.4,0.6,1.0) on the velocity profile are shown in Fig. 2. The ion permeability in the PEL grows together with the permittivity ratio of PEL and electrolyte region ϵ_r. As a result the velocity amplitude increases and it attain maximum value when $\epsilon_r = 1$. Here, the other parameters are Reynolds number $Re_w = 8$, Debye-Hückel parameter $\kappa h = 20$, softness parameter $\lambda_s = 1$, scaled charge density $q_{fix} = 6$ and PEL thickness p = 0.2.

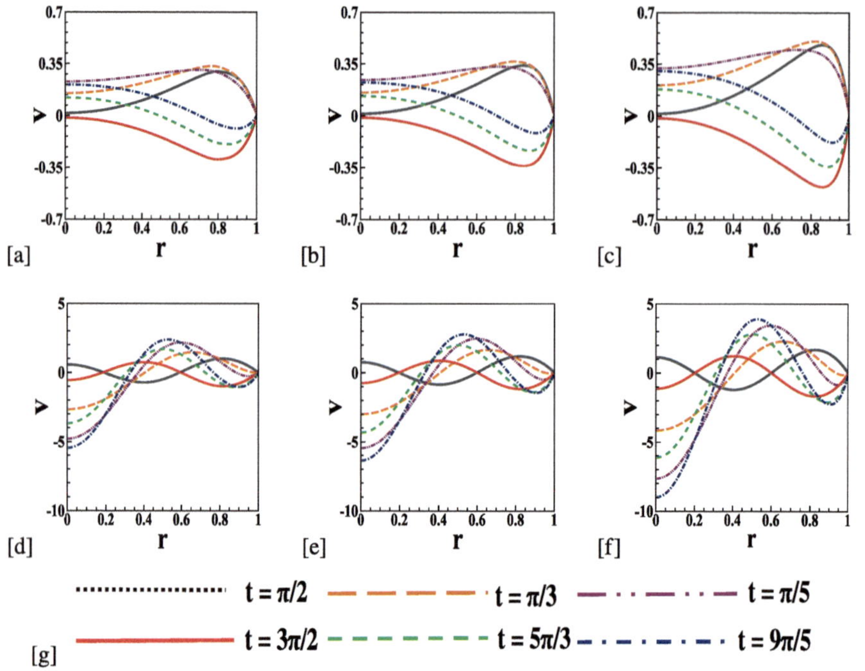

Fig. 2 Dimensionless velocity distribution at different non-dimensional time of Newtonian fluid in first row ($\lambda = 0$) and Maxwell fluid in second row ($\lambda\omega = 5$) for different permittivity ratio ϵ_r ($= 0.4, 0.6, 1.0$). Here, $Re_w = 8$, $\lambda\omega = 6$, $q_{fix} = 6$, $\sigma = -10\,\text{mC/m}^2$, $\kappa h = 20$, p = 0.2 and $\lambda_s = 1$

The Fig. 3 illustrates the effect of altering the softness parameter ($= 3, 7, 12$) on the axial velocity for both generalised Maxwell with $\lambda\omega = 5$ and Newtonian with $\lambda = 0$ fluids while the scaled charge density $q_{fix}(= 6)$, Debye-Hückel parameter $\kappa h(= 20)$, oscillating Reynolds number $Re_w(= 8)$, permittivity ratio $\epsilon_r(= 0.8)$ and PEL thickness (p=0.2) are fixed. As the softness parameter λ_s grows, the drug force in the PEL increases, and as a result, velocity amplitude decreases. Therefore with a rise in the softness parameter λ_s velocity amplitude for both Maxwell and Newtonian fluid decreases.

The oscillatory Reynolds number of a fluid has a significant dominance on the viscoelastic behaviour of the fluid. It has been stated in Fig. 4 that the oscillating Reynolds number Re_w has an effect on the velocity field. As the diffusion time scale is substantially longer than the period of oscillation, the flow oscillation increases as the Re_w increases, whereas the flow amplitude decreases as the Re_w increases. Other factors that we take into account in the figure include Debye-Hückel parameter $\kappa h = 20$, softness parameter $\lambda_s = 1$, PEL charge density $q_{fix} = 6$ and the permittivity ratio $\epsilon_r = 0.8$ and PEL thickness p = 0.2. When considering the impacts on both the Newtonian and Maxwell fluids, it is discovered that the influence on the Maxwell fluid is far greater.

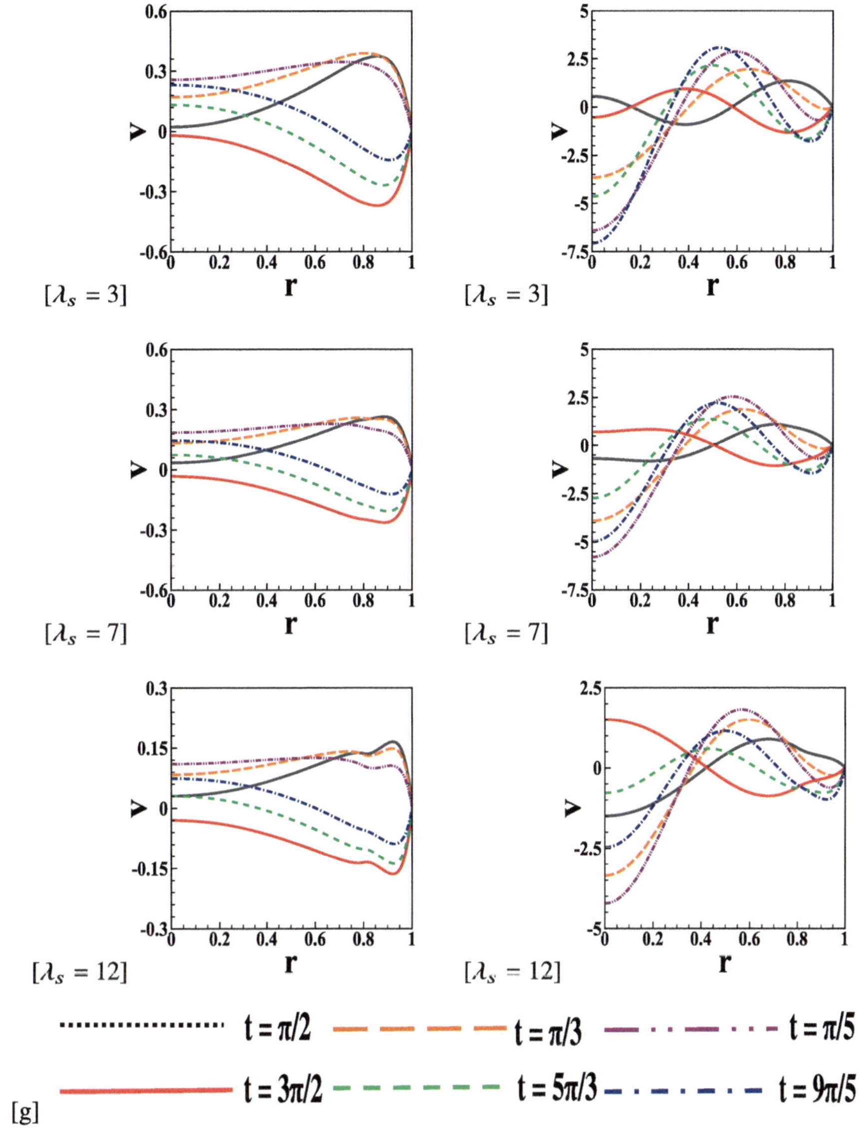

Fig. 3 Dimensionless velocity distribution at different non-dimensional time of Newtonian fluid in first column ($\lambda = 0$) and Maxwell fluid in second column ($\lambda\omega = 5$) for different softness parameter λ_s ($= 3, 7, 12$). Here, $\sigma = -10\,\mathrm{mC/m^2}$, $Re_w = 8$, $\lambda\omega = 6$, $\kappa h = 20$, $\epsilon_r = 0.7$, p = 0.2 and $q_{fix} = 6$

In Fig. 5 we have shown the impacts of the thickness of PEL p ($= 0.01, 0.1$ and p $=$ 0.2) on the Maxwell ($\lambda\omega = 5$) and Newtonian fluids ($\lambda = 0$).The PEL has positive impact on the OEOF. With the higher values of the PEL thickness the resistance

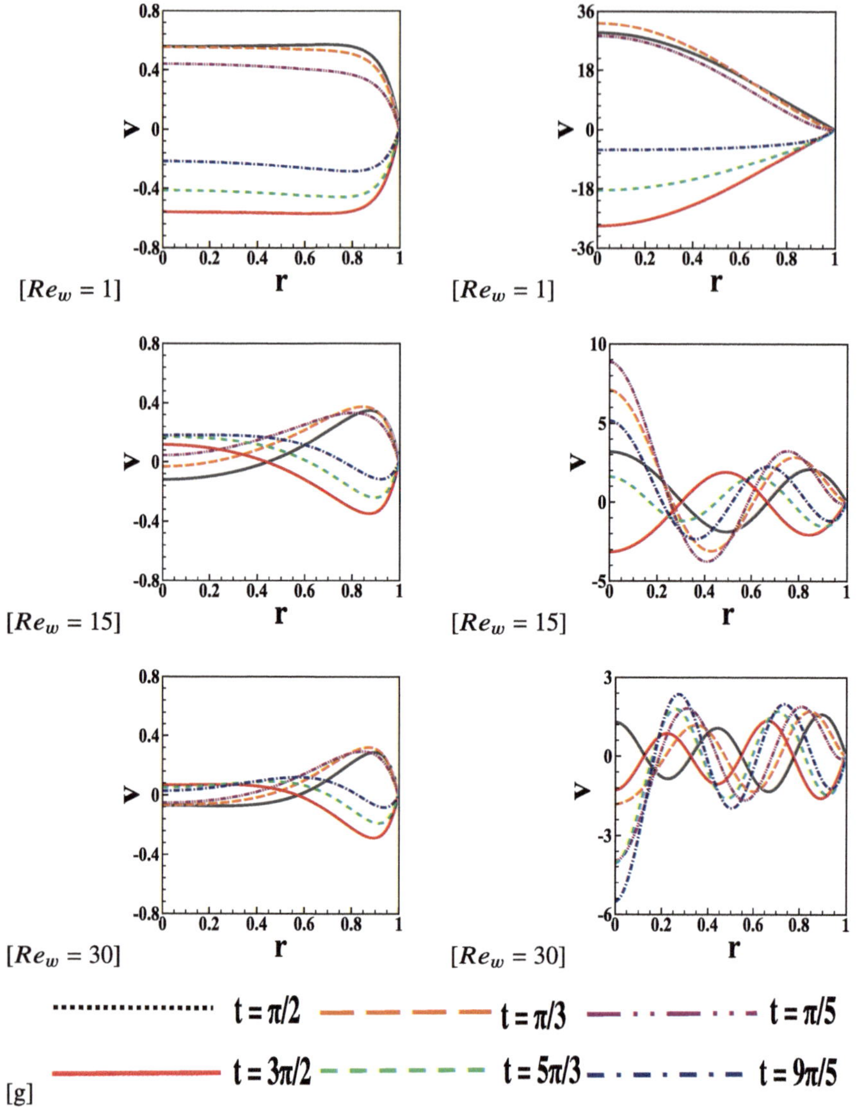

Fig. 4 Dimensionless velocity distribution at different non-dimensional time of Newtonian fluid in first column ($\lambda = 0$) and Maxwell fluid in second column ($\lambda\omega = 5$) for different Reynolds number Re_w ($= 1, 15, 30$). Here, $\sigma = -10\,\text{mC/m}^2$, $\lambda\omega = 6$, $\kappa h = 20$, $q_{fix} = 6$, $\epsilon_r = 0.7$, p = 0.2 and $\lambda_s = 1$

force in the PEL increases the ion permeability decreases and as a result the velocity amplitude decreases. It is clear from the figure that the velocity profiles correspond well with rigid nanopore when PEL is thin. It deviates from rigid channel when PEL thickness is larger. For Newtonian fluids the deviation in velocity distributions

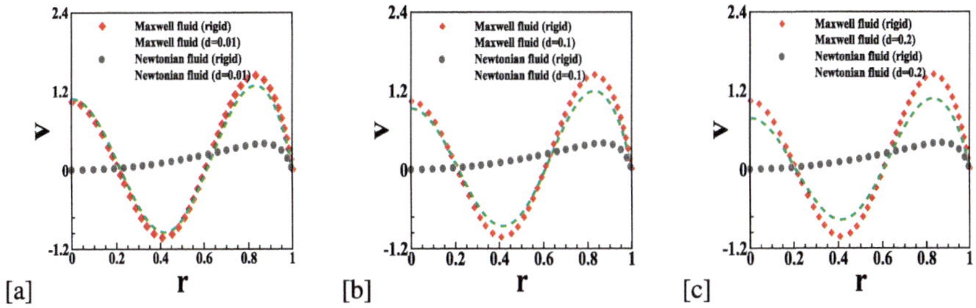

Fig. 5 Dimensionless velocity distribution for different PEL thickness **a** p = 0.01 **b** p = 0.1 and **c** p = 0.2 for Maxwell fluid ($\lambda\omega = 5$) and Newtonian fluid ($\lambda = 0$) at time t=$\pi/2$. Here, $\sigma = -10\,\text{mC/m}^2$, $Re_w = 8$, $\lambda\omega = 6$, $\epsilon_r = 0.7$, $\kappa h = 20$, $q_{fix} = 6$ and $\lambda_s = 1$

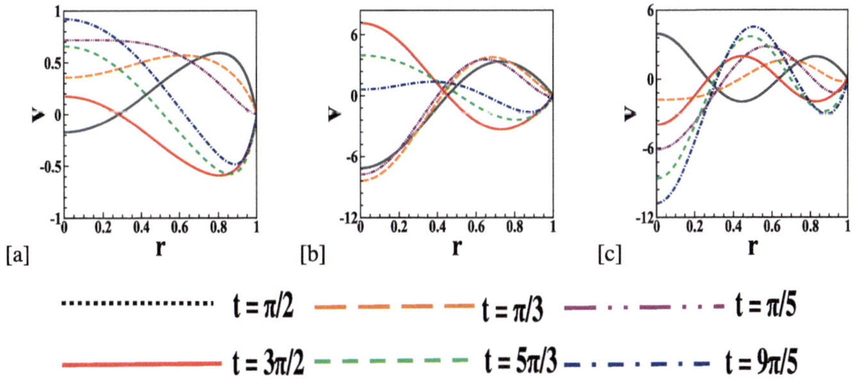

Fig. 6 Dimensionless velocity distribution at different non-dimensional time corresponding to different normalised relaxation time **a** $\lambda\omega = 1$ **b** $\lambda\omega = 4$ and (*c*) $\lambda\omega = 6$. Here, $\sigma = -10\,\text{mC/m}^2$, $Re_w = 8$, $\kappa h = 20$, $q_{fix} = 6$, $\epsilon_r = 0.7$, p = 0.2 and $\lambda_s = 1$

between the rigid nanopore and PEL grafted nanopore are smaller in comparison to Newtonian fluids.

The elasticity of solute rises when the normalised relaxation time $\lambda\omega$ is increased. For shear thinning effect, higher values of relaxation time $\lambda\omega$ result in a noticeable rise in flow profile. We demonstrated how the relaxation time $\lambda\omega$ (= 1,4,6) works for the viscoelastic features of fluid in Fig. 6 when the other parameters are fixed. It is visible from Fig. 6 that the velocity amplitude enlarge along with flow oscillation with the increases relaxation time.

We have used the formulation Eq. 19 to calculate the average flow rate in a cylindrical cross-section. In Fig. 7a–c we have displayed the variation of average flow corresponding to normalized relaxation time for different permittivity ratio, softness parameter and PEL charge density. In Fig. 7a, we calculate the average flow for different permittivity ratio ϵ_r. It's visible that the average flow rate enlarges

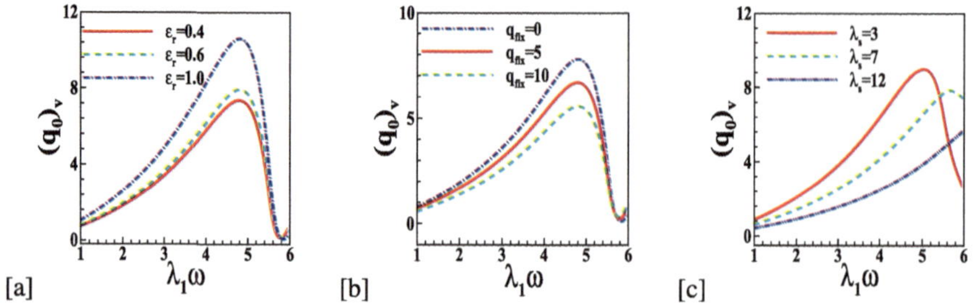

Fig. 7 Profile of average flow rate for Maxwell fluid with normalized relaxation time $\lambda\omega$ for **a** permittivity ratio ϵ_r, with $\lambda_s = 1$, $q_{fix} = 6$, **b** charge density of PEL q_{fix}, with $\lambda_s = 1$, $\epsilon_r = 0.8$ and **c** softness parameter λ_s, with $\epsilon_r = 0.8$, $q_{fix} = 6$. Here, $Re_w = 8$, $\kappa h = 10$, p $= 0.2$ and $\sigma = -10\,\text{mC/m}^2$

with permittivity ratio. With the increasing PEL charge density q_{fix}, average flow rate decreases as indicated in Fig. 7b. In Fig. 7c variation of average flow rate for different softness parameter λ_s is illustrated. As the softness parameter λ_s increases, average flow rate decreases.

3 Conclusions

In this paper we have disused about EOF of viscoelastic Maxwell fluid considering ion partitioning coefficient under time dependent electric field across the PEL surrounded nanopore. The expression of electric potential is deduced using Debye-Hückel linearized technique and the analytic solution is presented. The velocity field of OEOF is obtained using the known induced potential by solving CM equation where the stress coefficient is represented by the Maxwell constitutive relation. We have disused the flow properties of Newtonian and Maxwell fluids by considering the suitable dimensionless parameters. The influences of the parameters on the OEOF are portrayed here with description. With the increasing scaled PEL charge density and softness parameter, flow amplitude and average flow rate decreases where as the with the increasing value of the permittivity ratio the both the flow amplitude and average flow increases. The viscoelastic behavior is amplified by the parameters oscillatory Reynolds number and normalized relaxation time. With the increasing Reynolds number the flow oscillation increases whereas the flow amplitude drops. The normalized relaxation time has a commendable impacts on non-Maxwell fluid behavior. With the higher values of this parameter the flow amplitude increases for both the fluids. The velocity distributions on the PEL surround nanpore are deviates from the rigid nanopore and this deviation increases with higher value of the PEL thickness.

Acknowledgements Author (S. Bera) wish to acknowledge the Science & Engineering Research Board, Government of India through the project grants ECR/2016/000771.

References

1. Cheng, L.-J., Jay Guo, L.: Nanofluidic diodes. Chem. Soc. Rev. **39**(3), 923–938 (2010)
2. Stone, H.A., Stroock, A.D., Ajdari, A.: Engineering flows in small devices: microfluidics toward a lab-on-a-chip. Ann. Rev. Fluid Mech. **36**, 381–411 (2004)
3. Wang, X., Cheng, C., Wang, S., Liu, S.: Electroosmotic pumps and their applications in microfluidic systems. Microfluid. Nanofluidics **6**(2), 145–162 (2009)
4. van der Heyden, F.H.J., Bonthuis, DJ., Stein, D., Meyer, C., Dekker, C.: Power generation by pressure-driven transport of ions in nanofluidic channels. Nano Lett. **7**(4), 022–1025 (2007)
5. Li, J., Chen, D., Ye, J., Zhang, L., Zhou, T., Zhou, Y.: Direct numerical simulation of seawater desalination based on ion concentration polarization. Micromachines **10**(9), 562 (2019)
6. Kwak, R., Kim, S.J., Han, J.: Continuous-flow biomolecule and cell concentrator by ion concentration polarization. Anal. Chem. **83**(19), 7348–7355 (2011)
7. Vlassiouk, I., Siwy, Z.S.: Nanofluidic diode. Nano Lett. **7**(3), 552–556 (2007)
8. Ajdari, A.: Electro-osmosis on in homogeneously charged surfaces. Phys. Rev. Lett. **75**(4), 755 (1995)
9. Bera, S., Bhattacharyya, S.: Electroosmotic flow in the vicinity of a conducting obstacle mounted on the surface of a wide microchannel. Int. J. Eng. Sci. **94**, 128–138 (2015)
10. Moghadam, A.J.: Effect of periodic excitation on alternating current electroosmotic flow in a microannular channel. European J. Mech.-B/Fluids **48**, 1–12 (2014)
11. Stroock, A.D., Weck, M., Chiu, D.T., Huck, W.T.S., Kenis, P.J.A., Ismagilov, R.F., Whitesides, G.M.: Patterning electro-osmotic flow with patterned surface charge. Phys. Rev. Lett. **84**(15), 3314 (2000)
12. Bhattacharyya, S., Bera, S.: Nonlinear electroosmosis pressure-driven flow in a wide microchannel with patchwise surface heterogeneity. J. Fluids Eng. **135**(2) (2013)
13. Loucaides, N.G., Ramos, A., Georghiou, G.E.: Dielectrophoretic and ac electroosmotic trapping of DNA: numerical simulation incorporating fluid dynamics and steric particle effects. J. Electrostati. **69**(2), 111–118 (2011)
14. Jian, Y., Yang, L., Liu, Q.: Time periodic electro-osmotic flow through a microannulus. Phys. Fluids **22**(4), 042001 (2010)
15. Luo, W.-J.: Transient electroosmotic flow induced by ac electric field in micro-channel with patchwise surface heterogeneities. J. Colloid Interface Sci. **295**(2), 551–561 (2006)
16. Chanda, S., Das, S.: Effect of finite ion sizes in an electrostatic potential distribution for a charged soft surface in contact with an electrolyte solution. Phys. Rev. E **89**(1), 012307 (2014)
17. Keh, H.J., Ding, J.M.: Electrokinetic flow in a capillary with a charge-regulating surface polymer layer. J. Colloid Interface Sci. **263**(2), 645–660 (2003)
18. Saha, S., Gopmandal, P.P., Ohshima, H.: Steady/unsteady electroosmotic flow through nanochannel filled with electrolyte solution surrounded by an immiscible liquid. Colloid Polym. Sci. **295**(12), 2287–2297 (2017)
19. Leunissen, M.E., Zwanikken, J., Van Roij, R., Chaikin, P.M., Van Blaaderen, A.: Ion partitioning at the oil–water interface as a source of tunable electrostatic effects in emulsions with colloids. Phys. Chem. Chem. Phys. **9**(48), 6405–6414 (2007)
20. Koner, P., Bera, S., Ohshima, H.: Ion-partitioning effects on electrokinetic flow of generalized maxwell fluids through polyelectrolyte layer-coated nanopore under ac electric field. Colloid Polym. Sci. **299**(11), 1777–1795 (2021)
21. Sin, J.-S.: Ion partitioning effect on the electrostatic interaction between two charged soft surfaces. Colloids Surf. A: Physicochem. Eng. Aspects **628**, 127296 (2021)
22. Liu, Y., Jian, Y., Liu, Q., Li, F.: Alternating current magnetohydrodynamic electroosmotic flow of maxwell fluids between two micro-parallel plates. J. Mol. Liq. **211**, 784–791 (2015)
23. Jian, Y., Liu, Q., Yang, L.: Ac electroosmotic flow of generalized maxwell fluids in a rectangular microchannel. J. Non-Newton. Fluid Mech. **166**(21–22), 1304–1314 (2011)
24. Peralta, M., Arcos, J., Méndez, F., Bautista, O.: Mass transfer through a concentric-annulus microchannel driven by an oscillatory electroosmotic flow of a maxwell fluid. J. Non-Newton. Fluid Mech. **279**, 104281 (2020)

25. Sun, L.-X., Jian, Y.-J., Chang, L., Zhang, H.-Y., Liu, Q.-S.: Alternating current electro-osmotic flow of the maxwell fluids through a circular micro-pipe. J. Mech. **29**(2), 233–240 (2013)
26. Ganjizade, A., Ashrafizadeh, S.N., Sadeghi, A.: Effect of ion partitioning on the electrostatics of soft particles with a volumetrically charged core. Electrochem. Commun. **84**, 19–23 (2017)
27. López-Garcıa, J.J., Horno, J., Grosse, C.: Suspended particles surrounded by an inhomogeneously charged permeable membrane. solution of the poisson–boltzmann equation by means of the network method. J. Colloid Interface Sci **268**(2), 371–379 (2003)
28. Born, M.: Volumen und hydratationswärme der ionen. Zeitschrift für Physik **1**(1), 45–48 (1920)
29. Byron Bird, R.B., Stewart, W.E., Lightfoot, E.N.: Transport Phenomena, vol. 1. Wiley (2006)
30. Duval, J.F.L., van Leeuwen, H.P.: Electrokinetics of diffuse soft interfaces. 1. Limit of low donnan potentials. Langmuir **20**(23), 10324–10336 (2004)
31. Yeh, L.-H., Zhang, M., Hu, N., Joo, S.W., Qian, S., Hsu, J.-P.: Electrokineticion and fluid transport in nanopores functionalized by polyelectrolyte brushes. Nanoscale **4**(16), 5169–5177 (2012)
32. Bird, R.B., Armstrong, R.C., Hassager, O.: Dynamics of polymeric liquids. Fluid Mech. **1** (1987)

Numerical Simulation of Parabolic Partial Differential Equation

Inderjeet and Rashmi Bhardwaj

Abstract This paper discusses the numerical approach used to solve the parabolic partial differential equation. The Schmidt method and the Crank-Nicolson method are covered in this paper. These numerical methods are employed to discover a partial differential equation's approximate solution. These methods are unquestionably effective and practically good for solving partial differential equations, and they are all used to assess the degree of accuracy of each method. The Crank Nicolson method is exposed to both analysis of stability and analysis of absolute error in order to help mathematicians and engineers to understand how well these numerical solutions actually perform approaches. With Boundary condition $P(0, t) = P(L, t) = 0$ and $P(x, 0) = f(x), 0 < x < L$, the heat equation

$$\frac{\partial P}{\partial t} = c^2 \frac{\partial^2 P}{\partial x^2}$$

Where c is constant, x is space variable, L is the length of the rod, t is the time variable, and $P(x, t)$ is the temperature function with time t, and position x. The above heat equation is analysed using method of variable separation. The findings are then compared to the analytical solution after the identical equation is solved using both the Schmidt and Crank Nicolson techniques. We create a comparison table of the approximate and exact solutions in order to get and assess the level of accuracy of the numerical findings. We note that the exact and approximative solutions agree well, and we compare the computational effort required by the various proposed solutions. Now we identify the error in the suggested techniques that accounts for the proposed method's superiority. According to the result of the study, the Crank Nicolson approach is ultimately more efficient than the Schmidt approach and also yields less error.

Inderjeet
USBAS, GGSIPU, Dwarka, Delhi, India

R. Bhardwaj (✉)
Fellow of Institute of Mathematics and Applications (UK), Non-Linear Dynamics Research Lab, Indraprastha University (GGSIPU), Dwarka, Delhi, India
e-mail: rashmib@ipu.ac.in

S. Bhattacharyya and H. S. Mahato (eds.), *Modeling, Analysis and Simulations of Multiscale Transport Phenomena*, Springer Proceedings in Mathematics & Statistics 491, https://doi.org/10.1007/978-981-96-3098-1_11

Keywords Partial differential equations · Boundary value problems · Schmidt method · Crank-Nicolson method · Error estimation

1 Introduction

The branch of numerical analysis that analyses the numerical solution of a partial differential problem is known as numerical partial differential. Since ancient times, numerical mathematics has helped mathematics and made difficult mathematical problems simple to solve. Nearly all areas of science, particularly mathematics, engineering, and medicine, have been affected by this.

Although practically all fields employ these algorithms now, it is important to examine them carefully so that decisions may be made about them that are well-informed. Although many mathematicians have made efforts to evaluate these algorithms, all of their efforts have, regrettably, been limited to analysing their stability.

Many issues in the fields of science, engineering, and technology can be resolved using the formulation of partial differential equation. So to get complete the numerical solution of partial differential equation that is constrained to a set of boundary conditions, this study will assess the accuracy of various approaches, including the Bender Schmidt and Crank Nicolson difference method [1–3, 7–9].

In two independent variables x and y and one dependent variable u, the general linear partial differential equation of order second has following type:

$$R(x, y)\frac{\partial^2 u}{\partial x^2} + S(x, y)\frac{\partial^2 u}{\partial x \partial y} + T(x, y)\frac{\partial^2 u}{\partial y^2} + F(x, y, \frac{\partial u}{\partial x^2}, \frac{\partial u}{\partial y}) = 0$$

It is claimed that such a partial differential equation

1. Elliptic if $S^2 - 4RT < 0$
2. Hyperbolic if $S^2 - 4RT > 0$
3. Parabolic if $S^2 - 4RT = 0$

In particular, the one-dimensional heat equation

$$\frac{\partial P}{\partial t} = \alpha^2 \frac{\partial^2 P}{\partial x^2}$$

where, $\alpha^2 = \frac{k}{pc}$ is an example of parabolic equation.

If $\alpha^2 = \frac{1}{a}$ the equation becomes

$$\frac{\partial^2 P}{\partial x^2} = a\frac{\partial P}{\partial t}, \; here \; R = 1, \; S = 0 \; \& \; T = 0$$

$$S^2 - 4RT = 0.$$

Hence it is parabolic throughout.

The temperature function, denoted by the notation $P(x, t)$, is answer to the heat equation and is specified for all values of $x\varepsilon(0, 1)$ and all values of time $t\varepsilon(0, \infty)$. The solution is developed from initial values satisfying the specified boundary conditions rather than being defined in a closed domain. For the purpose of solving periodic partial differential equations, the finite difference approach is examined in this research [2].

The paper's final layout is as follows: The problem is presented in Sect. 2, followed by numerical examples in Sect. 3, discussion of the results in Sect. 4, and study conclusion in Sect. 5.

2 Problem Formulation

The approximate solution of the one-dimensional heat equation $\frac{\partial P}{\partial t} = \alpha^2 \frac{\partial^2 P}{\partial x^2}$, with Boundary condition $P(0, t) = T_0$, $P(l, t) = T_l$ and initial condition $P(x, 0) = f(x)$; $0 < x < l$, is derived in this section of the study using two numerical approaches.

Consider the X-T plane's rectangular region R. Rectangularize this area by using the formulas $x = h$ and $T = k$. Mesh or grid points are the locations where the dividing lies intersect. i, j is used to represent the mesh point $(x, t) = (ih, ik)$.

Bender Schmidt Method

To solve the parabolic equation

$$\frac{\partial^2 P}{\partial x^2} = a \frac{\partial^2 P}{\partial t} \tag{1}$$

with Boundary condition $P(0, t) = T_0$, $P(l, t) = T_l$ and with the Initial condition $P(x, 0) = f(x); 0 < x < l$.

At $P_{i,j}$ we have

$$\frac{\partial P}{\partial t} = \frac{P_{i,j+1} - P_{i,j}}{k} \; and \frac{\partial^2 P}{\partial x^2} = \frac{P_{i+j,j} - 2P_{i,j} + P_{i-1}, j}{h^2}$$

Put these values in Eq. (1), we get

$$\frac{P_{i+j,j} - 2P_{i,j} + P_{i-1,j}}{h^2} = a \frac{P_{i,j+1} - P_{I,j}}{k}$$

$$P_{i,j+1} - P_{i,j} = \frac{k}{ah^2}[P_{i+1,j} - 2P_{i,j} + P_{i-1,j}]$$

where $\lambda = \frac{k}{ah^2}$

$$P_{i,j+1} - P_{i,j} = \lambda[P_{i+1,j} - 2P_{i,j} + P_{i-1,j}]$$

$$P_{i,j+1} = \lambda P_{i+1,j} + (1 - 2\lambda)P_{i,j} + \lambda P_{i-1,j}$$

This is called Bender-Schmidt formula.

Put $\lambda = \frac{1}{2}$ the Bender Schmidt formula reduce to

$$P_{i,j+1} = \frac{1}{2}[P_{i+j,j} + P_{i-1,j}]$$

Crank Nicolson Difference Method

To solve the parabolic equation

$$\frac{\partial^2 P}{\partial x^2} = a\frac{\partial P}{\partial t} \tag{2}$$

with Boundary condition $P(0, t) = T_0$, $P(l, t) = T_l$ and with the Initial condition $P(x, 0) = f(x); 0 < x < l$.

At $P_{i,j}$ we have

$$\frac{\partial^2 P}{\partial x^2} = \frac{P_{i+j,j} - 2P_{i,j} + P_{i-1}, j}{h^2}$$

And at $P_{i,j+1}$

$$\frac{\partial^2 P}{\partial x^2} = \frac{P_{i+j,j+1} - 2P_{i,j+1} + P_{i-1,j+1}}{h^2}$$

Taking average of these two, we get

$$\frac{\partial^2 P}{\partial x^2} = \frac{P_{i+j,j} - 2P_{i,j} + P_{i-1}, j + P_{P_{i+1,j+1}} - 2Pi, j + 1 + P_{i-1,j+1}}{2h^2} \tag{3}$$

And also at $P_{i,j}$ we have

$$\frac{\partial P}{\partial t} = \frac{P_{i,j+1} - P_{i,j}}{k} \tag{4}$$

Put the values of equation (3) and (4) in equation (2), we get

$$\frac{P_{i+1,j} - 2P_{i,j} + P_{i+1,j+1} - 2P_{i,j+1} + P_{i-1,j+1}}{2h^2} = a\frac{P_{i,j+1} - P_{i,j}}{k}$$

$$P_{i,j+1} - P_{i,j} = \frac{\lambda}{2}[P_{i+1,j} - 2P_{i,j} + P_{i-1,j} + P_{i+1,j+1} - 2P_{i,j+1} + P_{i-1,j+1}]$$

where $\lambda = \frac{k}{ah^2}$

$$\lambda(P_{i+1,j+1} + P_{i-1,j+1}) - 2(\lambda + 1)P_{i,j+1} = 2(\lambda - 1)P_{i,j} - \lambda(P_{i+1,j} + P_{i-1,j})$$

This equation is called Crank-Nicolson difference scheme.
Put $\lambda = 1$, the Crank-Nicolson formula reduced to

$$P_{i,j+1} = \frac{1}{4}[P_{i-1,j+1} + P_{i+1,j+1} + u_{i-1,j} + P_{i+1,j}]$$

3 Numerical Examples

Using the suggested strategies, we solve two numerical examples in this section of the study. We also identify which method converges most quickly to the correct answer. The error for the suggested approach is then discovered, and the output is represented. The numerical method's mathematical convergence is defined as $\lim_{h\to 0} \max_{1\le n\le N} |y(x_n) - y_n| = 0$ and the error is defined as as $\lim_{h\to 0} \max_{1\le n\le N} |y(x_n) - y_n|$ where $y(x_n)$ is the approximation and y_n is the actual solution.

Examples (1)

Solve the Boundary value problem

$$\frac{\partial P}{\partial t} = \frac{\partial^2 P}{\partial x^2} \tag{5}$$

under the condition $P(0, t) = P(1, t) = 0$ and $P(x, 0) = \frac{3}{2}\sin \pi x$
$0 \le x \le 1$ and by taking $h = 0.2$.

The genaral solution of one-dimensional heat equation is

$$P(x, t) = e^{-c^2 m^2 t}(c_1 \cos mx + c_2 \sin mx)$$

Here $c^2 = 1$ so the above equation becomes

$$P(x,t) = e^{-m^2 t}(c_1 \cos mx + c_2 \sin mx)$$

By using the boundary conditions, we get $c_1 = 0$ and $m = n\pi$; where $n \in z$.

So $P(x,t)$ becomes $P(x,t) = e^{-n^2\pi^2 t}(c_2 \sin n\pi x)$; where $n \in z$.

The one-dimensional heat equation's most prevalent answer is

$$P(x,t) = \sum_{n=1}^{\infty} a_n \sin n\pi x e^{-n^2\pi^2 t}$$

Now using the Boundary condition $P(x,0) = \frac{3}{2}\sin \pi x$, we get $a_1 = \frac{3}{2}$.

Consequently, the one-dimensional heat equation's final solution is

$$P(x,t) = \frac{3}{2}\sin \pi x e^{-\pi^2 t}$$

Now, we make the table for different values of x and t (Table 1).

Now, we determine the approximate solution of above partial differential equation (5) by using the Bender Schmidt methods by $h = 0.2$ and then determine the error estimation of Bender Schmidt method.

Now, we determine the approximate solution of the above partial differential equation (5) by using the Crank-Nicolson methods by taking $h = 0.2$ and then determine the error estimation of Crank-Nicolson method (Table 4).

Table 1 Using the variable separation method, partial differential equation (4) is actually solved

	$x \rightarrow$	0	0.2	0.4	0.6	0.8	1
$t \downarrow$	J/i	0	1	2	3	4	5
0	0	0	88167783×10^{-8}	$1426584774 \times 10^{-9}$	$1426584774 \times 10^{-9}$	881677878×10^{-9}	0
0.02	1	0	$72374178925 \times 10^{-11}$	$117103881408 \times 10^{-11}$	$117103881408 \times 10^{-11}$	$72374178925 \times 10^{-11}$	0
0.04	2	0	$59409699428 \times 10^{-11}$	$96126912936 \times 10^{-11}$	$96126912936 \times 10^{-11}$	$59409699428 \times 10^{-11}$	0
0.06	3	0	$48767563771 \times 10^{-11}$	$78907575731 \times 10^{-11}$	$78907575731 \times 10^{-11}$	$48767563771 \times 10^{-11}$	0
0.08	4	0	$40031767525 \times 10^{-11}$	$64772760485 \times 10^{-11}$	$64772760485 \times 10^{-11}$	$40031767525 \times 10^{-11}$	0
0.10	5	0	$32860825638 \times 10^{-11}$	$53169932822 \times 10^{-11}$	$53169932822 \times 10^{-11}$	$32860825638 \times 10^{-11}$	0

Now we find the values of $E_1, E_2, E_3, \ldots, E_{22}, E_{24}, E_{24}$ for Bender Schmidt and Crank Nicolson methods as below.

Examples (2)

If initial temperature is $\frac{1}{2}\sin\frac{\pi x}{2}+\frac{3}{2}\sin\frac{5\pi x}{2}$ determine the temperature of a bar of length 2 whose ends are held at 0 and lateral surface is insulated. Given that $P(0,t)=0=P(2,t)$ for any t.

Solution

We know that one dimensional heat equation is

$$\frac{\partial P}{\partial t}=\frac{\partial^2 P}{\partial x^2} \tag{6}$$

where $c^2=1$.

And given the Boundary condition $P(0,t)=0=P(2,t)$ with initial condition

$$P(x,0)=\frac{1}{2}\sin\frac{\pi x}{2}+\frac{3}{2}\sin\frac{5\pi x}{2}$$

We know that general solution of one - dimensional heat equation is

$$P(x,t)=e^{-c^2m^2t}(c_1\cos mx+c_2\sin mx)$$

Here $c^2=1$, so the above equation becomes

$$P(x,t)=e^{-m^2t}(c_1\cos mx+c_2\sin mx)$$

By using the boundary conditions, we get $c_1=0$ and $m=\frac{n\pi}{2}$; where $n\in z$.

So $P(x,t)$ becomes

$$P(x,t)=e\frac{-n^2\pi^2t}{4}(c_2\sin\frac{n\pi x}{2})$$

where $n\in z$.

Most general solution of one - dimensional heat equation is

$$P(x,t) = \sum_{n=1}^{\infty} b_n e^{\frac{-n^2\pi^2 t}{4}} \sin\frac{n\pi x}{2}$$

Now using the boundary condition

$$P(x,0) = \frac{1}{2}\sin\frac{\pi x}{2} + \frac{3}{2}\sin\frac{5\pi x}{2}$$

we get $b_1 = \frac{1}{2}$.

Consequently, the one-dimensional heat equation's final solution is

$$P(x,t) = \frac{1}{2}(\sin\frac{\pi x}{2} e^{\frac{-\pi^2 t}{4}} + 3\sin\frac{5\pi x}{2} e^{\frac{-25\pi^2 t}{4}})$$

Now, we make the table for different values of x and t (Table 6).

Now, we determine approximate solution of the above partial differential equation (6) by using the Bender Schmidt methods by $h = 0.2$ and then determine the error estimation of Bender Schmidt method.

Now, we determine approximate solution of the above partial differential equation (6) by using the Crank-Nicolson methods by $h = 0.2$ and then determine the error estimation of Crank-Nicolson method (Table 9).

Now we find the values of E_1, E_2, $E_3, \ldots, E_{22}$, E_{24}, E_{24} for Bender Schmidt and Crank Nicolson methods as below.

4 Result Discussion

The approximate solutions from the two cases are shown in Tables 2, 3, 7, and 8, while error evaluations of the two procedures in comparison to the exact solution are shown in Tables 5 and 10. The results table shows that the Crank Nicolson method converges to the exact solution more frequently than the Bender Schmidt method. The Bender Schmidt method introduces the most error, whereas the Crank Nicolson method introduces the least error. The Crank Nicolson method, which converges exactly and quickly relative to Bender Schmidt method, has been determined to be the most effective technique for solving the Boundary value problems of Partial differential equations.

Table 2 Approximate solution of partial differential equation (5) by Bender Schmidt method

	$x \rightarrow$	0	0.2	0.4	0.6	0.8	1
$t \downarrow$	J/i	0	1	2	3	4	5
0	0	0	881677878×10^{-9}	$1426584774 \times 10^{-9}$	$1426584774 \times 10^{-9}$	881677878×10^{-9}	0
0.02	1	0	713292387×10^{-9}	$1154131326 \times 10^{-9}$	$1154131326 \times 10^{-9}$	713292387×10^{-9}	0
0.04	2	0	577065663×10^{-9}	$9337118565 \times 10^{-10}$	$9337118565 \times 10^{-10}$	577065663×10^{-9}	0
0.06	3	0	$4668559275 \times 10^{-10}$	755388759×10^{-9}	755388759×10^{-9}	$4668559275 \times 10^{-10}$	0
0.08	4	0	$3776943795 \times 10^{-10}$	$6111223425 \times 10^{-10}$	$6111223425 \times 10^{-10}$	$377694379 \times 10^{-10}$	0
0.10	5	0	$3055611705 \times 10^{-10}$	494408361×10^{-9}	494408361×10^{-9}	$3055611705 \times 10^{-10}$	0

Table 3 Approximate solution of partial differential equation (5) by Crank-Nicolson method

	$x \rightarrow$	0	0.2	0.4	0.6	0.8	1
$t \downarrow$	j / i	0	1	2	3	4	5
0	0	0	881677878×10^{-9}	$1426584774 \times 10^{-9}$	$1426584774 \times 10^{-9}$	881677878×10^{-9}	0
0.02	1	0	$72797016724 \times 10^{-11}$	$117788047345 \times 10^{-11}$	$117788047345 \times 10^{-11}$	$72797016724 \times 10^{-11}$	0
0.04	2	0	$60105915967 \times 10^{-11}$	$97253414953 \times 10^{-11}$	$97253414953 \times 10^{-11}$	$60105915967 \times 10^{-11}$	0
0.06	3	0	$49627323988 \times 10^{-11}$	$80298696978 \times 10^{-11}$	$80298696978 \times 10^{-11}$	$49627323988 \times 10^{-11}$	0
0.08	4	0	$40975522066 \times 10^{-11}$	$66299787393 \times 10^{-11}$	$66299787393 \times 10^{-11}$	$40975522066 \times 10^{-11}$	0
0.10	5	0	$33832035905 \times 10^{-11}$	$54741384031 \times 10^{-11}$	$54741384031 \times 10^{-11}$	$33832035905 \times 10^{-11}$	0

Table 4 The proposed two approaches Error estimation are compared to the exact answer

	x	0	0.2	0.4	0.6	0.8	1
t	j / i	0	1	2	3	4	5
0	0	0	E_1	E_2	E_3	E_4	0
0.02	1	0	E_5	E_6	E_7	E_8	0
0.04	2	0	E_9	E_{10}	E_{11}	E_{12}	0
0.06	3	0	E_{13}	E_{14}	E_{15}	E_{16}	0
0.08	4	0	E_{17}	E_{18}	E_{19}	E_{20}	0
0.10	5	0	E_{21}	E_{22}	E_{23}	E_{24}	0

Table 5 Error table

	Bender Schmidt value	Crank Nicolson value
E_1	0.0000000000000000	0.0000000000000000
E_2	0.0000000000000000	0.0000000000000000
E_3	0.0000000000000000	0.0000000000000000
E_4	0.0000000000000000	0.0000000000000000
E_5	1044940225 E -11	422837799 E -11
E_6	1690748808 E -11	684165937 E -11
E_7	1690748808 E -11	684165937 E -11
E_8	1044940225 E -11	422837799 E -11
E_9	1703133328 E -11	696216539 E -11
E_{10}	2755727286 E -11	1126502018 E -11
E_{11}	2755727286 E -11	1126502018 E -11
E_{12}	1703133328 E -11	696216539 E -11
E_{13}	2081971021 E -11	859760219 E -11
E_{14}	3368699831 E -11	1391121247 E -11
E_{15}	3368699831 E -11	1391121247 E -11
E_{16}	2081971021 E -11	859760219 E -11
E_{17}	2262329575 E -11	943754541 E -11
E_{18}	3660526235 E -11	1527026909 E -11
E_{19}	3660526235 E -11	1527026909 E -11
E_{20}	2262329575 E -11	943754541 E -11
E_{21}	2304708614 E -11	971210241 E -11
E_{22}	3729096722 E -11	1571451209 E -11
E_{23}	3729096722 E -11	1571451209 E -11
E_{24}	2304708614 E -11	971210241 E -11

Table 6 The actual solution of partial differential problem (6) using the variable separation method

	$x \to$	0	0.2	0.4	0.6	0.8	1
$t \downarrow$	j/i	0	1	2	3	4	5
0	0	0	$29389262615 \times 10^{-11}$	$47552825815 \times 10^{-11}$	$47552825815 \times 10^{-11}$	$29389262615 \times 10^{-11}$	0
0.02	1	0	$26627202575 \times 10^{-11}$	$43083718792 \times 10^{-11}$	$43083718792 \times 10^{-11}$	$26627202575 \times 10^{-11}$	0
0.04	2	0	$24124726308 \times 10^{-11}$	$39034627136 \times 10^{-11}$	$39034627136 \times 10^{-11}$	$24124726308 \times 10^{-11}$	0
0.06	3	0	$21857437626 \times 10^{-11}$	$35366076986 \times 10^{-11}$	$35366076986 \times 10^{-11}$	$21857437626 \times 10^{-11}$	0
0.08	4	0	$19803233143 \times 10^{-11}$	$32042304312 \times 10^{-11}$	$32042304312 \times 10^{-11}$	$19803233143 \times 10^{-11}$	0
0.10	5	0	$1794208679 \times 10^{-10}$	$29030906256 \times 10^{-11}$	$29030906256 \times 10^{-11}$	$1794208679 \times 10^{-10}$	0

Table 7 Approximate solution of partial differential equation (6) by Bender Schmidt method

	$x \rightarrow$	0	0.2	0.4	0.6	0.8	1
$t \downarrow$	j/i	0	1	2	3	4	5
0	0	0	$29389262615 \times 10^{-11}$	$47552825815 \times 10^{-11}$	$47552825815 \times 10^{-11}$	$29389262615 \times 10^{-11}$	0
0.02	1	0	$26582837761 \times 10^{-11}$	$43011935015 \times 10^{-11}$	$43011935015 \times 10^{-11}$	$26582837761 \times 10^{-11}$	0
0.04	2	0	$24044402634 \times 10^{-11}$	$38904660701 \times 10^{-11}$	$38904660701 \times 10^{-11}$	$24044402634 \times 10^{-11}$	0
0.06	3	0	$21748366492 \times 10^{-11}$	$35189596185 \times 10^{-11}$	$35189596185 \times 10^{-11}$	$21748366492 \times 10^{-11}$	0
0.08	4	0	$19671582292 \times 10^{-11}$	$31829288761 \times 10^{-11}$	$31829288761 \times 10^{-11}$	$19671582292 \times 10^{-11}$	0
0.10	5	0	$17793113337 \times 10^{-11}$	$28789862144 \times 10^{-11}$	$28789862144 \times 10^{-11}$	$17793113337 \times 10^{-11}$	0

Table 8 Approximate solution of partial differential equation (6) by Crank-Nicolson method

	$x \rightarrow$	0	0.2	0.4	0.6	0.8	1
$t \downarrow$	j/i	0	1	2	3	4	5
0	0	0	$29389262615 \times 10^{-11}$	$47552825815 \times 10^{-11}$	$47552825815 \times 10^{-11}$	$29389262615 \times 10^{-11}$	0
0.02	1	0	$26712721481 \times 10^{-11}$	$43068863311 \times 10^{-11}$	$43068863311 \times 10^{-11}$	$26712721481 \times 10^{-11}$	0
0.04	2	0	$24076312016 \times 10^{-11}$	$39079897964 \times 10^{-11}$	$39079897964 \times 10^{-11}$	$24076312016 \times 10^{-11}$	0
0.06	3	0	$22013769981 \times 10^{-11}$	$35549929749 \times 10^{-11}$	$35549929749 \times 10^{-11}$	$22013769981 \times 10^{-11}$	0
0.08	4	0	$19750878932 \times 10^{-11}$	$32096239684 \times 10^{-11}$	$32096239684 \times 10^{-11}$	$19750878932 \times 10^{-11}$	0
0.10	5	0	$18025260407 \times 10^{-11}$	$29089090792 \times 10^{-11}$	$29089090792 \times 10^{-11}$	$18025260407 \times 10^{-11}$	0

Table 9 The proposed two approaches Error estimation are compared to the exact solution

	x	0	0.2	0.4	0.6	0.8	1
t	j/i	0	1	2	3	4	5
0	0	0	E_1	E_2	E_3	E_4	0
0.02	1	0	E_5	E_6	E_7	E_8	0
0.04	2	0	E_9	E_{10}	E_{11}	E_{12}	0
0.06	3	0	E_{13}	E_{14}	E_{15}	E_{16}	0
0.08	4	0	E_{17}	E_{18}	E_{19}	E_{20}	0
0.10	5	0	E_{21}	E_{22}	E_{23}	E_{24}	0

Table 10 Error table

	Bender Schmidt value	Crank Nicolson value
E_1	0.00000000000000	0.00000000000000
E_2	0.00000000000000	0.00000000000000
E_3	0.00000000000000	0.00000000000000
E_4	0.00000000000000	0.00000000000000
E_5	44364814 E -11	85518906 E -11
E_6	71783777 E -11	14855481 E -11
E_7	71783777 E -11	14855481 E -11
E_8	44364814 E -11	85518906 E -11
E_9	80323674 E -11	48414292 E -11
E_{10}	129966345 E -11	45270828 E -11
E_{11}	129966345 E -11	45270828 E -11
E_{12}	80323674 E -11	48414292 E -11
E_{13}	109071134 E -11	156332355 E -11
E_{14}	176480802 E -11	183852763 E -11
E_{15}	176480802 E -11	183852763 E -11
E_{16}	109071134 E -11	156332355 E -11
E_{17}	13165085 E -10	51854211 E -11
E_{18}	21301555 E -10	53935372 E -11
E_{19}	21301555 E -10	53935372 E -11
E_{20}	13165085 E -10	52354211 E -11
E_{21}	13165085 E -10	83173617 E -11
E_{22}	241044112 E -11	58184536 E -11
E_{23}	241044112 E -11	58184536 E -11
E_{24}	148973454 E -11	83173617 E -11

5 Conclusion

The Boundary value problem of a differential equation is solved in this study using Bender Schmidt and Crank Nicolson, and the level of accuracy of each of these suggested approaches is also determined. The error tables generate a clear picture of which approach is more successful. Even if the numerical answers of all the approaches are in excellent agreement with the actual solution, the Crank Nicolson method was shown to be more reliable and to converge to the exact solution more quickly than the other methods. Because it is believed to be more precise, stable, consistent, and convergent than any other methods that have been proposed, the Crank Nicolson methodology is widely used.

References

1. Agyeman, E., Folson, D.: Int. J. Comput. Appl. **79**(5), 11
2. Quarteroni, A., Sacco, R., Saleri, F.: Numerical Mathematics. Springer, New York (2000)
3. Dahlquist, G., Bjorck, A.: Numerical methods in scientific computing. Soc. Ind. Appl. Math. (2008)
4. Richards, D., Abrahamsen, A.: The One Dimensional Heat Equation (2002)
5. Ledder, G.: Differential Equations: A Modeling Approach. McGraw-Hill (2005)
6. Golub, G.H., Ortega, J.M.: Scientific Computing and Differential Equation: An Introduction to Numerical Methods. Academic Press, San Diego (1992)
7. Grewal, B.S.: Numerical Methods in Engineering and Science: with Programs in C and C++. Khanna Publisher, Nai Sarak, Delhi (2013)
8. Iyengar, S.R.K., Jain, R.K.: Numerical Methods, pp. 182–184. New Age International Publishers (2009)
9. Jain, M.K., Iyengar, S.R.K., Jain, R.K.: Numerical Methods for Scientific and Engineering Computation (2005)
10. Jaiswal, A.K., Khandelwal, A.: A Text Book of Computer Based Numerical and Statistical Techniques, p. 618 (2009)
11. Matthews, J.H., Fink, K.D.: Numerical Methods using MATLAB. Pearson Education, Inc. (2004)
12. Kandasamy, P., Thilagavathy, K., Gunavathi, K., Numerical Methods, S. Chand
13. Liu, T.: Int. J. Hybrid Inf. Techonol. **8**(8), 91 (2015)
14. Neumerier, A.: Introduction to Numerical Analysis, Cambridge, UK, 1st ed. (2001)
15. Pandey, K., Verma, L.: Appl. Math. **2**(7), 883 (2011)
16. Dawkins, P.: Lecture Notes on Differential Equations (2007). http://tutorial.math.lamar.edu
17. Rao, K.S.: Numerical methods for Scientists and Engineers, 3rd edn. PHI Learning Private Ltd., New Delhi (2006)
18. Robinson, J.C.: Infinite-dimensional Dynamical Systems: An Introduction to Dissipative Parabolic PDEs and the Theory of Global Attractors. Cambridge University Press, Cambridge (2001)
19. Sastry, S.S.: Introduction Methods of Numerical Analysis, 5th edn. PHI Learning Private Ltd., New Delhi (2012)
20. Scott, L.R.: Numerical Analysis. Princeton, USA (2011)
21. Smith, G.D.: Numerical Solutions of Partial Differential Equations. Oxford University Press, Oxford (1965)
22. Suli, E., Mayers, D.F.: An Introduction to Numerical Analysis. Cambridge, New York (2003)
23. Wani, S.S., Thakar, S.H.: Int. J. Appl. Phys. Math. **3**(5), 324 (2013)

FVM Simulation Study for Dispersion Pattern of Indoor Thoron Gas Using Computational Fluid Dynamics (CFD) Modeling: Effect of Configuration

Neetika Chauhan

Abstract Thoron (Rn-220) is a radioactive gas which is present in our indoor environment. Long-term inhalation of short-lived decay products of thoron may produce radiological effects on human health. The reliable assessment of radiation dose due to inhaled thoron and its decay products requires knowledge about its distribution pattern and concentration levels in an indoor environment. Indoor concentration levels and dispersion of radioactive nuclides mainly influenced by magnitude of source of thoron and turbulent effect due to different living conditions. The present study includes the Finite Volume Method (FVM) for CFD simulations to study the dispersion pattern of indoor thoron gas. The physical problem involves incompressible indoor airflow with fixed position of inlet and outlets. Fluidyn-MP 5.2.1 is general-purpose Multi-Physics software which is used for modeling and simulation purpose. CFD simulation results show a significant effect of different occupation conditions on indoor thoron gas dispersion patterns.

Keywords Computational fluid dynamics modeling · Finite Volume Method (FVM) · Multi-physics software · Indoor thoron gas · Vertical and horizontal distribution

1 Introduction

Thoron gas (Rn-220), is a radioactive noble gas that belongs to the same family as Radon (Rn-222). Like Rn-222, thoron is colorless, odorless, and tasteless, making it unnoticeable to human senses. Rn-220 is part of the natural decay chain of thorium-232, found in varying concentrations in soil, rocks, and building materials. Thorium-232 undergoes a series of radioactive decay, eventually leading to the formation of thoron gas. Rn-220 is released from the ground or building materials; it can enter

N. Chauhan (✉)
Department of Physics, National Institute of Technology, Kurukshetra, Haryana, India
e-mail: neetika.nit@gmail.com

S. Bhattacharyya and H. S. Mahato (eds.), *Modeling, Analysis and Simulations of Multiscale Transport Phenomena*, Springer Proceedings in Mathematics & Statistics 491, https://doi.org/10.1007/978-981-96-3098-1_12

indoor environments [1]. Once inhaled, thoron and its decay products can attach to lung tissues and emit alpha particles, damaging lung cells and increasing the risk of lung cancer. The health risks associated with thoron exposure are higher to those of radon gas. Prolonged exposure to elevated thoron levels can pose a potential hazard, particularly in poorly ventilated or enclosed spaces [2–4].

Rn-220 has a short half-life of about 55.6 s, leading to fast decay to produce progeny, some of which are solid particles that can be inhaled in the indoor air. Indoor Rn-220 concentrations can vary greatly depending on geological conditions, soil composition, building materials, ventilation rates, etc. Mitigating thoron exposure involves strategies similar to those used for radon gas. These may include improving indoor ventilation, sealing cracks and openings in buildings, using ventilation systems with proper filtration, and considering construction materials that emit lower levels of thoron. The study of indoor Rn-222 gas has been extensively explored through CFD-based modeling, while relatively fewer investigations into the complex behavior of Rn-220 [5–9]. These limited studies on thoron gas behavior reveal the exponential decay of Rn-220 gas within enclosed indoor environments [10–12]. The presence of turbulence, airflow dynamics, and diffusion phenomena significantly influences the distribution of Rn-220 gas. These findings emphasize the need for further research and comprehensive CFD modeling to enhance the understanding of thoron gas behavior and its implications for indoor air quality. The complex behavior of Rn-220 gas refers to its unpredictable dispersion, concentration, and diffusion patterns within a given indoor environment due to turbulence [13–15]. In the present study, Computational Fluid Dynamics (CFD) modeling and simulation are performed to study the distribution of indoor Rn-220 in different room configurations, particularly concerning open rooms. The CFD study provides valuable insights into its dispersion patterns. CFD modeling allows to simulate and visualize airflow patterns, gas concentration distribution, and other factors affecting indoor thoron gas behavior.

2 CFD Modeling and Simulation

CFD simulation that was conducted to study the distribution of Rn-220 inside a room using Fluidyn-MP 5.2.1 is general purpose Multi-Physics software. The finite volume method (FVM), a common numerical approach used in CFD to discretize and solve fluid flow and transport equations, was implemented for simulations. Fluidyn-MP has two components: Graphic User Interface (GUI) and a command line-based solver. GUI helps define the problem to be studied, specify the properties of the different substances or materials involved, and select the numerical schemes and solution parameters. The second component solves the appropriate governing equations involved.

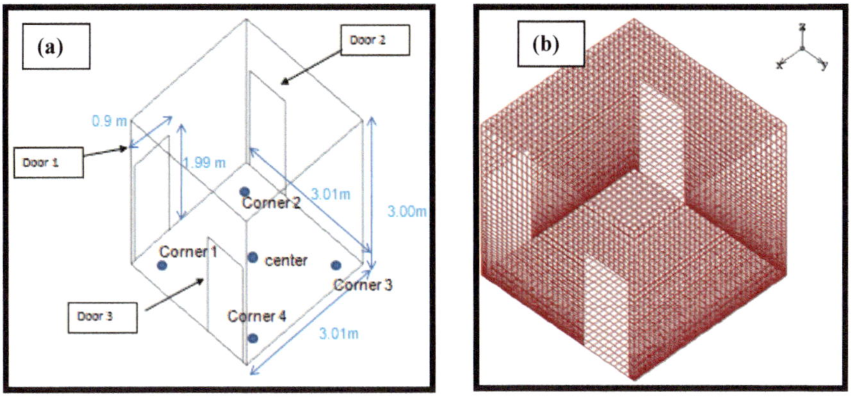

Fig. 1 **a** Model Room Geometry and **b** Mesh of Computational Domain

2.1 Geometry of Model Room and Mesh Generation

The geometry of the model room, depicted in Fig. 1a, possesses dimensions of 3.01 m × 3.01 m × 3.00 m and served as the controlled environment for investigating the dispersion of indoor Rn-220 concentration [6]. The model room configuration includes three doors, each measuring 0.9 m in width and 1.99 m in height. A narrow gap measuring 0.9 m in width and 0.02 m in height is situated between the door and the floor. During the simulation, three doors and a narrow gap were implemented to inlet and outlet boundary conditions. The model room's geometry is defined and discretized into a computational grid using the finite volume method, where the governing equations are solved. The generated mesh of the computational domain is shown in Fig. 1b.

2.2 Steady-State Flow Field Establishment and Thoron Dispersion Simulation

The simulation involved establishing a steady-state flow field followed and simulating Rn-220 dispersion in the room. The governing equations for the simulation were used to describe the physical behavior of the system. The fundamental equations used in CFD modeling are the Navier–Stokes equations for fluid flow. The incompressible Navier–Stokes equations in this study for air-flow describe the conservation of momentum. Indoor airflow was assumed incompressible. Once the steady-state flow field was established, the simulation focused on the dispersion of Rn-220 within the room. The distribution was driven by the fluid flow established. Diffusion processes governed the transport and dispersion of thoron. The simulation calculates how thoron particles disperse and mix within the room's air.

2.3 Governing Equations

A steady-state flow field was established within the room using the mass and momentum conservation equations (Eqs. 1 and 2). The $k - \varepsilon$ model is implemented within the Fluidyn_MP_5.2.3 software to capture the influence of turbulence. This model is a widely used two-equation closure approach for handling turbulence. Subsequently, the dispersion of Rn-220 was simulated within this established flow field using Eq. 3 [7].

$$\rho(\nabla.u_i) = 0, \tag{1}$$

$$\rho\left(\frac{\partial(u_i)}{\partial t} + \nabla.(u_i.u_j)\right) = -\nabla P + \nabla.(\mu_e \nabla u_i) + S, \tag{2}$$

$$\frac{\partial C_{Th}}{\partial t} = \nabla.(D\nabla C_{Th}) - \nabla.(\vec{u}C_{Th}) - \lambda_{Th}C_{Th} + S \tag{3}$$

where ρ: Air density (kg m^{-3}); u: Velocity vector (ms^{-1}); P: Pressure (Nm^{-2}); μ_e: Effective viscosity (Ns m^{-2}), S: Radioactive indoor source term; C_{Th}: Rn-220 concentration in the room (Bqm^{-3}), D: Rn-220 diffusion coefficient (m^2 s^{-1}); λ_{Th}: Rn-220 decay constant (0.01246 s^{-1}).

2.4 Material and Boundary Conditions

The computational domain was assigned air as the material, with its corresponding properties outlined in Table 1.

Four distinct CFD cases, designated as Case I, Case II, Case III, and Case IV, were formulated to exemplify a variety of potential layouts within an open room. The specific boundary conditions employed for each scenario are outlined in detail in Table 2. The visual representation of Cases I, II, III, and IV is presented in Fig. 2.

Table 1 Air properties used in the simulation

S. no.	Parameter	Value
1	Specific heat	1005 J kg^{-1} K^{-1}
2	Density	1.2 kg m^{-3}
3	Temperature	300 K

Table 2 Boundary conditions for Case I, Case II, Case III, and Case IV

Name	Position	Remark
Boundary Conditions (BC)1–4 are common for Case I–IV		
BC-1	At walls	No slip boundary condition with Rn-220 flux 0.81 $Bqm^{-2}\ s^{-1}$
BC-2	At floor	No slip boundary condition with Rn-220 flux 0.84 $Bqm^{-2}\ s^{-1}$
BC-3	At ceiling	No slip boundary condition with Rn-220 flux 0.84 $Bqm^{-2}\ s^{-1}$
BC-4	At doors	No slip boundary condition with Rn-220 flux 0.00 $Bqm^{-2}\ s^{-1}$
Case I		
BC-5	At door 1	Velocity inflow boundary condition with 0.02 ms^{-1}
BC-6	Narrow gap area under door 2	Pressure static boundary condition
BC-7	Narrow gap area under door 3	Pressure static boundary condition
Case II		
BC-5	At door 1	Velocity inflow boundary condition with 0.02 ms^{-1}
BC-6	At door 2	Pressure static boundary condition
BC-7	Narrow gap area under door 3	Pressure static boundary condition
Case III		
BC-5	At door 1	Velocity inflow boundary condition with 0.02 ms^{-1}
BC-6	Narrow gap area under door 2	Pressure static boundary condition
BC-7	At door 3	Pressure static boundary condition
Case IV		
BC-5	At door 1	Velocity inflow boundary condition with 0.02 ms^{-1}
BC-6	At door 2	Pressure static boundary condition
BC-7	At door 3	Pressure static boundary condition

2.5 *Post-processing and Result Analysis*

Post-processing involves visualizing thoron concentration contours that provide insights into how thoron disperses and concentrates in different room areas. With GUI, the results file loaded to demonstrate the vertical and horizontal distribution profile of Rn-220 in contour plots (see Table 2).

Fig. 2 The representation of **a** Case I, **b** Case II, **c** Case III, and **d** Case IV

3 Result and Discussion

The complex nature of Rn-220 gas behavior includes its unpredictable dispersion, concentration, and diffusion patterns within an enclosed indoor setting, primarily attributed to turbulence. Rn-220 gas demonstrates non-uniform distributions, subject to influences including airflow dynamics and diffusion phenomena. Predicting the behavior of Rn-220 gas presents a challenging task, demanding comprehensive investigation assisted by CFD modeling to illuminate the details of Rn-220 gas dispersion dynamics. Contour plots illustrating the indoor Rn-220 distribution pattern within the model room at horizontal XY planes (corresponding to $z = 1.5$ m heights) are presented in Fig. 3. These illustrations provide insight into how the thoron is distributed horizontally plane within the open room for each of the four cases. These representations offer a visual understanding of how thoron disperses within the horizontal plane at the specified height for each defined cases. The distribution of indoor

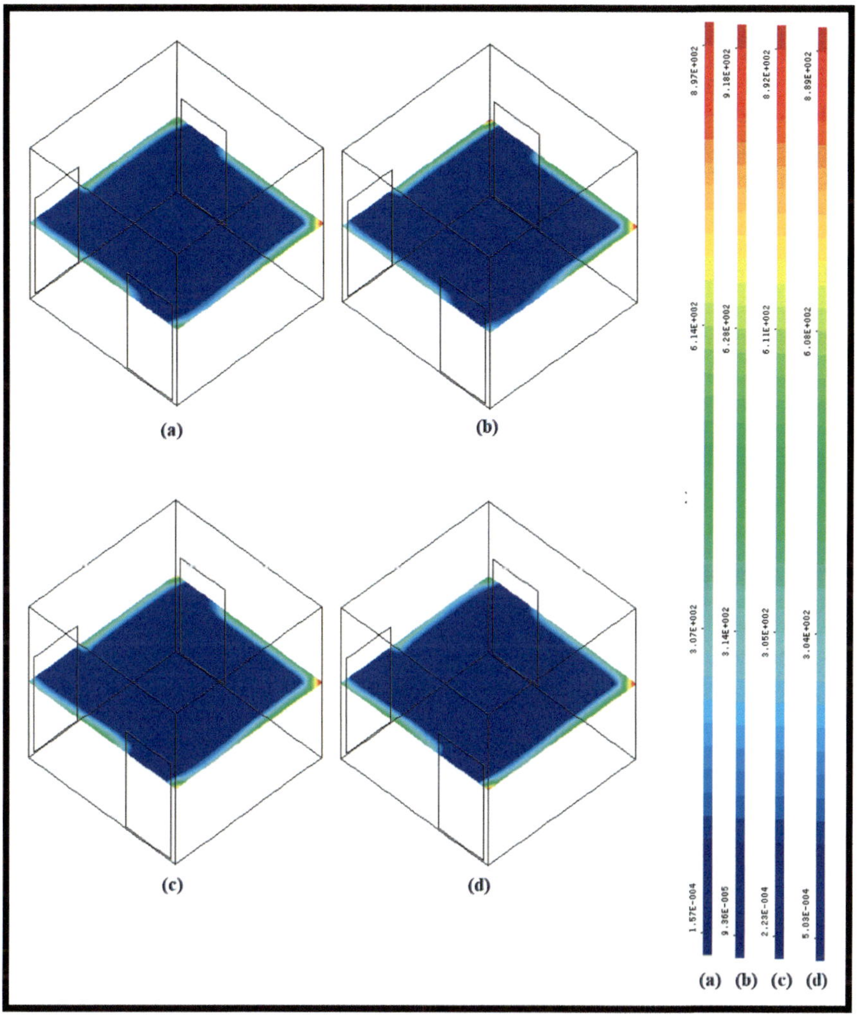

Fig. 3 Thoron dispersion in the horizontal plan at a height of z = 1.5 m for **a** Case I, **b** Case II, **c** Case III, and **d** Case IV

thoron exhibits complex behavior that is particularly sensitive to open-room configuration. Across all scenarios, the distribution of indoor Rn-220 consistently displays a characteristic non-uniform pattern [10–15]. The CFD simulation yielded the average Rn-220 concentration of horizontal distribution at the central domain of model room of 96 Bqm^{-3}, 97 Bqm^{-3}, 94 Bqm^{-3}, and 94 Bqm^{-3}, for Case I, Case II, Case III, and Case IV, respectively. Regarding the dispersion of Rn-220, the effectiveness of gas distribution appears to be comparable for Case III and Case IV. As illustrated in Fig. 3b, the dispersion of thoron gas for Case II reveals a lower concentration of Rn-220 gas near the wall facing Door 1. Even with the exhalation of Rn-220 gas from the door maintained at zero values, air circulation facilitates the transport of Rn-220 gas to the area in front of the closed door. Similarly, Fig. 4 represents the contours of Rn-220 gas distribution in vertical plans at the y = 1.5 m distance for each four cases. The average Rn-220 concentration in vertical plans at the central domain of the model room was found to be 144 Bqm^{-3}, 160 Bqm^{-3}, 120 Bqm^{-3}, and 146 Bqm^{-3}, for each four cases, respectively. Also, Fig. 5 depicts contour plots illustrating the distribution of Rn-220 gas in vertical planes at a distance of x = 1.5 m for each of the four cases. Moreover, the average Rn-220 concentration in the vertical planes at the central domain of the model room was determined to be 150 Bqm^{-3}, 152 Bqm^{-3}, 153 Bqm^{-3}, and 151 Bqm^{-3} for each of the four cases, respectively. The observed variation in the average Rn-220 concentration within horizontal and vertical distributions can be attributed to surfaces with dissimilar source terms. Specifically, the walls exhibit lower thoron exhalation rates than the current model room's roof and floor. This difference in exhalation rates across surfaces contributes to the differences in thoron concentration observed.

4 Conclusions

Based on the CFD investigation, the following conclusions are drawn:

1. Sensitivity to Room Configuration: The distribution of Rn-220 gas is subject to indoor space configuration. Visualizations of Rn-220 gas distribution patterns within the model room's horizontal XY planes demonstrate the varying impact of room configurations on Rn-220 gas dispersion.
2. Non-Uniform Patterns: Regardless of the specific scenario or room configuration (Case I, Case II, Case III, or Case IV), the distribution of indoor Rn-220 gas consistently exhibits a characteristic non-uniform pattern.
3. Thoron Concentration Variability: The average Rn-220 gas concentrations differ among the investigated four room configurations. This variation in concentration levels reflects each room configuration's distinct gas dispersion and distribution effectiveness.
4. Role of Air Circulation: Thoron gas dispersion pattern highlights the significant influence of air circulation gas transport. The air circulation facilitates the movement of Rn-220 gas to areas in front of closed doors with assuming no thoron is emitting from the door.
5. Vertical Distribution Insights: Vertical distribution plots within the room provide additional insights into Rn-220 gas behavior. These plots further highlight the complexity of gas dispersion patterns and their sensitivity to room configuration.
6. Surface Source Term Impact: The observed variations in average Rn-220 gas concentrations across horizontal and vertical distributions are attributed to surfaces with differing thoron source terms.

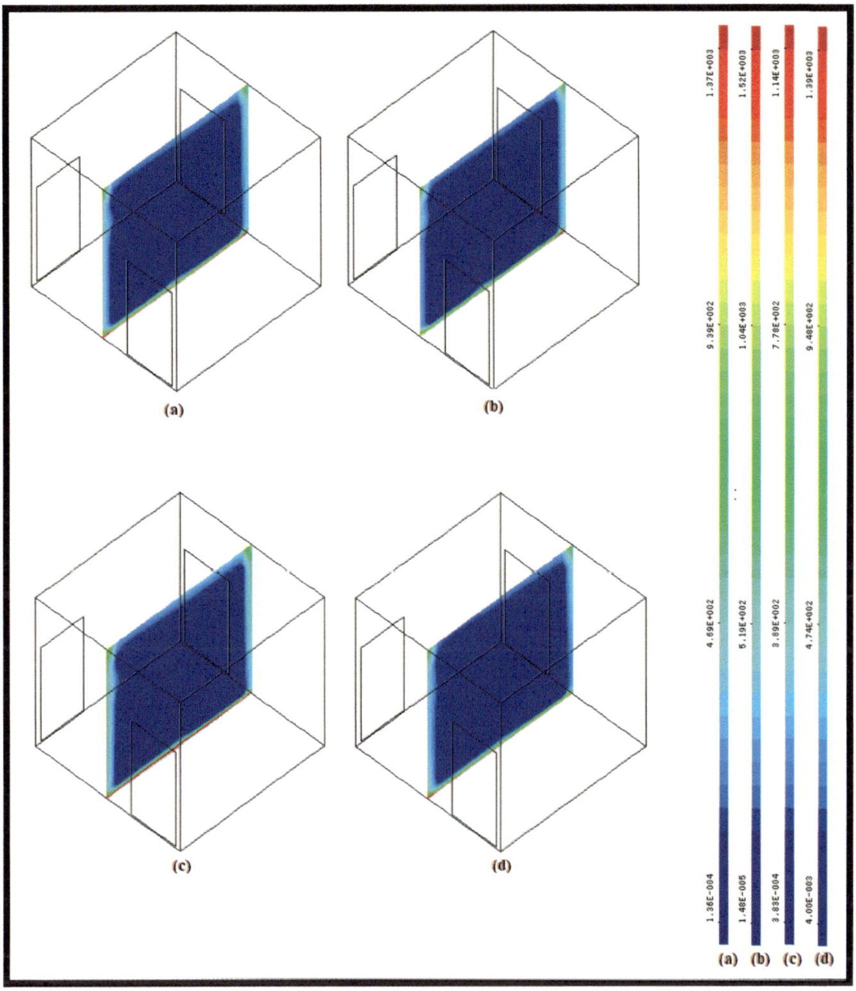

Fig. 4 Thoron dispersion in the vertical plan at the distance of y = 1.5 m for **a** Case I, **b** Case II, **c** Case III, and **d** Case IV

The CFD investigation provides knowledge about the behavior of Rn-220 gas within indoor environments. The study emphasizes the importance of comprehensive CFD modeling, which can incorporate the improvements to analyze the more factors affecting the dispersion patterns of Rn-220 gas.

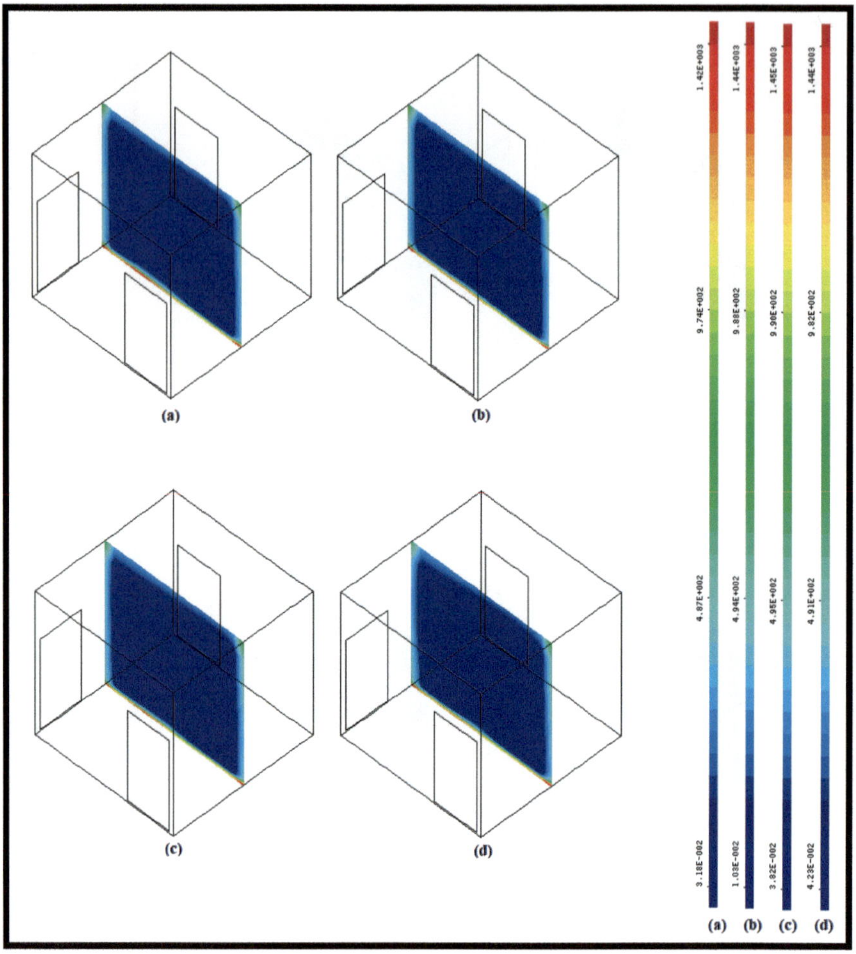

Fig. 5 Thoron dispersion in the vertical plan at the distance of x = 1.5 m for **a** Case I, **b** Case II, **c** Case III, and **d** Case IV

References

1. Nazaroff, W.W., Nero, A.V.: Radon and Its Decay Products in Indoor Air. Wiley, New York (1988)
2. UNSCEAR: United Nations scientific committee on the effects of atomic radiation. Report A/AC.82/-644, exposures of workers and the public from various sources of radiation. United Nations, New York (2000)
3. World Health Organization: WHO Handbook on Indoor Radon: a Public Health Perspective. World Health Organization (2009)
4. International Atomic Energy Agency: Radiation protection against radon in workplaces other than mines. Internat. Atomic Energy Agency (2003)
5. Akbari, K., Mahmoudi, J., Ghanbari, M.: Influence of indoor air conditions on radon concentration in a detached house. J. Environ. Radioact. **116**, 166–173 (2013)

6. Chauhan, N., Chauhan, R.P., Joshi, M., Agarwal, T.K., Aggarwal, P., Sahoo, B.K.: Study of indoor radon distribution using measurements and CFD modeling. J. Environ. Radioact. **136**, 105–111 (2014)
7. Chauhan, N., Chauhan, R.P.: Active-passive measurements and CFD based modelling for indoor radon dispersion study. J. Environ. Radioact. **144**, 57–61 (2015)
8. Rabi, R., Oufni, L.: Study of radon dispersion in typical dwelling using CFD modeling combined with passive-active measurements. Radiat. Phys. Chem. **139**, 40–48 (2017)
9. Rabi, R., Oufni, L., Amrane, M.: Modeling of indoor Rn-222 distribution in ventilated room and resulting radiation doses measured in the respiratory tract. J. Radiat. Res. Appl. Sci. **10**(3), 273–282 (2017)
10. Meisenberg, O., Tschiersch, J.: Thoron in indoor air: modeling for a better exposure estimate. Indoor Air **21**(3), 240–252 (2011)
11. Urosevic, V., Nikezic, D., Vulovic, S.: A theoretical approach to indoor radon and thoron distribution. J. Environ. Radioact. **99**(12), 1829–1833 (2008)
12. de With, G., de Jong, P.: CFD modelling of thoron and thoron progeny in the indoor environment. Radiat. Prot. Dosim., 1–7 (2011)
13. Zhuo, W., Iida, T., Moriizumi, J., Aoyagi, T., Takahashi, I.: Simulation of the concentrations and distributions of indoor radon and thoron. Radiat. Prot. Dosim. **93**(4), 357–368 (2001)
14. Chauhan, N., Chauhan, R.P., Joshi, M., Agarwal, T.K., Sapra, B.K.: Measurements and CFD modeling of indoor thoron distribution. Atmos. Environ. **105**, 7–13 (2015)
15. Chauhan, R.P., Kumar, A., Chauhan, N.: Distribution of indoor thoron in dwellings under normal and turbulent flow conditions using CFD simulation technique. Nucl. Technol. Radiat. Prot. **32**(2), 180–184 (2017)

Painlevé Analysis, Symmetry Reduction and Analytic Solutions for Variable Coefficients STOB Equation

Shailendra Singh, Vinita, and S. Saha Ray

Abstract The integrable models have the various applications in the numerous areas of science and engineering. It is an interesting fact to check that these integrable models retain the integrable property while the constant coefficients are replaced by the variable coefficients. In order to investigate such problem, this article considers the variable coefficients well posed nonlinear dynamics Sharma-Tasso-Olver-Burgers equation. The Painlevé analysis technique is implemented to retain the integrable property of the considered equation without any constraints condition on variable coefficients. To derive the analytical solutions, the auto-Bäcklund transformation approach is adopted. The 3D graphs have been drawn for the obtained solutions. These graphs depict the physical behaviour of the equation under consideration in the form of periodic, kink and anti-kink wave surfaces. At last, the Lie symmetry approach is used in order to derive the infinitesimal transformation and point symmetries for this equation.

Keywords Sharma-Tasso-Olver-Burgers equation · Integrability · Painlevè analysis · Auto-Bäcklund transformation · Lie symmetry technique · Analytical solutions

S. Singh (✉)
Department of Mathematics, Institute of Technical Education and Research, Siksha 'O' Anusandhan (Deemed to be University), Bhubaneswar, India
e-mail: shailendra.chauhan012@gmail.com

Vinita
Department of Mathematics, National Institute of Technology, Rourkela, Odisha, India

S. S. Ray
Department of Mathematics, Indian Institute of Engineering Science and Technology Shibpur, Howrah, West Bengal, India

S. Bhattacharyya and H. S. Mahato (eds.), *Modeling, Analysis and Simulations of Multiscale Transport Phenomena*, Springer Proceedings in Mathematics & Statistics 491, https://doi.org/10.1007/978-981-96-3098-1_13

1 Introduction

The study of nonlinear dynamic models plays an important role in various fields of engineering, mathematical physics, biological molecules and chemical systems [1]. Analysis of such nonlinear models allows us to explore more and to predict the possible behavior of the corresponding physical system. Applications of the solutions to these models are quite promising in the areas of applied mathematics, engineering sciences, and theoretical physics [2]. Several mathematical models have been investigated by various classical methods such as the Darboux transformation, the inverse scattering transformation, the variable separation method, the Hirota bilinear method, $\left(\frac{G}{G'}\right)$−method, the homogeneous balance method, the sin-cosine method and many others [3–6].

In this article, the well posed nonlinear dynamics model with variable coefficients known as the Sharma-Tasso-Olver-Burgers (STOB) equation [7] has been investigated. The (1+1)-dimensional STOB equation has the following form:

$$\gamma(t)w_t - \alpha(t)\left(2ww_x + w_{xx}\right) - \beta(t)\left(3w^2w_x + 3w_x^2 + 3ww_{xx} + w_{xxx}\right) = 0, \tag{1}$$

where γ, α and β are the differentiable functions of t. Equation (1) will reduces to usual Burgers system by substituting $\gamma = 1, \beta = 0$ and α as a constant and (1) will degenerate to the STO system by taking $\gamma = 1, \alpha = 0$ and β as a constant. In [8], super extension for the Burgers equation and, the soliton fission, fusion, topological, non-topological and rogue waves for the STO equation are derived by the solitary wave ansatz approach. In [9], the existence of diverse lump and interaction solutions to the STOB equation has been explored. Ref. [10] is concerned with the formation of soliton molecules by the resonant mechanism of the STOB equation. In this paper, the STOB equation has been investigated through Painlevé analysis, auto-Bäcklund transformation and Lie symetry approach.

The Painlevé analysis [11] is used for the integrability test of Eq. (1). To investigate the various exact solutions for Eq. (1), an auto-Bäcklund transformation (ABT) method and Lie symmetry approach have been used. Both the techniques are computational and algorithmic. The key idea behind the Lie symmetry analysis is taken into account the tangent structural equations under one or more parametric transformation groups [12]. The ABT technique and the Lie symmetry method are more reliable, effective, and straightforward to implement for finding the exact closed form solutions of nonlinear PDEs.

This article is summarized as follows: In Sect. 2, Painlevé analysis [13] for variable coefficients STOB equation (1) has been discussed. In Sect. 3, exact solutions for Eq. (1) are obtained via the ABT method [14, 15]. The Lie symmetry analysis [16, 17] for the aforementioned PDE was discussed in Sect. 4. The physical interpretation of these solutions is the insights of Sect. 5. Finally, the conclusion is drawn in Sect. 6.

2 Painlevé Analysis for Variable Coefficients STOB Equation

Integrability is one of the conventional procedures which assure the numerous kinds of analytical solutions for a NLPDE. Painlevé integrability analysis is often employed in order to test the integrability. One may refer to [14] for detailed procedure of this method. Considering a singular manifold which is given by $g(x, t) = 0$. The Laurent series defines the solution to Eq. (1) about the manifold $g(x, t) = 0$ as

$$w = g^{\sigma} \sum_{j=0}^{\infty} w_j g^j, \tag{2}$$

where σ is an integer value and w, w_j, g are the functions of x and t ($u_0 \neq 0$).

To get the nature of leading orders, $w = w_0 g^{\sigma}$ is substituted into Eq. (1). This generates

$$\sigma = -1, \qquad w_0 = g_x, \qquad w_0 = 2g_x. \tag{3}$$

The results show the two leading order coefficients. Therefore, it can be proceeded by considering two cases:

Case 1:

In this instance, we select $\alpha = -1$, $w_0 = g_x$. This will turn Eq. (2) as

$$w = g_x g^{-1} + \sum_{j=1}^{\infty} w_j g^{j-1}. \tag{4}$$

Next, Eq. (4) is substituted into Eq. (1) to produce a recursion relation. In this relation, the coefficient of g^{j-4} generates the resonance values at $j = -1$, 1 and 3. The Laurent series will always be terminated in the Painlevé test at the highest resonance value. Therefore, Eq. (4) becomes

$$w = g_x g^{-1} + w_1 + w_2 g + w_3 g^2. \tag{5}$$

Now, Eq. (5) is substituted into Eq. (1) and collecting the coefficients of various powers of g. One gets

$$g^{-3}: \quad 0.$$

$$\begin{aligned} g^{-2}: \quad w_2(x, t) =& \frac{1}{3\beta g_x^2}(\gamma g_t - 2\alpha u_1 g_x - 3\beta u_1^2 g_x - 3\beta g_x u_{1,x} - \alpha g_{xx} \\ & - 3\beta u_1 g_{xx} - \beta g_{xxx}). \end{aligned}$$

$$g^{-1}: \quad 0.$$

The negative resonance values can be neglected because of the consistent nature at these values. The absence of w_1 and w_3 with respect to the resonances at $j = 1$ and 3 respectively, show that w_1 and w_3 are the arbitrary functions. This remarks that the compatibility at the resonances is satisfied. Thus, Eq. (1) is known to be integrable by the Painlevé analysis.

Case 2:

In this case, we choose $\alpha = -1$, $w_0 = 2g_x$. This will turn Eq. (2) as

$$w = 2g_x g^{-1} + \sum_{j=1}^{\infty} w_j g^{j-1}. \tag{6}$$

Equation (6) is substituted into Eq. (1) to give a recursion relation. In this relation, the coefficient of g^{j-4} generates the resonance values at $j = -2, -1$ and 3. The Laurent series will always be terminated in the Painlevé test at the highest resonance value. Therefore, Eq. (6) becomes

$$w = g_x g^{-1} + w_1 + w_2 g + w_3 g^2. \tag{7}$$

Now, Eq. (7) is substituted into Eq. (1) and collecting the coefficients of various powers of g. One gets

$$g^{-3}: \quad w_1(x,t) = \frac{-\alpha g_x - 3\beta g_{xx}}{3\beta g_x}.$$

$$g^{-2}: \quad w_2(x,t) = \frac{1}{18\beta^2 g_x^3}\left(3\beta\gamma g_t g_x + \alpha^2 g_x^2 - 9\beta^2 g_{xx}^2 + 6\beta^2 g_x g_{xxx}\right).$$

$$g^{-1}: \quad 0.$$

The negative resonance values can be neglected because of the consistent nature at these values. The absence of w_3 with respect to the resonances at $j = 3$ declares that w_1 is the arbitrary function. This remarks that the compatibility at the resonances is satisfied. Thus, Eq. (1) is also known to be integrable by the Painlevé analysis in this case.

3 Analytic Solutions for Variable Coefficients STOB Equation via ABT Method

The Laurent series, which appears in Painlevé approach at a constant level in g, may be truncated in this method to get an ABT. From the previous section, two Laurent series have been derived. Therefore, it can be proceeded via considering two cases:

3.1 First ABT

From Eq. (5), the truncated series is given as

$$w = g_x g^{-1} + w_1. \tag{8}$$

In order to get the coefficients for various powers of g, Eq. (8) is substituted into Eq. (1). One gets

$$g^{-3}: \quad 0.$$

$$g^{-2}: -\gamma g_t + \alpha\left(2w_1 g_x + g_{xx}\right) + \beta\left(3w_1^2 g_x + 3g_x w_{1,x} + 3w_1 g_{xx} + g_{xxx}\right) = 0. \tag{9}$$

$$\begin{aligned} g^{-1}: \gamma g_{xt} &- \alpha\left(2g_x w_{1,x} + 2w_1 g_{xx} + g_{xxx}\right) - \beta\big(3w_1^2 g_{xx} + 6w_{1,x} g_{xx} + 3g_x w_{1,xx} \\ &+ 3w_1\left(2g_x w_{1,x} + g_{xxx}\right) + g_{xxxx}\big) = 0. \end{aligned} \tag{10}$$

$$\begin{aligned} g^{0}: \gamma w_{1,t} &- \alpha\left(2w_1 w_{1,x} + w_{1,xx}\right) - \beta\left(3w_1^2 w_{1,x} + 3w_{1,x}^2 + 3w_1 w_{1,xx} + w_{1,xxx}\right) \\ &= 0. \end{aligned} \tag{11}$$

It is clear from Eq. (11) that w_1 is obtained as an exact solution of Eq. (1) with g and w_1 satisfying Eqs. (9) and (10) (where $g_x \neq 0$). Thus, Eq. (8) is taken as the first ABT of Eq. (1). More precisely, it is re-written as

$$w = \frac{\partial}{\partial x}(\ln g) + w_1. \tag{12}$$

Equation (12) will generate several analytical solutions of Eq. (1) by giving appropriate values of g and w_1.

In a particular instance, the values of g and w_1 are selected as

$$g(x,t) = 1 + e^{p(t)x+q(t)}, \tag{13}$$
$$w_1(x,t) = k(t), \tag{14}$$

where $p(\neq 0), q$ and k are some arbitrary functions of t. The following system of equations is produced as Eq. (12) is plugged into Eqs. (9)–(11).

$$\alpha p\,(2k+p) + \beta p\left(3k^2 + 3kp + p^2\right) - \gamma\left(xp' + q'\right) = 0. \tag{15}$$

$$\alpha p^2\,(2k+p) + \beta p^2\left(3k^2 + 3kp + p^2\right) - \gamma\left((1+xp)\,p' + pq'\right) = 0. \tag{16}$$

$$\gamma k' = 0, \qquad \text{where } \gamma \neq 0 \tag{17}$$

Equation (17) produces the value of k as

$$k = \nu_1,$$

where ν_1 is an integrating constant. Next, by solving Eqs. (15) and (16), the values of p and q are found as

$$p = \nu_2,$$
$$q = \nu_3 + \int \frac{\nu_2}{\gamma}\left(2\nu_1\alpha + 3\nu_1^2\beta + \nu_2\alpha + 3\nu_1\nu_2\beta + \nu_2^2\beta\right)dt,$$

where ν_2 and ν_3 are real constants.

Thus, the first analytic solution of Eq. (1) is produced using Eq. (12) as

$$w(x,t) \tag{18}$$
$$= \nu_1 + \frac{\nu_2 \exp\left(x\nu_2 + \nu_3 + \displaystyle\int \frac{\nu_2\left((2\nu_1+\nu_2)\,\alpha + \left(3\nu_1^2 + 3\nu_1\nu_2 + \nu_2^2\right)\beta\right)}{\gamma}dt\right)}{1 + \exp\left(x\nu_2 + \nu_3 + \displaystyle\int \frac{\nu_2\left((2\nu_1+\nu_2)\,\alpha + \left(3\nu_1^2 + 3\nu_1\nu_2 + \nu_2^2\right)\beta\right)}{\gamma}dt\right)}.$$

3.2 Second ABT

From Eq. (7), the truncated series is given as

$$w = 2g_x g^{-1} + w_1. \tag{19}$$

In order to get the coefficients for various powers of g, Eq. (19) is substituted into Eq. (1). One gets

$$g^{-3} : \alpha g_x + 3\beta \left(w_1 g_x + g_{xx}\right) = 0. \tag{20}$$

$$g^{-2} : \gamma g_t g_x + \alpha g_x \left(-2w_1 g_x + g_{xx}\right) + \beta(-3w_1^2 g_x^2 + 3w_1 g_x g_{xx} + 3g_{xx}^2 + 2g_x g_{xxx}) = 0. \tag{21}$$

$$g^{-1} : -\gamma g_{xt} + \alpha \left(2g_x w_{1,x} + 2w_1 g_{xx} + g_{xxx}\right) + \beta\left(3w_1^2 g_{xx} + 6w_{1,x} g_{xx} + 3g_x w_{1,xx} + 3w_1\left(2g_x w_{1,x} + g_{xxx}\right) + g_{xxxx}\right) = 0. \tag{22}$$

$$g^{0} : \gamma w_{1,t} - \alpha \left(2w_1 w_{1,x} + w_{1,xx}\right) - \beta \left(3w_1^2 w_{1,x} + 3w_{1,x}^2 + 3w_1 w_{1,xx} + w_{1,xxx}\right) = 0. \tag{23}$$

It is clear from Eq. (23) that w_1 is obtained as an exact solution of Eq. (1) with g and w_1 satisfying Eqs. (20)–(22) (where $g_x \neq 0$). Thus, Eq. (19) is taken as the second ABT of Eq. (1), and it is re-written as

$$w = 2\frac{\partial}{\partial x}(\ln g) + w_1. \tag{24}$$

Equation (24) will generate several analytical solutions of Eq. (1) by giving appropriate values of g and w_1.

In a particular instance, the values of g and w_1 are selected as

$$g(x, t) = 1 + e^{p(t)x+q(t)}, \tag{25}$$

$$w_1(x, t) = k(t), \tag{26}$$

where $p(\neq 0), q$ and k are some arbitrary functions of t. The following system of equations is produced as Eq. (24) is plugged into Eqs. (20)–(23).

$$\alpha + 3\beta \left(k + p\right) = 0. \tag{27}$$

$$\alpha p \left(-2k + p\right) + \beta p \left(-3k^2 + 3kp + 5p^2\right) + \gamma \left(xp' + q'\right) = 0. \tag{28}$$

$$\alpha p^2 \left(2k + p\right) + \beta p^2 \left(3k^2 + 3kp + p^2\right) - \gamma \left(\left(1 + xp\right) p' + pq'\right) = 0. \tag{29}$$

$$\gamma k' = 0, \qquad \text{where } \gamma \neq 0 \tag{30}$$

Equation (30) produces the value of k as

$$k = \nu_1,$$

where ν_1 is an integrating constant. Equation (27) gives the value of p as

$$p = \frac{-\alpha - 3\nu_1\beta}{3\beta}, \qquad \text{where } \beta \neq 0.$$

Next, by solving Eq. (28), the value of q is found as

$$q = \nu_2 + \int \frac{2\alpha^3 - 27\nu_1^2\alpha\beta^2 - 27\nu_1^3\beta^3 + 9x\beta\gamma\alpha' - 9x\alpha\gamma\beta'}{27\beta^2\gamma} dt,$$

where ν_2 is the real constant.

Here, the values of p, q and k do not satisfy Eq. (29). A constraint condition $\beta(t) = \alpha(t)$ takes place to make Eq. (29) satisfy. Thus, the second analytic solution of Eq. (1) is produced using Eq. (24) with the condition $\beta(t) = \alpha(t)$ as

$$w(x, t) = \tag{31}$$

$$\frac{3\nu_1 \exp\left(x\left(\frac{1}{3} + \nu_1\right)\right) - (2 + 3\nu_1)\exp\left(\nu_2 - \int \frac{\alpha\left(-2 + 27\nu_1^2 + 27\nu_1^3\right)}{27\gamma} dt\right)}{3\left(\exp\left(x\left(\frac{1}{3} + \nu_1\right)\right) + \exp\left(\nu_2 - \int \frac{\alpha\left(-2 + 27\nu_1^2 + 27\nu_1^3\right)}{27\gamma} dt\right)\right)}.$$

4 Lie Symmetry Analysis for the STOB Equation

In this section, Lie symmetry approach has been used in order to derive the infinitesimal transformation and point symmetries for the STOB equation. Suppose that, $\gamma(t) = 1, \alpha(t) = t$, $\beta(t) = t^2$, by substituting these values in Eq. (1.1), the following particular form of STOB equation is obtained as:

$$w_t - t\left(2ww_x + w_{xx}\right) - t^2\left(3w^2w_x + 3w_x^2 + 3ww_{xx} + w_{xxx}\right). \tag{32}$$

Assume that the Lie group of point transformations with one parameter say (ε), given by

$$\begin{aligned}\tilde{x} &= x + \varepsilon\xi^x(x, t, w) + O(\varepsilon^2),\\ \tilde{t} &= t + \varepsilon\xi^t(x, t, w) + O(\varepsilon^2),\\ \tilde{w} &= w + \varepsilon\chi(x, t, w) + O(\varepsilon^2),\end{aligned} \tag{33}$$

where ξ^x, ξ^t and χ are the infinitesimals of the above transformation and ε is small parameter.

Suppose, the vector field in the tangent space determined by the transformation (4.2) is

$$\rho = \xi^x \frac{\partial}{\partial x} + \xi^t \frac{\partial}{\partial t} + \chi \frac{\partial}{\partial w}. \tag{34}$$

If STOB equation (4.1) remains invariant under the transformation (4.2), then it must fulfill the following invariance criteria:

$$pr^{(3)}\rho[\Delta(x, t)]|_{\Delta=0} = 0, \tag{35}$$

where, $\Delta = w_t - t\left(2ww_x + w_{xx}\right) - t^2\left(3w^2w_x + 3w_x^2 + 3ww_{xx} + w_{xxx}\right) = 0$, $pr^{(3)}$ is the fourth order prolongation of vector field ρ, which is formulated as:

$$\begin{aligned}pr^{(3)}\rho =& \rho + \chi^x \frac{\partial}{\partial w_x} + \chi^t \frac{\partial}{\partial w_t} + \chi^{xx} \frac{\partial}{\partial \psi_{xx}} + \chi^{xt} \frac{\partial}{\partial \psi_{xt}} + \chi^{tt} \frac{\partial}{\partial \psi_{tt}} + \chi^{xxx} \frac{\partial}{\partial \psi_{xxx}}\\ &+ \ldots\end{aligned} \tag{36}$$

By operating (4.5) in Eq. (4.4) together with $\Delta(x, t)$, one obtains

$$\begin{aligned}pr^{(4)}\rho =& \rho + \gamma(t)\chi^t - \alpha(t)(2w\chi^x + \chi^{xx}) - \beta(t)(3w^2\chi^x + 6w_x\chi^x + 3w\chi^{xx}\\ &+ \chi^{xxx}),\end{aligned} \tag{37}$$

here, χ^x, χ^t, χ^{xx}, χ^{xxx}, are known as the extended invariant coefficients, given by

$$\begin{aligned}\chi^x &= D_x\chi - u_x D_x\xi^x - u_t D_x\xi^t,\\ \chi^t &= D_t\chi - u_x D_t\xi^x - u_t D_t\xi^t,\\ \chi^{xx} &= D_x\chi_x - u_{xx} D_x\xi^x - u_{xt} D_x\xi^t,\end{aligned}$$

where D_x and D_t represent total derivative operators. By substituting these invariant coefficients in Eq. (4.6), and extracting coefficients of w and their derivatives, a system of determining equations have been obtained. On solving those determining equations, the following infinitesimals of the transformation (4.2), are obtained as:

$$\begin{aligned}\xi^x &= c_1 x + c_2,\\ \xi^t &= c_1 t,\\ \chi &= -c_1 w,\end{aligned} \tag{38}$$

where, $c_1, c_2, ..., c_6$ are arbitrary constants. Therefore, corresponding infinitesimal generators of the transformation (4.2) are given by:

$$\rho_1 = x\frac{\partial}{\partial x} + t\frac{\partial}{\partial t} - w\frac{\partial}{\partial w}, \quad \rho_2 = \frac{\partial}{\partial x}. \tag{39}$$

Now, the relation between these generators can be defined by using Lie bracket operation i.e.,

$$[\rho_1, \rho_1] = 0, [\rho_1, \rho_2] = \rho_2, [\rho_2, \rho_1] = -\rho_2, [\rho_2, \rho_2] = 0. \tag{40}$$

Now, characteristic equation corresponding to ρ_1 is

$$\frac{dx}{x} = \frac{dt}{t} = -\frac{dw}{w}. \tag{41}$$

On solving (4.10), the similarity variable and group invariant solution, is given by

$$\begin{aligned}X &= \frac{x}{t},\\ w &= \frac{1}{t}F(X).\end{aligned} \tag{42}$$

By using them (4.1) reduces to

$$3F^2F' + 2FF' + 3FF'' + 3F'^2 + F' + XF' + F'' + F''' = 0, \tag{43}$$

which is an ordinary differential equation and can be solved analytically. Similarly, by generator ρ_2 trivial solution has been obtained.

5 Physical Interpretation of Solutions

The variable coefficients STOB Eq. (1) is examined in this work. Two analytic solutions have been obtained using two ABTs as given in Sects. 3.1 and 3.2. These analytic solutions are given by Eqs. (18) and (31).For each of the solutions, two 3D graphs have been plotted. These graphs represent the periodic, kink and anti-kink solution surfaces. The descriptions of the graphs are given as follows:

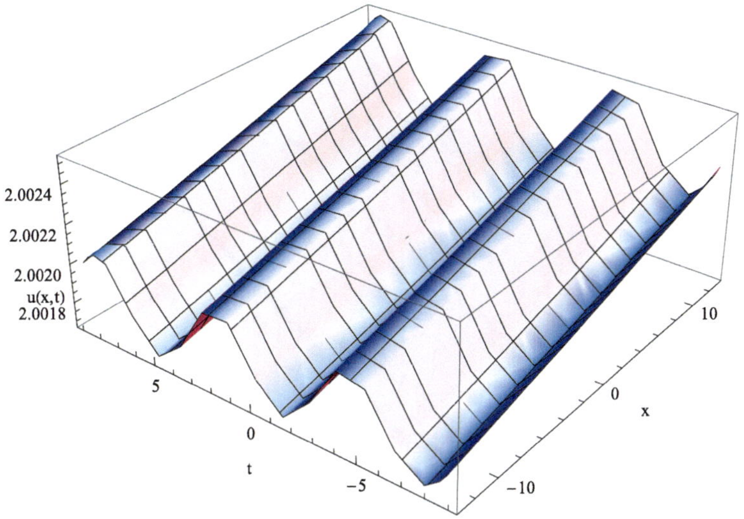

Fig. 1 Periodic solution surface of Eq. (18)

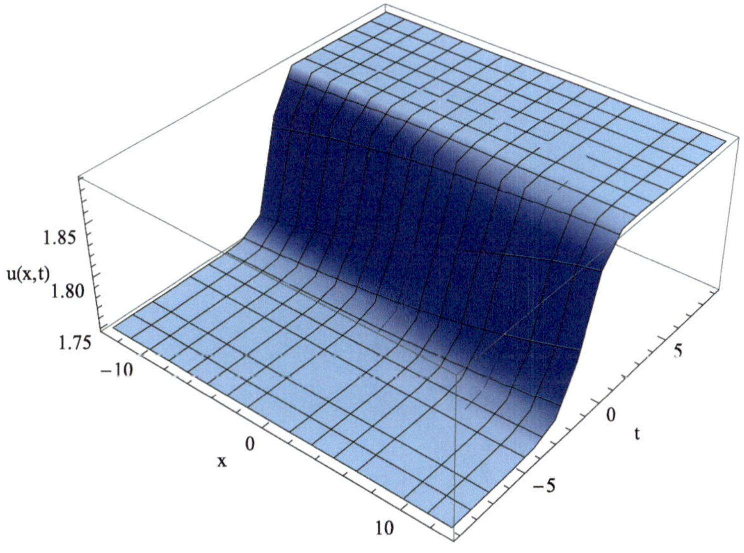

Fig. 2 Kink solution surface of Eq. (18)

The three dimensional surfaces for Eq. (18) where Fig. 1 represents the periodic wave solution for $\alpha = \sin(t)$, $\beta = \cos(t)$, $\gamma = 1$, $\nu_1 = 2$, $\nu_2 = 0.01$ and $\nu_3 = -1.26$, while Fig. 2 shows the kink wave solution for $\alpha = \exp(t)$, $\beta = 1$, $\gamma = 1$, $\nu_1 = 1.89$, $\nu_2 = -0.14$ and $\nu_3 = -2$.

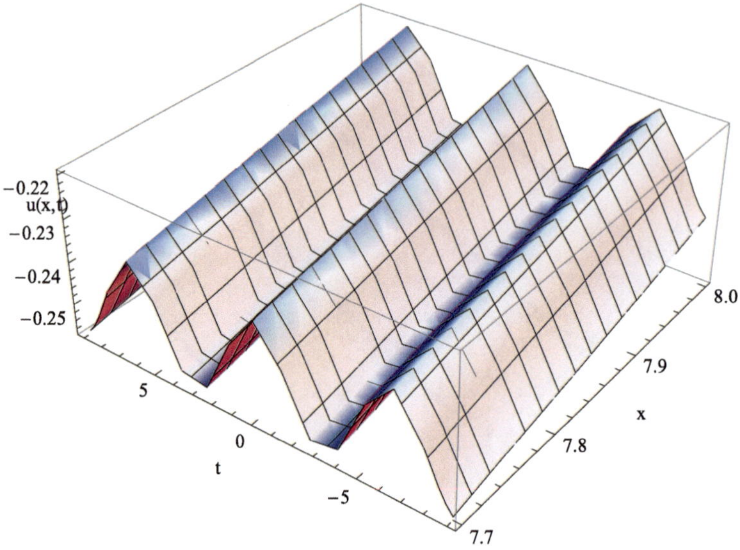

Fig. 3 Periodic solution surface for Eq. (31)

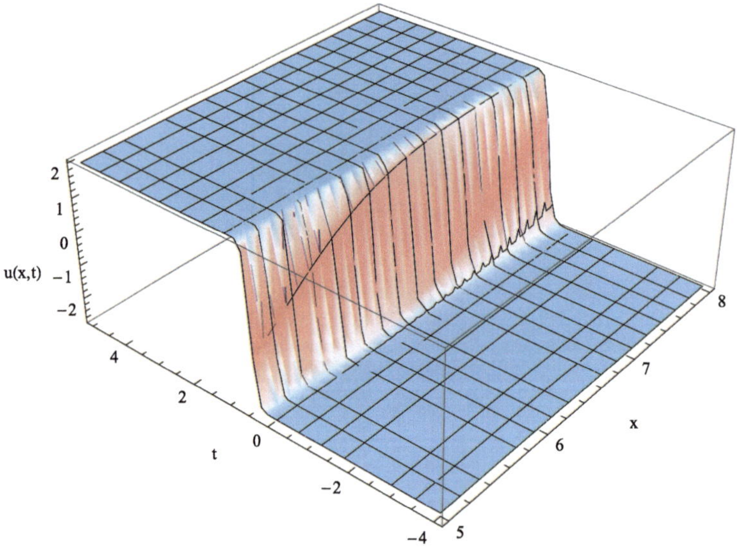

Fig. 4 Anti-kink solution surface for Eq. (31)

Moreover, the three dimensional surfaces for Eq. (31) where Fig. 3 represents the periodic wave solution for $\alpha = \sin(t)$, $\gamma = 1$, $\nu_1 = 0.001$ and $\nu_2 = 2$, while Fig. 4 shows the anti-kink wave solution for $\alpha = \exp(t)$, $\gamma = 1$, $\nu_1 = 2$ and $\nu_2 = -0.33$.

6 Conclusion

This article investigated the well posed nonlinear dynamics with variable coefficients STOB equation (1). In this paper, the Painlevé analysis technique is implemented to retain the integrable property of Eq. (1) without any constraints condition on variable coefficients. Secondly, two ABTs are generated with the help of Laurent series arising in the Painlevé analysis method. Using these ABTs, two analytical solutions have been derived successfully. The exact solution (18) is obtained with the free variables, whereas the solution (31) is generated with the condition $\beta(t) = \alpha(t)$. By taking suitable functions and parametric values, the 3D graphs have been drawn for each of the solutions. These graphs depict the physical behaviour of Eq. (1) in the form of periodic, kink and anti-kink wave surfaces. Finally, the Lie symmetry approach is used in order to derive the infinitesimal transformation and point symmetries for the STOB equation (1) by choosing specific variable coefficients $\gamma(t) = 1, \alpha(t) = t, \beta(t) = t^2$.

References

1. Logan, J.D.: An Introduction to Nonlinear Partial Differential Equations. Wiley(2008)
2. Bai, C.: Exact solutions for nonlinear partial differential equation: a new approach. Phys. Lett. A **288**(3–4), 191–195 (2001)
3. Niwas, M., Kumar, S., Kharbanda, H.: Symmetry analysis, closed-form invariant solutions and dynamical wave structures of the generalized (3+ 1)-dimensional breaking soliton equation using optimal system of Lie subalgebra. J. Ocean Eng. Sci. **7**(2), 188–201 (2022)
4. Saha, R.S., Singh, S.: New various multisoliton kink-type solutions of the (1+ 1)-dimensional Mikhailov-Novikov-Wang equation. Math. Methods Appl. Sci. **44**(18), 14690–14702 (2021)
5. Arora, R., Chauhan, A.: Lie symmetry analysis and some exact solutions of (2+ 1)-dimensional KdV-Burgers equation. Int. J. Appl. Comput. Math. **5**(1), 1–13 (2019)
6. Raza, N., Jhangeer, A., Rezazadeh, H., Bekir, A.: Explicit solutions of the (2+ 1)-dimensional Hirota-Maccari system arising in nonlinear optics. Int. J. Mod. Phys. B **33**(30), 1950360 (2019)
7. Yan, Z., Lou, S.: Soliton molecules in Sharma-Tasso-Olver-Burgers equation. Appl. Math. Lett. **104**, 106271 (2020)
8. Yu, J., Ren, B., Wang, W.L.: Symmetry Reductions, Cte Method and Interaction Solutions for Sharma-Tasso-Olver-Burgers Equation. Symmetry **14**(8), 1690 (2022)
9. Miao, Z., Hu, X., Chen, Y.: Interaction phenomenon to (1+ 1)-dimensional Sharma-Tasso-Olver-Burgers equation. Appl. Math. Lett. **112**, 106722 (2021)
10. Hu, X., Miao, Z., Lin, S.: Solitons molecules, lump and interaction solutions to a (2+ 1)-dimensional Sharma-Tasso-Olver-Burgers equation. Chinese J. Phys. **74**, 175–183 (2021)
11. Singh, S., Saha Ray, S.: Painlevé analysis, auto-Bäcklund transformation and analytic solutions for modified KdV equation with variable coefficients describing dust acoustic solitary structures in magnetized dusty plasmas. Mod. Phys. Lett. B **35**(30), 2150464 (2021)
12. Vinita, S.S.R.: Invariant analysis, optimal system, power series solutions and conservation Laws of Kersten-Krasil'shchik coupled KdV-mKdV equations. J. Geom. Phys. 104677 (2022)
13. Conte, R.: Invariant Painlevé analysis of partial differential equations. Phys. Lett. A **140**(7–8), 383–390 (1989)
14. Singh, S., Saha Ray, S.: Painlevé integrability, auto-Bäcklund transformations, new abundant analytic solutions including multi-soliton solutions for time-dependent extended KdV8 equation in nonlinear physics. J. Ocean Eng. Sci. (2022)

15. Wahlquist, H.D., Estabrook, F.B.: Bäcklund transformation for solutions of the Korteweg-de Vries equation. Phys. Rev. Lett. **31**(23), 1386 (1973)
16. Saha Ray S.: Invariant analysis and conservation laws for the time fractional (2+ 1)-dimensional Zakharov-Kuznetsov modified equal width equation using Lie group analysis. Comput. Math. Appl. **76**(9), 2110–2118 (2018)
17. Vinita, Saha Ray, S.: Symmetry analysis with similarity reduction, new exact solitary wave solutions and conservation laws of (3+ 1)-dimensional extended quantum Zakharov-Kuznetsov equation in quantum physics. Mod. Phys. Lett. B **35**(09), 2150163 (2021)

Delay Modelling of Selected Trains in Indian Railways

Shyamal Sundar Banerjee, Avijit Bhattacharyya, Sunil Kumar Sharma, and Subhash Chandra Panja

Abstract India operates the fourth largest railway network in the world, transporting over 8 billion passengers annually. Despite this extensive network, passengers frequently encounter delays, causing significant inconvenience. Given the vast number of passengers, improving the punctuality of trains in India is crucial. This study aims to model the delays of selected trains to understand their patterns over time. Train delay data, collected at various points over a period, has been used for this analysis. Researchers have developed functions to extract and process this delay data from HTML content obtained from a list of URLs. The study focuses on delays in the arrival of 20 trains at their destinations, employing various tools and machine learning algorithms, particularly regression-based methods, to model this data. Findings indicate that the Extra Trees Regression model outperforms other regression models in predicting train delays.

Keywords Train delay · RandomForest · ExtraTrees · Regression · XGBoost

S. S. Banerjee (✉) · S. C. Panja
Department of Mechanical Engineering, Jadavpur University, Kolkata, India
e-mail: shyamalsb.me.rs@jadavpuruniversity.in

S. C. Panja
e-mail: subhash.panja@jadavpuruniversity.in

A. Bhattacharyya
Founder Director, Radiant Research & Academics Pvt Ltd., Kolkata, India
e-mail: avijit.bhattacharyya@gmail.com

Professor of Practice, St. Xavier's University, (XBS), New Town, Kolkata, India

S. K. Sharma
Freight, South Eastern Railways, Kolkata, India
e-mail: crsefrser@gmail.com

S. Bhattacharyya and H. S. Mahato (eds.), *Modeling, Analysis and Simulations of Multiscale Transport Phenomena*, Springer Proceedings in Mathematics & Statistics 491, https://doi.org/10.1007/978-981-96-3098-1_14

1 Introduction

Trains have long been a key mode of long-distance travel, both in terms of freight tonnage and passenger numbers, outperforming other transport systems. This is particularly evident in countries like India, with its vast land area and a large population of 1.4 billion, where railways are a crucial and reliable transport system [1]. Indian Railways (IR) operates nearly 20,000 trains, including superfast trains, express trains, suburban trains, local short-distance trains, freight trains, and metro railways [2]. An estimated 80 million people travel by train daily in India [3]. These services, provided by the government, cater to passengers from all socio-economic backgrounds. Over time, maintaining the expected level of service has posed numerous challenges for IR, one of the most significant being train delays. Train cancellations, often due to bad weather, political unrest, strikes, and other issues, force passengers to cancel their tickets. In January 2024 alone, 4.586 million tickets were canceled, resulting in earnings of 430 million INR for the Railways due to non-availability of berths or cancellations caused by delays [4]. Addressing these delays is crucial for improving service quality and passenger satisfaction. In India, the same railway tracks are used by all the aforementioned types of trains, with no separate tracks for 90–95% of freight and passenger trains [5]. This single degree of freedom means that a delay in one train can lead to subsequent delays for others on the same track. Other significant causes of train delays include delays at the origin, engine breakdowns, waiting times at overtaking points, and varying weather conditions (such as temperature fluctuations, unpredictable wind speeds, heavy rain, irregular snowfall, and humidity). Additional factors include aging infrastructure that takes time to replace due to overcrowding, internal policies, and spare parts availability [6]. These issues can disrupt smooth train operations, affecting manpower and spare parts scheduling, overall railway efficiency, and organizational productivity. Hence, studying train delays and finding solutions is essential to meet passengers' expectations of on-time arrivals.

The numerous components of the railways, exposed to the open atmosphere, exhibit varying behavior patterns, making it challenging to embed data acquisition systems, especially across diverse terrains. Capturing and storing delay data is complex and crucial. One potential solution is to measure the actual positions of trains within the railway network [7]. Below is a brief summary of the literature reviewed to establish the foundation of this study.

1.1 Literature Review

In the Netherlands, train progress over a railway network is tracked by train describer systems using track occupation detection messages from safety systems. Similarly, in Iranian Railways, passenger train delay data is recorded daily and consolidated into weekly and monthly reports [8]. The Dutch train describer system, known as the train observation and tracking system (TROTS), requires an algorithm that efficiently processes the large TROTS log files by visiting each line only once, thereby reducing computation time [9]. An object-oriented approach stores relevant data from log files in infrastructure and train number objects, allowing the algorithm to efficiently revisit, use, and update information [9]. The State Railway of Thailand provides train location information on their website, displaying the name of the last station each train arrived at or departed from, along with timestamps and cumulative delay minutes [10].

For Indian Railways (IR) delay analysis, raw data is captured from multiple platforms such as runningstatus.in, the IR website, and railyatri.in through user-defined functions. These functions iterate through each URL and train number pair, send GET requests to the URLs, retrieve and process HTML content, extract text content, perform data preprocessing (such as splitting columns and renaming), and design the data into a consumable form, making it a significant research topic.

A Dutch Railways delay model, incorporating train path conflicts and dispatching decisions, uses 1-factor simple linear regression on historical data to help dispatchers predict running and arrival times, especially for trains with substantial running time supplements. However, it doesn't cover all delay ranges [7]. In Iranian Railways, a neural network model demonstrates high accuracy by exploring different input approaches [8]. An online traffic prediction model in the Netherlands employs a dynamic algorithm but requires consideration of additional factors [9]. For Thailand's trains, two algorithms estimate arrival times, with researchers aiming to understand more about delay factors [10]. A Backpropagation Neural Network (BPNN) simulates biological decision systems [11]. In Iranian railway delay prediction, a hybrid Adaptive Neuro-Fuzzy Inference System (ANFIS) combined with a clustering method addresses delay propagation impacts [12]. Stockholm Railways' delay analysis compares neural networks and decision trees, suggesting further research incorporating weather data [11]. Indian train delays are analyzed using various models, indicating scalability despite limited data [13]. Delhi's delay data is analyzed with Extreme Learning Machines, highlighting the need for pan-India coverage and online Recurrent Neural Network (RNN) algorithms [5, 14].

1.2 Approach to the Problem

Keeping all the above related work in mind, there is a need to model delay data from the pan-India network, encompassing various categories of trains running daily, weekly, and bi-weekly/tri-weekly. This model should consider different factors such as the type of train, travel distance, route information (including halts), day of the week, and region of the journey. It is also essential to establish the statistical significance of these factors in relation to delay times for a significant number of selected Indian Railways (IR) trains (20 in this study), which originate outside New Delhi and are long-distance express trains traversing diverse terrains.

To analyze delays in IR, researchers gathered raw data on delay times and train movements from various platforms. This process involves using custom functions to systematically traverse each Uniform Resource Locator (URL) and corresponding train number combination from provided lists. These functions send GET requests to the URLs, retrieve and analyze HTML content directly from the input element, extract textual information, and perform necessary data preprocessing tasks such as column splitting and renaming. The data is then meticulously structured into a usable format through thoughtful design.

The processed data is utilized for performance analysis of predictive machine learning models on the data collected from IR. This analysis employs two broad categories of supervised learning methods: regression and four tree-based algorithms, with 8-9 factors considering different bagging and boosting techniques.

This research paper is divided into five broad sections:

- Section 1 establishes the background and the need to solve the problem.
- Section 2 discusses the learning tools and techniques adopted to address the problem.
- Section 3 demonstrates and discusses the model results in detail.
- Section 4 provides conclusions and outlines the scope for future work.

2 Learning Tools and Techniques

The present study deals with the supervised learning methods mainly regression.From exploratory data analysis through delay histograms and scatter diagrams and barplots,the patterns in the train delay data and the assumptions which are necessary to be fulfilled for linear regression model suggests that it might be a good fit to explain the relationship between the factors laid out in the previous section. It is important to find the variables which shall affect train delay and how they are used to model it.This step involves Feature Extraction and scaling and selecting the informative features through standard feature selection techniques like backward elimination and recursive feature elimination. In the present study, below methods have been used for modelling the delay.

2.1 Regression

Simple linear regression models the relationship between two variables by fitting a linear equation to the observed data. One variable is considered to be an explanatory variable, and the other is considered to be a dependent variable.

2.2 Hypothesis Testing

The null hypothesis in the context of linear regression is that the slope of the relationship between the predictor and the response variable is zero:

$$H_0 : \beta = 0 \tag{1}$$

Failure to reject the null hypothesis is equivalent to concluding that there is no linear relationship between x and y.

2.3 Confidence Interval of β_0, β_1, and σ^2

Confidence interval is a measure of overall quality of the regression line. If the error terms ϵ_i in the regression model are NID (0, σ^2), then it is already shown that:

$$\frac{\hat{b} - b}{\sqrt{\sigma^2 / S_{xx}}} \tag{2}$$

follows t-distribution with (n-2) degrees of freedom. Thus, a 100(1-α) percent confidence interval on the slope b in simple linear regression is:

$$\hat{b} - t_{\alpha/2,n-2}\sqrt{\sigma^2/S_{xx}} \leq b \leq \hat{b} + t_{\alpha/2,n-2}\sqrt{\sigma^2/S_{xx}} \tag{3}$$

2.4 Multiple Linear Regression Equation Form

Instead of one predictor variable, when there are at least two predictor variables, multiple linear regressions are used. In the case of p regressor variables, the multiple linear regression model is given by:

$$y_i = b_0 + \sum_{j=1}^{p} b_j x_{ij} + \epsilon_i, \quad i = 1, 2, 3, \ldots, n \tag{4}$$

Errors $\epsilon_i, i = 1, 2, 3, \ldots, n$ are assumed independent and identically distributed $N(0, \sigma^2)$, as in simple linear regression.

The least square estimators, $\hat{b}_0, \ldots, \hat{b}_p$, must satisfy:

$$\left.\frac{\partial L}{\partial b_0}\right|_{b_0, b_1, \ldots, b_p} = -2 \sum_{i=1}^{n} (y_i - \hat{b}_0 - \sum_{j=1}^{p} \hat{b}_j x_{ij}) = 0 \tag{5}$$

$$\left.\frac{\partial L}{\partial b_j}\right|_{\hat{b}_0,\ldots,\hat{b}_p} = -2\sum_{i=1}^{n}(y_i - \hat{b}_0 - \sum_{j=1}^{p}\hat{b}_j x_{ij})x_{ij} = 0 \tag{6}$$

2.5 Model Assumptions

- Independence: y_i are independent of one another.
- Normality: For any fixed combination of $x_{1i}, \ldots, x_{ki}$, $y_i \sim N(b_0 + b_1 x_{1i} + \ldots + b_k x_{ki}, \sigma^2)$ since $\epsilon \sim NID(0, \sigma^2)$.
- Homoscedasticity: var(y_i) is the same for any fixed combination of $x_{1i}, \ldots, x_{ki}$. That is, var$(y_i|x_{i1}, \ldots, x_{ip}) = \sigma^2$.
- Linearity: $E(y_i)$ is a linear function of $x_{1i}, \ldots, x_{ki}$. That is, $E(y_i|x_{1i}, \ldots, x_{ki}) = b_0 + b_1 x_{1i} + \ldots + b_k x_{ki}$. The effect of the regressors on the mean of the response is additive.

2.6 Estimation of Model Parameters (Least Squares and Maximum Likelihood)

The least-squares solution gives the best fitting regression plane by minimizing the sum of squares of the distances between the observed outcomes y_i and the corresponding expected outcome values.

2.7 Test for Significance of Overall Regression

Hypothesis:

$$H_0 : \beta_1 = \beta_2 = \cdots = \beta_k = 0 \tag{7}$$

Versus

$$H_1 : \beta_j \neq 0 \text{ for at least one } j \tag{8}$$

Total Sum of Squares (SST) = Regression Sum of Squares (SSR) + Residual Sum of Squares (SSE)

Total Sum of Squares (SST) = Total variation in Y. (May also be interpreted as the Total Residual Variation (Sum of Squares) when the predictor is a Constant)

Regression Sum of Squares (SSR) = Portion of the Total variation in Y explained by the variations in X's

Residual Sum of Squares (SSE) = Portion of the Total variation in Y that still remained unexplained.

2.8 Confidence Intervals on Regression Coefficients

By assumption errors ϵ_i are distributed iid $N(0, \sigma^2)$. So the observations $\{y_i\}_{i=1}^n$ are normally and independently distributed with mean $b_0 + \sum_{j=1}^p b_j x_{ij}$ and variance σ^2. Since the least squares estimator $\hat{b}$ is a linear combination of the observations, it follows that $\hat{b}$ is normally distributed with mean vector b and the variance covariance matrix $\sigma^2(X^T X)^{-1}$. Thus, each of the statistics:

$$\frac{\hat{b}_i - b_i}{\sqrt{MSEC_{jj}}}, \quad j = 0, 1, \ldots, p \tag{9}$$

has a t-distribution with $n - p - 1$ degrees of freedom, where C_{jj} and MSE are the jj-th element of $(X^T X)^{-1}$ matrix and estimate of error variance respectively. This leads to the following 100(1-α)

$$\hat{b} - t_{\alpha/2,n-p-1}\sqrt{MSEC_{jj}} \leq b \leq \hat{b} + t_{\alpha/2,n-p-1}\sqrt{MSEC_{jj}} \tag{10}$$

2.9 Diagnostics Concept of Influential Point and Outliers

- A point that makes a lot of difference in estimation of regression coefficients is called 'an influential point'.
- A point where the observed and the fitted values of the regression differ considerably, is called an outlier.

Based on the problem researchers have considered, three methods namely Linear Regression, Random Forest and Extra Trees are applied. There are numerous regression algorithms available, and it is not always obvious which one is optimal for the dataset which is being studied. The only way to find the best model is to test a number of them and compare the results. So, in this study, two algorithms in addition to the above, i.e., in total five algorithms, viz. Multiple Linear Regression, Decision Tree, Random Forest, Extra Trees, Gradient Boosting with a combination of linear and tree-based algorithms have been tested.

1. **Multiple Linear Regression (MLR)**: This method models the relationship between the dependent variable and one or more independent variables by fitting a linear equation to the observed data. The linear regression model is given by:

$$y = \beta_0 + \beta_1 x_1 + \beta_2 x_2 + \ldots + \beta_n x_n + \epsilon \tag{11}$$

where y is the dependent variable, $x_1, x_2, \ldots, x_n$ are the independent variables, $\beta_0, \beta_1, \ldots, \beta_n$ are the coefficients, and ϵ is the error term.

2. **Decision Tree (DT) Regression**: Employs a tree-like model of decisions, where each internal node represents a decision based on the value of an attribute, and each leaf node represents the predicted value. The decision tree regression model is built by recursively splitting the data based on feature values that minimize the mean squared error.
3. **Random Forest (RF) Regression**: An ensemble learning method that constructs multiple decision trees and merges their results to improve accuracy and prevent overfitting. The random forest algorithm is given by:

$$\hat{f}(\mathbf{x}) = \frac{1}{B} \sum_{b=1}^{B} f_b(\mathbf{x}) \tag{12}$$

where B is the number of trees and f_b is the b-th decision tree.
4. **Extra Trees (ET) Regression**: Similar to random forest but with a focus on creating unpruned decision trees and using random splits of all candidate features. The extra trees algorithm is given by:

$$\hat{f}(\mathbf{x}) = \frac{1}{T} \sum_{t=1}^{T} h_t(\mathbf{x}) \tag{13}$$

where T is the number of trees and h_t is the t-th decision tree.
5. **Gradient Boosting (GB) Regression**: An ensemble technique that builds multiple decision trees sequentially, where each tree attempts to correct the errors of the previous one. The gradient boosting algorithm is given by:

$$F_m(\mathbf{x}) = F_{m-1}(\mathbf{x}) + \eta \sum_{i=1}^{N} w_i g_i(\mathbf{x}) \tag{14}$$

where F_m is the m-th model, η is the learning rate, w_i are the weights, and g_i are the base learners.

2.10 Performance Evaluation Metrics

After prediction of delay is obtained on test dataset, the following measures are used to evaluate the performance:

1. **Mean Absolute Error (MAE)**: The MAE is given by:

$$\text{MAE} = \frac{1}{n} \sum_{i=1}^{n} |y_i - \hat{y}_i| \tag{15}$$

where y_i is the actual value, $\hat{y}_i$ is the predicted value, and n is the number of observations.

2. **Root Mean Square Error (RMSE)**: The RMSE is given by:

$$\text{RMSE} = \sqrt{\frac{1}{n}\sum_{i=1}^{n}(y_i - \hat{y}_i)^2} \tag{16}$$

3. **Mean Absolute Percent Error (MAPE)**: The MAPE is given by:

$$\text{MAPE} = \frac{100\%}{n}\sum_{i=1}^{n}\left|\frac{y_i - \hat{y}_i}{y_i}\right| \tag{17}$$

4. **R-square** (R^2): The R-square value is given by:

$$R^2 = 1 - \frac{\sum_{i=1}^{n}(y_i - \hat{y}_i)^2}{\sum_{i=1}^{n}(y_i - \bar{y})^2} \tag{18}$$

where $\bar{y}$ is the mean of the actual values.

3 Results and Discussion

This section primarily consists of the below Sects. 3.1 Data Collection Approach, 3.2 Exploratory Data Analysis, 3.3 Solution Approach and Results, and 3.4 Analysis of Research Findings.

3.1 Data Collection Approach

At the outset, different sets of data on the railway systems and train movements have been studied for pilot-level analysis. Several discussions have been arranged with different levels of officials to understand different types of operations. Based on these approaches and studies, some of the data elements have been considered important and different websites and platforms have been visited to fetch the relevant data for analysis. Data resides in staging table(s) in the public domain, and web scraping functions are applied to extract data from the public domain to some readable format. Overall, these functions work together to extract and process train delay data by fetching information from a webpage, picking out the relevant information parts, organizing them neatly, and then using them. 2.6 MB of data is stored in 5-10 s. Veracity has been evaluated by checking data consistency, completeness, correctness, and whether it aligns with known facts or expectations. Data validation processes, data profiling, and data quality assessments aid in determining veracity. The final

Table 1 Description of train data collected for each train

Variable	Description
x_1	Actual starting date of a train at a station (e.g., 15th Oct 2018)
x_2	Distance between source station and destination (e.g., 639 km)
x_3	Number of days to complete journey (e.g., 3 days)
x_4	Train running days (e.g., Sunday, Monday, Tuesday, Wednesday, Thursday, Friday, Saturday)
x_5	Delay in departing
x_6	Number of stopping stations (e.g., 32 halts)
x_7	Time taken to complete the journey in minutes (e.g., 1260 mins)
x_8	Rajdhani, Superfast, Express, local passenger, etc. in categorical form
Y	Final predicted delay in minutes (e.g., 20 min)

Table 2 Data statistics for complete data of 20 trains

Statistic	Value
Total number of trains considered	20
Total number of trains running daily	5
Total number of trains running weekly	13
Total number of trains running bi-weekly/tri-weekly	2
Longest journey	4286 km (Vivek express)
Smallest journey	1531 km (Poorva express)

dataset thus obtained consists of 9192 records of 20 trains running daily, weekly, or bi-weekly from 1st Jan 2015 to 31st Dec 2019, referring to the longest-running trains from the Indian Railways website. In this analysis, 20 trains have been chosen, out of which 5 trains run daily, 13 trains run weekly, and 2 trains run bi-weekly/tri-weekly. Out of these, 2 trains have been chosen, namely Vivek Express which runs for 4286 km in one trip and Poorva Express which runs for 1531 km in one trip. For a better understanding of the dataset, the independent variables, types, and range along with descriptions are given in Tables 1 and 2.

In the current delay prediction system, the objective variable (Y) is the trains' delay in minutes behind its scheduled arrival at the terminal station, whereas the explanatory variables ($x_1, x_2, \ldots, x_8$) are the following.

3.2 Exploratory Data Analysis

Exploratory data analysis aims to address the issue of missing values and erroneous data from the observed data. Feature engineering is another crucial step in comprehending the data and selecting a suitable algorithm. Identifying the variables that influence train delays and leveraging them for prediction is a crucial task. When dealing with an extensive dataset, data filtration and clustering techniques become valuable tools to manage and extract relevant information efficiently, and some visuals about the data specifications are demonstrated below.

A pairplot is used to explore relationships between pairs of variables in a dataset. It creates a grid of scatterplots and histograms where each variable in the dataset is compared with every other continuous variable, viz. Delay (Mins.), Delay in Departure, NoDTCJ, Halt, and Distance. The diagonal of the pairplot typically consists of histograms or kernel density estimation (KDE) plots for each variable in the dataset. These plots show the distribution of each variable individually. The shape of each histogram or KDE plot is examined to understand the distribution of the corresponding variable. Common distributions include normal (bell-shaped), skewed (asymmetric), or multimodal (having multiple peaks). The off-diagonal plots in a pairplot are scatterplots that visualize the relationship between pairs of variables. Each point on the scatterplot represents a single data point, with one variable plotted against another. By examining the scatterplots, patterns or relationships between pairs of variables are identified. When points in the scatterplot follow a roughly straight line pattern, it indicates a linear correlation between the variables. When points in the scatterplot follow a curved or non-linear pattern, it suggests a non-linear relationship between the variables. When points in the scatterplot appear randomly scattered, it indicates little to no relationship between the variables. When groups or clusters of points are identified in the scatterplot, subpopulations or patterns within the data are identified. Individual points that deviate significantly from the overall pattern may indicate outliers or anomalies in the data. After careful observation of the above delay static plots and boxplots, the specification of the data is clear and this fundamental information helps the researchers in model selection discussed in Sect. 2.9 (Figs. 1 and 2).

The above plot provides a visual representation of the distribution of railway stations in India considered for the study. It helps in understanding the geographical coverage of railway infrastructure and assists in assessing the representativeness of railway services across different regions of the country considered in the present study.

3.3 Solution Approach and Results

To divide the data into training and testing sets, typically, the data is randomly split into 70% for training the model and 30% for testing the model's performance. To evaluate machine learning models, k-Fold Cross-Validation, where 'k' represents an integer (e.g., k = 10) is employed for a robust assessment of model performance by repeatedly splitting the data into training and testing subsets. In the present study, several classes of methods for addressing regression problems have been compared to understand their performance and effectiveness in predicting delays. It is to be noted that Extra Trees Regressor gives the best results since MAPE of 0.25 has been achieved compared to 0.28 achieved by Random Forest Regressor with the same R-squared for both (refer Table 3). Feature Importance Plot is seen in Fig. 3 and day-wise actual delay and predicted delay is seen in Fig. 5.

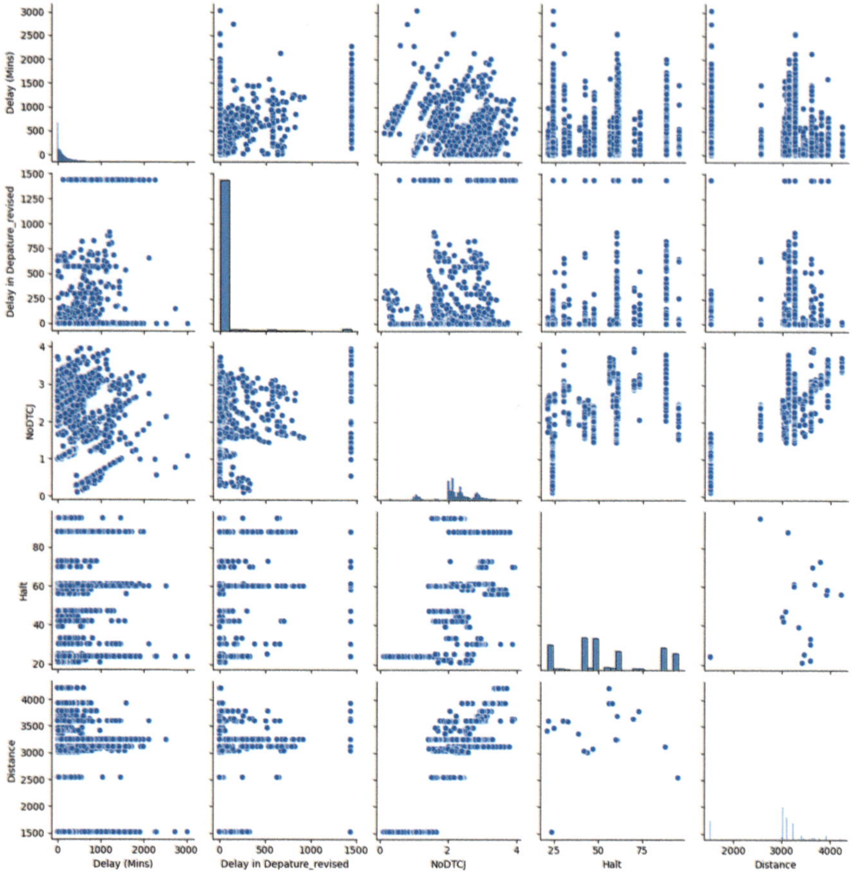

Fig. 1 Pairplots of delay (mins.) versus the independent (X) variables/features

3.4 Analysis of Research Findings

The complete analysis is demonstrated for the ET regressor as it performs the best as shown in Table 3. Feature importance plots have been used to understand which features have the most influence on the model's predictions. The feature importance plot highlights the most influential features in predicting the target variable. Features with higher importance scores contribute more to the model's predictive performance.

By examining the plot in Fig. 3, the following top 3 features have been identified as the factors which are most critical in determining the outcome of interest.

1. NoDTCJ = No. of Days Taken to Complete the journey
2. Delay in Departure
3. If train originates from North Eastern Division it gets delayed

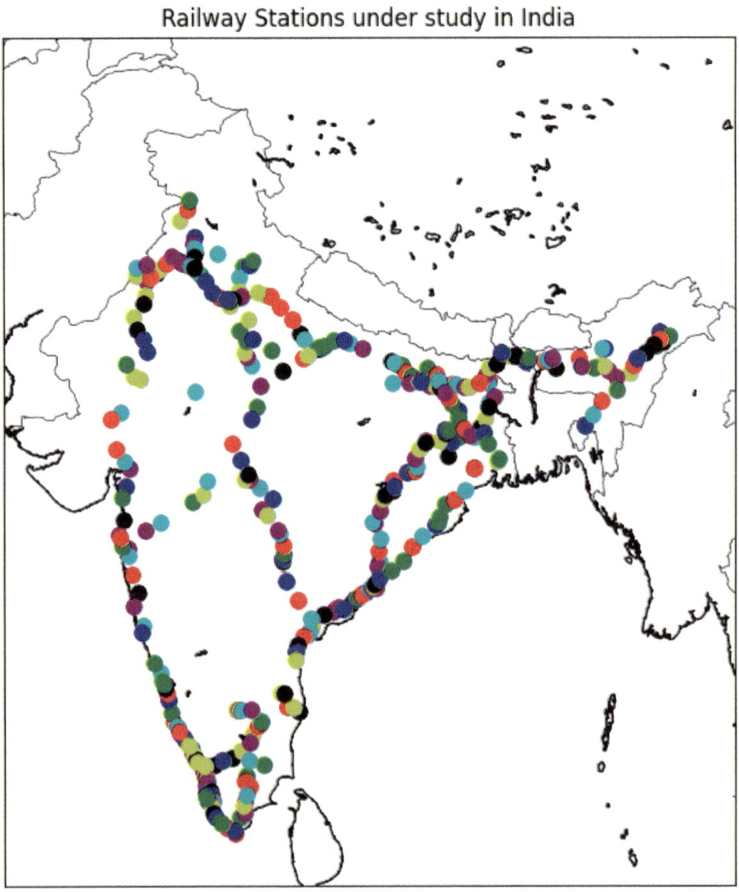

Fig. 2 Geospatial distribution of the railroad network under study

The plot provides insights into how the ET model makes predictions and which variables it relies on the most. Features with high importance indicate that they play a significant role in shaping the model's decision boundaries or predictions. Understanding the importance of features helps gain insights into the underlying relationships between predictors and the target variable. Feature importance analysis provides a form of model validation by highlighting which variables are consistently influential across different model runs or iterations. It is verified whether the identified important features align with domain knowledge or prior hypotheses about the problem domain i.e., Railways. Discrepancies between expected and observed feature importance may indicate areas for further investigation or potential model biases. Feature importance analysis may reveal unexpected or previously overlooked factors that significantly impact the target variable, thus identifying novel insights, which are used to generate new hypotheses or directions for further research. By uncovering

Table 3 Performance comparison of the MAE, RMSE, R-squared, and MAPE of the model for ETR, RFR, GBR, and DTR

Machine learning algorithm name	MAE	RMSE	R-squared	MAPE
Extra trees regressor	21.78	117.8	0.71	0.25
Random forest regressor	23.8	116.89	0.71	0.28
Gradient boosting regressor	51.15	126.41	0.67	1.13
Decision tree regressor	24.45	132	0.62	0.27

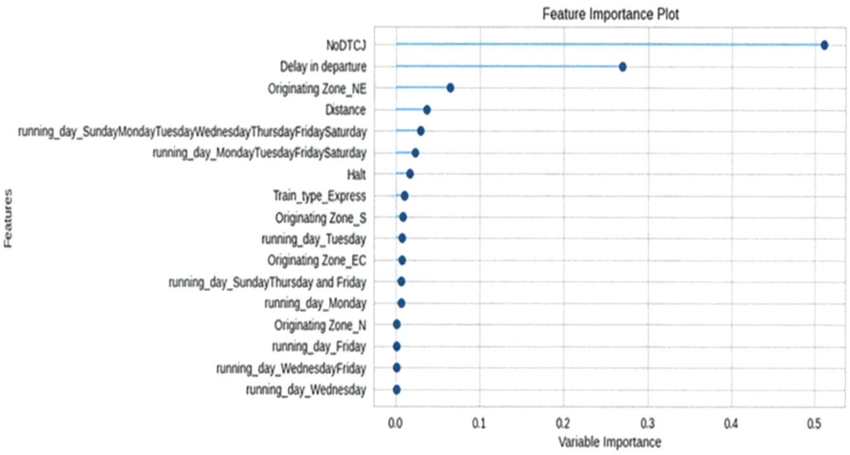

Fig. 3 Feature importance plot for the ET model

hidden relationships or patterns, feature importance plots lead to a deeper understanding of the underlying phenomena being studied. The train and test residuals distribution in Fig. 4 suggest minimal overfitting and model generalizes best on the Indian Railways dataset.

A barplot comparing actual and predicted values is a useful visualization for assessing the performance of the ET model. The plot typically consists of bars representing the actual values and the corresponding predicted values generated by the model for each day of the week for the train Avadh Assam Express (Fig. 5).

By comparing the actual and predicted values for each data point, we can assess how well the model's predictions align with the true outcomes. Ideally, the predicted values should closely match the actual values, indicating that the model is making accurate predictions. It is evident that the model performs better for Monday, Wednesday, Friday, and Sunday where the error is 5 mins or lower. Factors contributing to prediction errors for Tuesday, Thursday, and Saturday may further be studied to identify potential areas for model improvement or feature engineering. In summary, a barplot comparing actual and predicted values allows us to visually assess the performance of the ET model, identify areas of strength and weakness, and gain insights into the model's predictive capabilities and reliability.

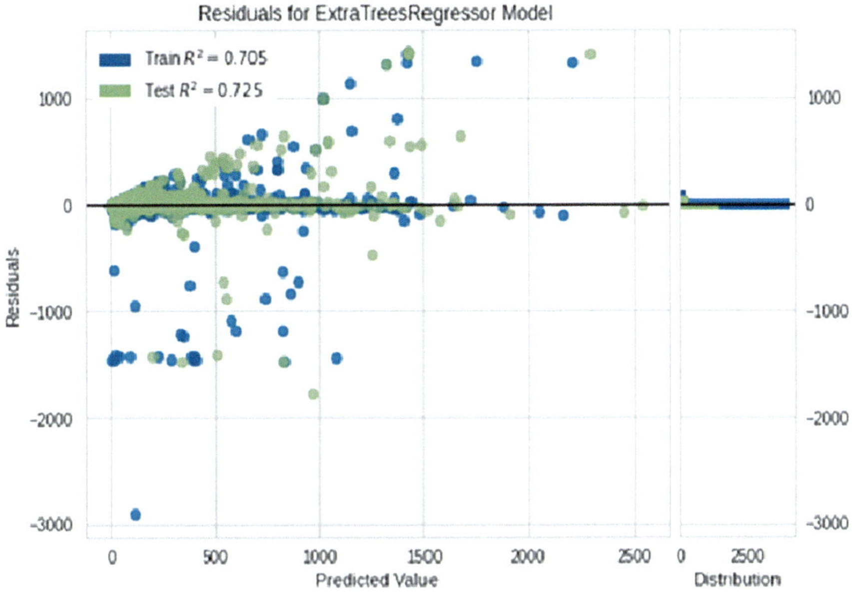

Fig. 4 Train and test error distribution

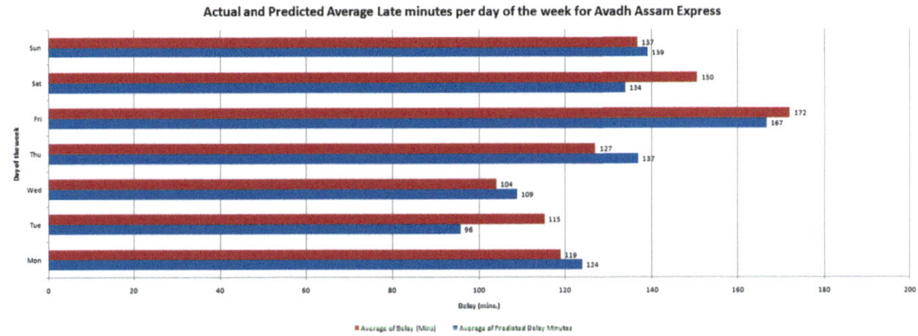

Fig. 5 Actual delay versus predicted delay

4 Conclusions

The primary aim of this paper is to model the delays of trains to understand their patterns over time. In this study, delay data at the destination for 20 trains is collected from Indian Railways across different zones over a period of five years, from 2015 to 2019. This extensive dataset is analyzed using five regression techniques to identify the most effective methods for delay prediction.The analysis reveals that the Extra Trees Regressor outperformed other regression methods in terms of Mean Absolute Percentage Error (MAPE), Mean Absolute Error (MAE), and R-squared values. This indicates that the Extra Trees Regressor is highly effective in providing accurate and reliable predictions for train delays. Further research could explore the application of other modeling techniques to enhance the accuracy and reliability of delay predictions. Specifically, time series models such as AutoRegressive Moving Average (ARMA) and AutoRegressive Integrated Moving Average (ARIMA) could be employed to predict future delays. These models are particularly well-suited for handling temporal data and could provide valuable insights into delay trends, allowing Indian Railways to proactively manage and mitigate potential delays.Moreover, incorporating additional data sources, such as weather conditions, maintenance schedules, and real-time train tracking information, could further improve the predictive models. Advanced machine learning techniques, including deep learning and ensemble methods, could also be investigated to capture complex patterns and interactions within the data. Collaborative efforts with domain experts in rail transportation could provide practical insights and help in refining the models to better suit the operational needs of Indian Railways.In conclusion, this study demonstrates the potential of using machine learning models to analyze and predict train delays, offering a pathway to more efficient and reliable railway operations. By leveraging advanced predictive analytics, Indian Railways can enhance its ability to anticipate delays and implement timely interventions, ultimately improving service quality and passenger satisfaction.

References

1. Ministry of Railways, Indian Railways year book 2015–2016, in Ministry of Railways (Railway Board) (2015)
2. Chengula, T.J., Mwakalonge, J., Comert, G., Siuhi, S.: Improving road safety with ensemble learning: detecting driver anomalies using vehicle inbuilt cameras. Mach. Learn. Appl. **14**, 100510 (2023). https://doi.org/10.1016/J.MLWA.2023.100510
3. Ministry of Railway, Indian Railways Statistical Publications 2016–2017: Passenger Business, p. 23. Archived (PDF) from the original on 3 Mar. 2018. Accessed 2 Mar. 2018
4. Desprez, M., Zawada, K., Ramp, D.: Overcoming the ordinal imbalanced data problem by combining data processing and stacked generalizations. Mach. Learn. Appl. **7**, 100241 (2022). https://doi.org/10.1016/J.MLWA.2021.100241
5. Indian Railways. https://indianrailways.gov.in/
6. Ari, B., Güvenir, H.A.: Clustered linear regression. Knowledge-Based Systems, vol. 15, no. 3, pp. 169–175 (2002). https://doi.org/10.1016/S0950-7051(01)00154-X
7. Hansen, I.A., Goverde, R., van der Meer, D.J.: Online train delay recognition and running time prediction. In: IEEE Conference on Intelligent Transportation Systems, Proceedings, ITSC, pp. 1783–1788 (2010). https://doi.org/10.1109/ITSC.2010.5625081
8. Masoud, Y., et al.: Railway passenger train delay prediction via neural network model. J. Adv. Transp. **47**, 355–368 (2013)
9. Aria, M., Cuccurullo, C., Gnasso, A.: A comparison among interpretative proposals for Random Forests. Mach. Learn. Appl. **6**, 100094 (2021). https://doi.org/10.1016/J.MLWA.2021.100094
10. Campbell, T.W., Roder, H., Georgantas III, R.W., Roder, J.: Exact Shapley values for local and model-true explanations of decision tree ensembles. Mach. Learn. Appl. **9**, 100345 (2022). https://doi.org/10.1016/J.MLWA.2022.100345
11. Hu, J., Noche, B.: Application of artificial neuron network in analysis of Railway Delays. Open J. Soc. Sci. **4**, 59–68 (2016)

12. Yaghini, M., Setayesh Sanai, M., Amin Sadrabady, H.: Passenger train delay classification. Int. J. Appl. Metaheuristic Comput. **4**, 21–31 (2015). https://doi.org/10.4018/jamc.2013010102
13. Che, D., Liu, Q., Rasheed, K., Tao, X.: Decision tree and ensemble learning algorithms with their applications in bioinformatics. Adv. Exp. Med. Biol. **696**, 191–199 (2011). https://doi.org/10.1007/978-1-4419-7046-6_19
14. Satyakrishna, J., et al.: Train delay prediction systems using big data analytics. Int. J. Innov. Res. Comput. Commun. Eng. **6**(3) (2018)

Analytical Solutions of Fractional Order Newell-Whitehead-Segel Equation

Sidheswar Behera

Abstract In this article, the $(\frac{G'}{G})$-expansion method along with the $(\frac{G'}{G}, \frac{1}{G})$-expansion method was considered to obtain analytical solutions of fractional order Newell-Whitehead-Segel(FNWS) equation. FNWS equation has paramount significance in nonlinear systems describing different stripe patterns in various surfaces(two-dimensional). The parametric condition influences the periodic solutions, kink solutions, and soliton solutions for the FNWS equation are extracted and classified into hyperbolic, trigonometric, and rational solutions. The numerical results and their graphical simulations are presented by the use of MATLAB.

Keywords Travelling wave solutions · Fractional newell-whitehead-segel equation · The $(\frac{G'}{G})$-expansion method · The $(\frac{G'}{G}, \frac{1}{G})$-expansion method

1 Introduction

The significance of fractional order partial differential equations(FPDEs) is presently grounded in pure and applied mathematics along with other scientific horizons, which are involved in many fields such as physics, applied mathematics, atmospheric science, oceanic sciences, fluid mechanics, population growth dynamic models, chemical kinetics and nonlinear optics. Historically, the fractional calculus is also classmate of classical calculus. Whereas, the former one is the generalization of the later one by a flexible approach to arbitrary orders of calculus. Considering fractional order differentials, FPDEs can model better than integer order differentials [1]. Therefore, searching for exact solutions to FPDEs is a crucial concern for researches and scientists. Since so many mathematical-physical models are governed by FPDEs, the analytical solutions to all these equations are of utmost important. FPDEs are notoriously difficult to find exact solutions to, however as symbolic computation has advanced, it is reasonable to use the test function method to create these exact

S. Behera (✉)
Department of Physics, Veer SurendraSai University of Technology, Burla, Odisha 768018, India
e-mail: sbehera_phy@vssut.ac.in

S. Bhattacharyya and H. S. Mahato (eds.), *Modeling, Analysis and Simulations of Multiscale Transport Phenomena*, Springer Proceedings in Mathematics & Statistics 491, https://doi.org/10.1007/978-981-96-3098-1_15

solutions [1–5]. Exact FPDE solutions, particularly travelling wave solutions, are crucial for understanding nonlinear physical phenomena. To reveal more information for a better comprehension of diverse nonlinear complex events as well as for further applications in real-time scientific disciplines, it is more important to design analytical solutions of FPDEs. The main aim of this study is to provide the $(\frac{G'}{G})$ method and the $(\frac{G'}{G}, \frac{1}{G})$ method for travelling wave analysis of the nonlinear fractional Newell-Whitehead-Segel equation(FNWS) [6].

$$\frac{\partial^\alpha u}{\partial t^\alpha} = \frac{\partial^{2\eta} u}{\partial x^{2\eta}} + au - bu^3,\ t > 0,\ 0 < \alpha \leq 1 \tag{1}$$

If we use $\alpha = \eta$=1 in the aforementioned equation, then the classical form of Newell-Whitehead-Segel equation will come into picture.

$$u_t = u_{xx} + au - bu^3. \tag{2}$$

Fractional models of differential equations have some great advantage that it will represent non-local property of nonlinear physical problems while the classical integer order derivatives are mostly local operators. Where a and b are real numbers. The term u_t on the left hand side of above FNWS (1) represents the time variation of $u(x, t)$ at a certain spatial location, whereas the first term u_{xx} on the right hand side of FNWS (1) represents the spatial variation of $u(x, t)$ at a specific time similarly the terms $(au - bu^3)$ on the right hand side FNWS (1) plays the role of the source term. The function $u(x, t)$ sometimes describes the nonlinear distribution of temperature gradient in a infinitely long wire and sometimes the fluid flow inside a capillary tube of infinite length as a function of velocity. In nature many non-equilibrium systems are shown in variety of forms: some uniform states, some oscillatory states, some chaotic states and in terms of some patterns. Amplitude equations play a significant role to model Many stripes patterns, for example nonlinear dynamics of sand ripples, seashell stripes can be understand in a great extent by implementing amplitude equations. FNWS is one among the amplitude equations defines stripe pattern in two dimensional systems. The dynamics near Bifurcation realm of Rayleigh-Benard convection of binary fluid mixtures modeled by FNWS [7]. The famous paper of Newell and Whitehead [8] attracts attention of mathematicians for Be'nard's problem. Be'nard's problem is an important hydrodynamic problem, in fact which describes the thermal heating of water encapsulated between two plates(heated from the bottom surface). Bifurcation parameters will display specific patterns, such as rolls where fluid stream lines form cylinders, when temperature variations between the two surfaces are over a specific threshold. By bending, these cylinders create additional spiral-like patterns. Similarly, hexagonal or rectangular patterns resembling honeycombs are produced when liquid flow is separated into cells. Some of the patterns, such as stripes and hexagons, are present in various nonlinear physical systems. Stripe patterns can be detected, for instance, in human fingerprints, in zebra skin, and in the visual cortex. In case of laser beam propagation in optical fiber and systems associated with chemical reaction and diffusion species [9] hexagonal patterns are noticed.

Many significant and effective techniques have recently been created in order to discover the properties of travelling wave solutions. For instance, variational iteration [10, 11], hirota bilinear transformation method [12], sub-equation method [13], backlund transform method [14], sine-cosine method [15–18], generalized tanh-coth method [19], jacobi elliptic function expansion method [20], adomianos decomposition method [21], the $(\frac{G'}{G})$ method [22, 23], the $(\frac{G'}{G}, \frac{1}{G})$ method [24], the modified $(\frac{G'}{G^2})$-expansion method [25, 26] modified trial function method [27], and so on.

So many researchers are devoted a great deal of their effort to the investigation of exact solution of NWS equation. E. Saberi et al. explored the conservation rules and similarity reductions of the NWS equation using the Lie symmetry approach [28]. Robert Graham studied the re-normalization group approach for an extended NWSE. Amit Prakash and Vijay Verma offered a helpful strategy based on the Adomian decomposition method in the case of time-fractional FNWS [30]. Considering Caputo fractional generalized FNWS different cases of the problem were investigated by Ali et al. [31].

The $(\frac{G'}{G})$-expansion approach, which Wang et al. developed in 2008 [32], has become widely employed to look for various exact solutions of integer order differentials. The $(\frac{G'}{G})$-expansion method is fantastic since it effectively treats nonlinear issues by transforming them to linear ones for travelling waves with constant coefficients. In essence, it converts a complex nonlinear equation into a much simpler algebraic equation. The one variable $(\frac{G'}{G})$-expansion approach really takes advantage of the fact that the exact solutions of nonlinear PDEs can be described by a polynomial in one variable $(\frac{G'}{G})$, where $G = G(\xi)$ solves the second order auxiliary equation $G'' + \lambda G' + \mu G = 0$, given λ and μ The fundamental principle of the two-variable $(\frac{G'}{G}, \frac{1}{G})$-expansion approach is that the exact solutions of nonlinear FPDEs can be described by a polynomial in two variables of $(\frac{G'}{G}, \frac{1}{G})$, where $G = G(\xi)$ satisfies the second order auxiliary equation $G'' + \lambda G' + \mu G = 0$, In a nonlinear FPDE, there is homogeneous balance between the highest order derivatives and the related nonlinear terms. is utilizes the fact for determining the degree of the polynomial. The coefficients of the resulting polynomial can then be found by applying the methods described in this work's process flow to a set of algebraic equations.

This paper presents the applicability and effectiveness of the $(\frac{G'}{G})$-expansion method and $(\frac{G'}{G}, \frac{1}{G})$-expansion method on FPDEs which model many nonlinear physical problems such as wave movement in fluid-dynamics, biological chains in population dynamics, current related problems in electrodynamics, particle vibrations in lattices in condensed matter physics, different chemical reactions in applied chemistry, and signal transmissions in optical communication system.

2 Succinct Overview of the Conformable Fractional Derivative Operator

Here we present some basic properties which will subsequently imposed for constructing the traveling wave solutions of fractional Newell-Whitehead-Segel equation.

Let a function $f : [0, \infty) \mapsto R$. The αth order space-time-fractional derivative of f structured as:

$$D_{\alpha} f(t) = \lim_{\varepsilon \mapsto 0} \frac{f(t + \varepsilon t^{1-\alpha} - f(t))}{\varepsilon}, t > 0\ ,\ \alpha \epsilon (0, 1), \tag{3}$$

Some basic properties fractional derivative operator.

1. $D^{\alpha}(cf + dg) = cD^{\alpha}(f) + dD^{\alpha}(f)$
2. $D^{\alpha}(t^{p}) = pt^{p-\alpha}$, for all $p \epsilon R$.
3. $D^{\alpha}(\lambda) = 0$, for all constant functions $f(t) = \lambda$
4. $D^{\alpha}(fg) = fD^{\alpha}(g) + gD^{\alpha}(f)$
5. $D^{\alpha}(fg)(t) = \frac{gD^{\alpha}(f) - fD^{\alpha}(g)}{g^{2}}$

3 Methodologies

This section provide the description about the $(\frac{G'}{G})$-expansion method and the $(\frac{G'}{G}, \frac{1}{G})$-expansion method in a concise way. We will have some common initial steps for both the methods, the travelling wave transformation approach is taken into account when integrating an ordinary differential equation (ODE) from a FPDEs. Given is a generalised FPDE.

$$S(u, D_{\text{t}}^{\alpha} u, D_{\text{x}}^{\alpha} u, D_{\text{tt}}^{2\alpha} u, D_{\text{xx}}^{2\alpha} u, D_{\text{xt}}^{2\alpha} u ...) = 0,\ \ 0 < \alpha \leq 1 \tag{4}$$

where $u = u(x, t)$ is an unidentified function, and S stands for a polynomial involving $u = u(x, t)$ and some its partial derivatives, which include highest order derivatives and nonlinear terms. We introduce the complex wave variable in order to determine the travelling wave solution of Eq. (4).

$$\xi = k(\frac{x^{\alpha}}{\Gamma(\alpha + 1)} - \frac{ct^{\alpha}}{\Gamma(\alpha + 1)}), \tag{5}$$

that is

$$u(x, t) = u(\xi), \tag{6}$$

we make the modifications listed below in light of this:

$$\frac{\partial}{\partial t} = -kc\frac{\partial}{\partial \xi},\quad \frac{\partial^{2}}{\partial t^{2}} = kc^{2}\frac{\partial^{2}}{\partial \xi^{2}},\quad \frac{\partial}{\partial x} = k\frac{\partial}{\partial \xi},\quad \frac{\partial^{2}}{\partial x^{2}} = k^{2}\frac{\partial^{2}}{\partial \xi^{2}}, \ldots\ldots.., \tag{7}$$

etc. for further derivatives. Using Eqs. (7), (4) changes to an ODE.

$$P(u, u_{\xi}, u_{\xi\xi}, u_{\xi\xi\xi}, ...) = 0, \tag{8}$$

where u_{ξ} denotes $\frac{du}{d\xi}$.

Now, integrate the ODE (8) to reduce it simpler form, and setting the constant of integration to zero.

3.1 The $(\frac{G'}{G})$-Expansion Method

The $(\frac{G'}{G})$-expansion method's primary steps are as follows:

Step 1: Let's look at how a polynomial in $(\frac{G'}{G})$ can be used to define the ODE (8) solution:

$$u(\xi) = \sum_{i=0}^{m} \alpha_i (\frac{G'}{G})^i,$$

When the second order linear ODE of the form $G = G(\xi)$ is satisfied.

$$G^{''} + \lambda G^{'} + \mu G = 0, \tag{9}$$

where $\alpha_i^{'}$s λ and μ are constants that must be a later time, $\alpha_m \neq 0$. The homogeneous balance between the two can be used to find the positive integer m. ODE terms that are nonlinear and have the highest order derivatives (8).

Step 2: Equation (9) is substituted into Eqs. (8) and (7), all terms with the same order of $(\frac{G'}{G})$ are collected, and the obtained polynomial's coefficients are all equaled to zero. A series of algebraic equations will then exist for the variables c, $lambda$, m, $\alpha_i^{'}$s etc.

Step 3: The well known general solutions of Eq. (8) along with α_m, α_{m-1},.....,α_0 and c values we have traveling wave solutions of FPDE (4) as follows:

$$\frac{G'(\xi)}{G(\xi)} = \begin{cases} \frac{\sqrt{\lambda^2-4\mu}}{2}(\frac{A_1 \sinh(\frac{\sqrt{\lambda^2-4\mu}}{2}\xi)+A_2\cosh(\frac{\sqrt{\lambda^2-4\mu}}{2}\xi)}{A_1\cosh(\frac{\sqrt{\lambda^2-4\mu}}{2}\xi)+A_2\sinh(\frac{\sqrt{\lambda^2-4\mu}}{2}\xi)}) - \frac{\lambda}{2}, & \lambda^2-4\mu > 0, \\ \frac{\sqrt{4\mu-\lambda^2}}{2}(\frac{-A_1 \sin(\frac{\sqrt{4\mu-\lambda^2}}{2}\xi)+A_2\cos(\frac{\sqrt{4\mu-\lambda^2}}{2}\xi)}{A_1\cos(\frac{\sqrt{4\mu-\lambda^2}}{2}\xi)+A_2\sin(\frac{\sqrt{4\mu-\lambda^2}}{2}\xi)}) - \frac{\lambda}{2}, & \lambda^2-4\mu < 0, \\ \frac{A_2}{A_1+A_2\xi} - \frac{\lambda}{2}, & \lambda^2-4\mu = 0. \end{cases} \tag{10}$$

here A_1 and A_2 are constants.

3.2 The $(\frac{G'}{G}, \frac{1}{G})$ Expansion Method

The main steps of the $(\frac{G'}{G}, \frac{1}{G})$-expansion method are given as:

Step 1: Assume that a polynomial in ϕ and ψ may be used to define the ODE (8) solution.

$$u(\xi) = \sum_{i}^{n} \alpha_i \phi^i + \sum_{i-1}^{n} \beta_i \phi^{i-1} \psi,$$

where the second order linear ordinary differential equation is satisfied by $G(\xi)$.

$$G^{''} + \lambda G - \mu = 0, \tag{11}$$

and we let

$$\phi = (\frac{G'}{G}), \psi = (\frac{1}{G}) \tag{12}$$

The positive integer n can be determined, and the constants $a_i^{\prime}$s bi's μ and λ are to be ascertained subsequently. Utilising homogenous balance between the highest order derivatives as well as the nonlinear terms in ODE (8) Because of Eq. (11) and in (12), we have,

$$\phi' = -\phi^2 + \mu\psi - \lambda, \psi' = -\phi\psi, \tag{13}$$

The general solutions of the LODE (11) can be classified into three cases as follows:

$$G(\xi) = \begin{cases} A\sinh(\sqrt{-\lambda}\xi) + B\cosh(\sqrt{-\lambda}\xi) - \frac{\mu}{\lambda}, & \psi^2 = \frac{-\lambda}{\lambda^2\nu+\mu^2}(\phi^2 - 2\mu\psi + \lambda), (\lambda < 0), \\ A\sin(\sqrt{\lambda}\xi) + B\cos(\sqrt{\lambda}\xi) + \frac{\mu}{\lambda}, & \psi^2 = \frac{\lambda}{\lambda^2\nu-\mu^2}(\phi^2 - 2\mu\psi + \lambda), (\lambda > 0), \\ \frac{\mu}{2}\xi^2 + A\xi + B, & \psi^2 = \frac{1}{A_1^2-2\mu A_2}(\phi^2 - 2\mu\psi), (\lambda = 0). \end{cases} \tag{14}$$

here constants A and B can be determined later, $\nu = A^2 + B^2$ for $(\lambda < 0)$ and $\nu = A^2 - B^2$ for $(\lambda > 0)$.

Step 2: Equation (11) can be substituted for Eq. (8) using Eqs. (14) and (13) , and then Eq. (8) can be stated in terms of a polynomial in ϕ and ψ, provided the degree of ψ is smaller. A system of algebraic equations involving $a_i^{\prime}$s, $b_i^{\prime}$s, A, B, c, μ , λ can be realised. the polynomial's coefficients can be created by setting each one to zero.

Step 3: Using the values of $a_i^{\prime}$s and $b_i^{\prime}$s where $(i = 1....n)$, solve the algebraic systems created in the previous step, we will have travelling wave solutions of three different forms such as: hyperbolic solutions, trigonometric solutions and rational solutions.

4 Application of the Methods on Governing Model

In this section, we will discuss the potential application of two proposed methods on FNWS equation to examine the closed form travelling wave type solutions.

4.1 Implementation of $(\frac{G'}{G})$-Expansion Method

We start with the FNWS Eq. (1). As described in Section. 2 we perform a traveling wave transformation u(x,t)= u(ξ) with the argument $\xi = k(\frac{x^\alpha}{\alpha} - \frac{ct^\alpha}{\alpha})$, generating the reduced ODE,

$$k^2u^{''} + kcu^{'} + au - bu^3 = 0, \tag{15}$$

Suppose, the exact solution of Eq. (15) can be expressed by a polynomial in $(\frac{G'}{G})$ as follows

$$u(\xi) = \sum_{i=0}^{m} \alpha_i (\frac{G'}{G})^i, \tag{16}$$

Where α_i (i = 0,1,2,3, ... m) are arbitrary constants, while G(ξ) satisfies the following second order linear ODE:

$$G^{''} + \lambda G^{'} + \mu G = 0, \tag{17}$$

where μ and λ are constants. We arrive to $m = 1$ by using the homogeneous balance between u^3 and $u^{''}$ in Eq. (15). As a result, we write Eq. (16) in the form

$$u(\xi) = \alpha_1 (\frac{G^{'}}{G}) + \alpha_0, \quad \alpha_1 \neq 0, \tag{18}$$

When Eqs. (18) and (17) are substituted into Eq. (15) and all terms with the same power of $\frac{G'}{G}$ are collected, the left hand side of Eq. (15) is transformed into a polynomial in $\frac{G'}{G}$ A series of algebraic equations for α_1, α_0, λ, μ and v are produced by equating the coefficients of powers of $\frac{G'}{G}$ separately to zero. solving the algebraic equations yields

$$\alpha_0 = \frac{\lambda}{\sqrt{2b}} \pm \frac{c}{3\sqrt{2b}}, \quad \alpha_1 = \pm\sqrt{\frac{2}{b}}. \tag{19}$$

where λ, b and c are arbitrary constant.

Using Eq. (19) in Eq. (18), we get the general solution of Eq. (15) as

$$u(\xi) = \pm\sqrt{\frac{2}{b}}(\frac{G'}{G}) + \frac{\lambda}{\sqrt{2b}} \pm \frac{c}{3\sqrt{2b}}, \tag{20}$$

Case 1: (hyperbolic solution), when $\lambda^2 - 4\mu > 0$:

when $A_2 = 0$, $A_1 > 0$, $\mu = 0$, we have travelling wave solution of Eq. (15),

$$u_1(\xi) = \sqrt{\frac{\lambda^2 - 4\mu}{2b}} \left(\frac{A_1 sinh(\frac{1}{2}\sqrt{\lambda^2 - 4\mu})\xi + A_2 cosh(\frac{1}{2}\sqrt{\lambda^2 - 4\mu})\xi}{A_1 cosh(\frac{1}{2}\sqrt{\lambda^2 - 4\mu})\xi + A_2 sinh(\frac{1}{2}\sqrt{\lambda^2 - 4\mu})\xi} \right) - \frac{\lambda}{\sqrt{2b}} - \frac{c}{3\sqrt{2b}}, \tag{21}$$

similarly when $A_1 = 0$, $A_2 > 0$, $\mu = 0$, we have travelling wave solution of Eq. (15),

$$u_2(\xi) = \sqrt{\frac{\lambda^2 - 4\mu}{2b}} \left(\frac{A_1 sinh(\frac{1}{2}\sqrt{\lambda^2 - 4\mu})\xi + A_2 cosh(\frac{1}{2}\sqrt{\lambda^2 - 4\mu})\xi}{A_1 cosh(\frac{1}{2}\sqrt{\lambda^2 - 4\mu})\xi + A_2 sinh(\frac{1}{2}\sqrt{\lambda^2 - 4\mu})\xi} \right) + \frac{\lambda}{\sqrt{2b}} + \frac{c}{3\sqrt{2b}}, \tag{22}$$

Case 2: (trigonometric solution),when $\lambda^2 - 4\mu < 0$:
when $A_1 = 0$, $A_2 > 0$, $\mu = 0$ we have travelling wave solution of Eq. (15) as

$$u_3(\xi) = -\sqrt{\frac{4\mu - \lambda^2}{2b}} \left(\frac{-A_1 sin(\frac{1}{2}\sqrt{4\mu - \lambda^2})\xi + A_2 cos(\frac{1}{2}\sqrt{4\mu - \lambda^2})\xi}{A_1 cos(\frac{1}{2}\sqrt{4\mu - \lambda^2})\xi + A_2 sin(\frac{1}{2}\sqrt{4\mu - \lambda^2})\xi} \right) + \frac{\lambda}{\sqrt{2b}} - \frac{c}{3\sqrt{2b}}, \tag{23}$$

similarly when $A_2 = 0$, $A_1 > 0$, $\mu = 0$ we have travelling wave solution of Eq. (15),

$$u_4(\xi) = -\sqrt{\frac{4\mu - \lambda^2}{2b}} \left(\frac{-A_1 sin(\frac{1}{2}\sqrt{4\mu - \lambda^2})\xi + A_2 cos(\frac{1}{2}\sqrt{4\mu - \lambda^2})\xi}{A_1 cos(\frac{1}{2}\sqrt{4\mu - \lambda^2})\xi + A_2 sin(\frac{1}{2}\sqrt{4\mu - \lambda^2})\xi} \right) - \frac{\lambda}{\sqrt{2b}} + \frac{c}{3\sqrt{2b}}, \tag{24}$$

Case 3: (rational solution), when $\lambda^2 - 4\mu = 0$:
we have travelling wave solution of Eq. (15),

$$u_5(\xi) = \sqrt{\frac{2}{b}} \left(\frac{A_1}{A_1 + A_{2\xi}} - \frac{\lambda}{2} \right) \pm \frac{c}{3\sqrt{2b}}, \tag{25}$$

4.2 *Implementation of* $(\frac{G'}{G}, \frac{1}{G})$*-Expansion Method*

Assume that a polynomial in $(\frac{G'}{G}, \frac{1}{G})$ may be used to define the exact solution of Eq. (15).

$$u(\xi) = \alpha_0 + \alpha_1 \phi + \beta_1 \psi, \tag{26}$$

where the second order LODE is satisfied by $G(\xi)$.

$$G'' + \lambda G - \mu = 0, \tag{27}$$

By combining Eq. (26) with Eqs. (27), (15), and carrying out the procedures already outlined in Subsection (3.2), We will be given a sequence of algebraic problems, and by resolving the algebraic problems,

$$\alpha_0 = \frac{-\mu}{3b_1 b}, \quad \alpha_1 = \sqrt{\frac{-2}{b}}, \quad \beta_1 = \frac{c}{3a_1}. \tag{28}$$

Case 1: (hyperbolic solution), when $\lambda < 0$:

$$u_{hyp}(\xi) = \frac{-\mu}{3b_1 b} + \frac{\sqrt{\frac{-2}{b}}(A_1 \cosh(\sqrt{-\lambda}\xi) + A_2 \sinh(\sqrt{-\lambda}\xi))}{A_1 \sinh(\sqrt{-\lambda}\xi) + A_2 \cosh(\sqrt{-\lambda}\xi) + \frac{\mu}{\lambda}} + \frac{c}{3\alpha_1(A_1 \sinh(\sqrt{-\lambda}\xi) + A_2 \cosh(\sqrt{-\lambda}\xi) + \frac{\mu}{\lambda})}, \tag{29}$$

when $A = 0$, $B > 0$, $\mu = 0$, we have travelling wave solution of Eq. (15) as

$$u_6(\xi) = \sqrt{\frac{-2}{b}}\sqrt{-\lambda}\tanh(\xi\sqrt{-\lambda}) + \frac{c}{3\alpha_1}B\sqrt{-\lambda}\,\mathrm{csch}(\xi\sqrt{-\lambda}), \tag{30}$$

similarly when $B = 0$, $A > 0$, $\mu = 0$, we have travelling wave solution of Eq. (15) as

$$u_7(\xi) = \sqrt{\frac{-2}{b}}\sqrt{-\lambda}\coth(\xi\sqrt{-\lambda}) + \frac{c}{3\alpha_1}A\sqrt{-\lambda}\,\mathrm{sech}(\xi\sqrt{-\lambda}), \tag{31}$$

Case 2: (trigonometric solution), when $\lambda > 0$:

$$u_{tri}(\xi) = \frac{-\mu}{3b_1 b} + \frac{\sqrt{\frac{-2}{b}}(A\cos(\sqrt{\lambda}\xi) - B\sin(\sqrt{\lambda}\xi))}{A\sin(\sqrt{\lambda}\xi) + B\cos(\sqrt{\lambda}\xi) + \frac{\mu}{\lambda}} + \frac{c}{3a_1(A\sin(\sqrt{\lambda}\xi) + B\cos(\sqrt{\lambda}\xi) + \frac{\mu}{\lambda})}, \tag{32}$$

when $A = 0$, $B > 0$, $\mu = 0$, we have travelling wave solution of Eq. (15),

$$u_8(\xi) = \sqrt{\frac{-2}{b}}\sqrt{\lambda}\tan(\xi\sqrt{\lambda}) + \frac{c}{3\alpha_1}B\sqrt{\lambda}\csc(\xi\sqrt{\lambda}), \tag{33}$$

similarly when $B = 0$, $A > 0$, $\mu = 0$, we have travelling wave solution of Eq. (15),

$$u_9(\xi) = \sqrt{\frac{-2}{b}}\sqrt{\lambda}\cot(\xi\sqrt{\lambda}) + \frac{c}{3\alpha_1}A\sqrt{\lambda}\sec(\xi\sqrt{\lambda}), \tag{34}$$

Case 3: (rational solution), when $\lambda = 0$
we have travelling wave solution of Eq. (15),

$$u_{10}(\xi) = \frac{-\mu}{3\beta_1 b} + \sqrt{\frac{-2}{b}}\frac{\mu\xi + A}{\frac{\mu\xi^2}{2} + A\xi + B} + \frac{c}{3\alpha_1}\frac{1}{\frac{\mu\xi^2}{2} + A\xi + B}. \tag{35}$$

5 Results and Discussion

Under various dynamic conditions, the analyzed solutions are described in the form of accurate periodic and solitary wave solutions. In Fig. 1a, b simulated 3D plots of the periodic travelling wave solutions of u_1 and u_2 are presented for the parameterized values of $A_1 = 2$, $A_2 = 3$, $\lambda = \sqrt{-1}$, $b = 2$, $c = 4$, $v = 1$ respectively. Figure 1c, d represents hybrid dark and bright soliton solutions for the parameterized values of $A_1 = 2$, $A_2 = 3$, $\lambda = 1$, $b = 2$, $c = 4$, $v = 1$ for u_3 and u_4 respectively. It is noticed from Fig. 2a, b that the wave solutions of u_6 and u_7 are propagating towards positive x- direction, progressive wave solutions and singular optical solitons solutions are represented for Eq. (30) and Eq. (31) respectively. In Fig. 2c, d the direction of propagation of waves for u_8 and u_9 are quite opposite to that of Fig. 2a and seems like an mirror image of Fig. 2a. Figures 1e and 2e display the dynamics of the traveling waves for u_5 and u_{10} respectively. To the best of our notice, some new solutions in the form of $u_1, u_2, u_3, u_4, u_6, u_7, u_8$ and u_9 are presented while u_5 and u_{10} are presented by Prakash et al. [30].

6 Conclusion

The $(\frac{G'}{G})$-expansion method and the $(\frac{G'}{G}, \frac{1}{G})$-expansion method have been provided and utilised to produce numerous precise travelling wave solutions utilising arbitrary FNWS equation parameters.We get the recognised solitary solutions as the two parameters $A1$ and $A2$ take on specific values. The $(\frac{G'}{G}, \frac{1}{G})$-expansion approach is used when $mu = 0$ in equations Eqs. (30) and (34) and bi in Eq. (26) for expansion. It is seen that the graphical simulation of Eqs. (25) and (35) are having similar nature and in full agreement with the assumptions. In fact the two variable $(\frac{G'}{G}, \frac{1}{G})$-expansion method is an extended version of the $(\frac{G'}{G})$-expansion method. In summary, the the two variable$(\frac{G'}{G}, \frac{1}{G})$-expansion method is have great advantage over the $(\frac{G'}{G})$-expansion method because the solutions discovered using the earlier technique recoup the discoveries made using the later technique. It's also interesting to note that, depending on the parametric parameters, the general traveling wave solutions can provide periodic or soliton solutions. The obtained solutions of FNWS equation may be helpful to understand different nonlinear systems describing stripe patterns in two dimensional systems.

Declarations

Ethical Approval

This article does not involve with any human or animal studies. I also confrm that, I have read and abided by the statement of ethical standards for the manuscript

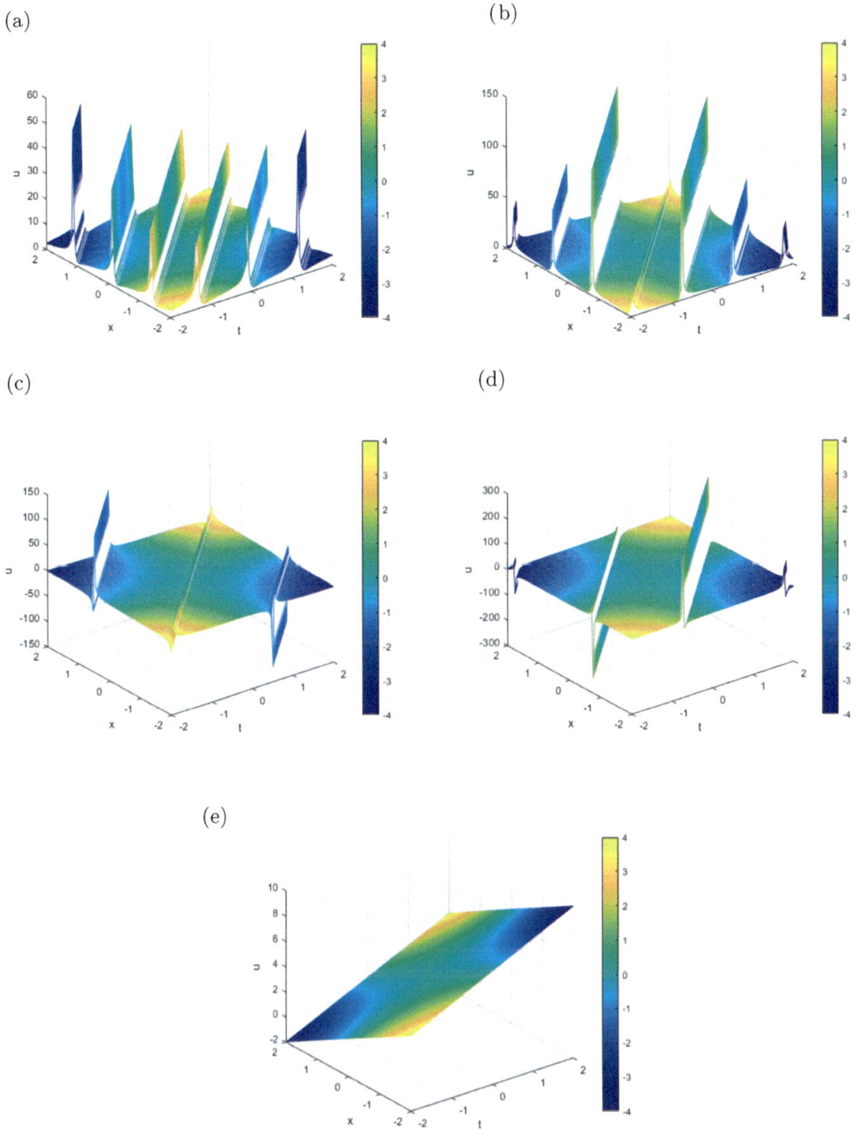

Fig. 1 Traveling wave solution corresponding to the fnws equation by $(\frac{G'}{G})$-expansion method. **a** Plot of u_1 with $A_1 = 2$, $A_2 = 3$, $\lambda = \sqrt{-1}$, $b = 2$, $c = 4$, $v = 1$. **b** Plot of u_2 with $A_1 = 2$, $A_2 = 3$, $\lambda = \sqrt{-1}$, $b = 2$, $c = 4$, $v = 1$. **c** Plot of u_1 with $A_1 = 2$, $A_2 = 3$, $\lambda = 1$, $b = 2$, $c = 4$, $v = 1$. **d** Plot of u_1 with $A_1 = 2$, $A_2 = 3$, $\lambda = 1$, $b = 2$, $c = 4$, $v = 1$. **e** Plot of u_1 with $\lambda = 1$, $b = 2$, $c = 4.5$, $v = 1$, $A_1 = 2$, $A_2 = 3$

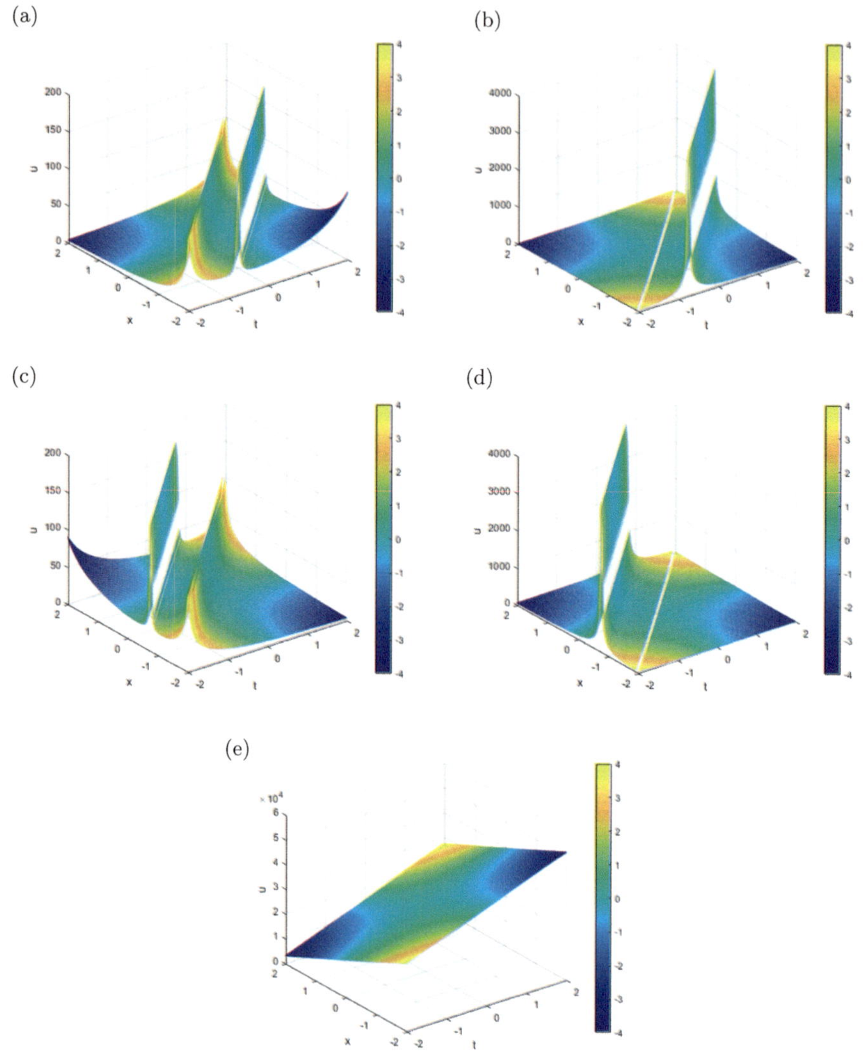

Fig. 2 Traveling wave solution corresponding to the FNWS equation by $(\frac{G'}{G}, \frac{1}{G})$-expansion method. **a** Plot of u_6 with $A = 5$, $B = 3$, $\lambda = -4$; $b = 2$; $\alpha_1 = 4$; $v = 1$; $c = 3.5$. **b** Plot of u_7 with $A = 5$, $B = 3$, $\lambda = -4$; $b = 2$; $\alpha_1 = 4$; $v = 1$; $c = 3.5$. **c** Plot of u_8 with $A = 5$, $B = 3$, $\lambda = 4$; $b = 2$; $\alpha_1 = 4$; $v = 1$, $c = 3.5$. **d** Plot of u_9 with $A = 5$, $B = 3$, $\lambda = 4$; $b = 2$; $\alpha_1 = 4$; $v = 1$, $c = 3.5$. **e** Plot of u_{10} with $\mu = 1$, $A = 4.5$, $B = 2.5$, $\alpha_1 = 5$, $\beta_1 = 2$, $v = 1$, $c = 1.5$

submission to this journal and that the manuscript has not been copyrighted, published, or submitted elsewhere.

Confict of Interest

I have no conflicts of interest to disclose.

Author Contributions

The author confirms sole responsibility for the following: study conception and design, data collection, analysis and interpretation of results, and manuscript preparation.

Funding

This research received no funding to acknowledge.

Data Availability

The data used to support the fndings of this study are included within the article.

References

1. Podlubny, I.: Fractional Differential Equations: An Introduction to Fractional Derivatives, Fractional Differential Equations, to Methods of their Solution and Some of their Applications. Academic Press, New York (1999)
2. Meerschaert, M.M., Tadjeran, C.: Finite difference approximations for two-sided space-fractional partial differential equations. Appl. Numer. Math. **56**(1), 80–90 (2006)
3. Barkai, E., Metzler, R., Klafter, J.: From continuous time random walks to the fractional Fokker-Planck equation. Phys. Rev. E **61**(1), 132–138 (2000)
4. Daftardar-Gejji, V., Bhalekar, S.: Solving fractional diffusion-wave equations using a new iterative method. Fract. Calc. Appl. Anal. **11**(2), 193–202 (2008)
5. Jaradat, O.K., Al-Banawi, K., Joudeh, A.A.: Solving fractional hyperbolic partial differential equations by the generalized differential transform method. World Appl. Sci. J. **23**(12), 89–96 (2013)
6. Elaheh Saberi, S.: Reza Hejazi, Ahmad Motamednezhad, Lie symmetry analysis, conservation laws and similarity reductions of Newell-Whitehead-Segel equation of fractional order. J. Geom. Phys. **135**, 116–128 (2019)
7. Getling, A.V.: Rayleigh-Benard Convection: Structures and Dynamics. World Scientific (1998). 978-981-02-2657-2
8. Newell, A., Whitehead, J.: Finite bandwidth, finite amplitude convection. J. FluidMech. **38**, 279–303 (1969)
9. Golovinb, N.A.: General Aspect of pattern formation, pattern formation and growth phenomena in Nano-System. Alxaander 1–54 (2007)
10. Guo-cheng Wu, A., Lee, E.W.M.: Fractional variational iteration method and its application. Phys. Lett. A **374**, 2506–2509. https://doi.org/10.1016/j.physleta.2010.04.034
11. Sakar, M.G., Ergören, H.: Alternative variational iteration method for solving the time-fractional Fornberg-Whitham equation. Appl. Math. Model. **39**, 3972–3979 (2015). http://dx.doi.org/10.1016/j.apm.2014.11.048
12. Jaradat, H.M.: New solitary wave and multiple soliton solutions for the time-space fractional boussinesq equation, Italian. J. Pure Appl. Math. **36**(367–376) (2016)

13. Zhang, S., Zhang, H.-Q.: Fractional sub-equation method and its applications to nonlinear fractional PDEs. Phys. Lett. A **375**, 1069–1073 (2011)
14. Lu, B.: Bäcklund transformation of fractional Riccati equation and its applications to nonlinear fractional partial differential equations. Phys. Lett. A **376**, 2045–2048 (2012). https://doi.org/10.1016/j.physleta.2012.05.013
15. Yao, S.W., Behera, S., Inc, M., Rezazadeh, H., Virdi Virdi, J.P.S., Mahmoud, W., Arqub, O.A., Osman, M.S.: Res. Phys. **42**, 105990 (2022). https://doi.org/10.1016/j.rinp.2022.105990
16. Behera, S., Virdi, J.S.: Solitons and Breather type solutions of some nonlinear equations by the Sine-Cosine method. Authorea 24 May 2020. https://doi.org/10.22541/au.159031622.22777311
17. Behera, S., Virdi, J.S.: Traveling wave solutions of some coupled nonlinear systems by sine-cosine approach. In: Giri, D., Raymond Choo, K.K., Ponnusamy, S., Meng, W., Akleylek, S., Prasad Maity, S. (eds.), Proceedings of the Seventh International Conference on Mathematics and Computing. Advances in Intelligent Systems and Computing, vol. 1412. Springer, Singapore (2022). https://doi.org/10.1007/978-981-16-6890-6_66
18. Behera, S., Virdi, J.P.S.: Analytical solutions of some fractional order nonlinear evolution equations by sine-cosine method. Discont. Nonlinearity Compl. **12**(2), 275–286 (2023). https://doi.org/10.5890/DNC.2023.06.004
19. Manafian, J., Lakestani, M.: A new analytical approach to solve some of the fractional-order partial differential equations, Indian. J. Phys. **91**, 243–258 (2017). https://doi.org/10.1007/s12648-016-0912-z
20. Tasbozan, O., Çenesiz, Y., Kurt, A.: New solutions for conformable fractional Boussinesq and combined KdV-mKdV equations using Jacobi elliptic function expansion method. Eur. Phys. J. Plus **131**, 244 (2016). https://doi.org/10.1140/epjp/i2016-16244-x
21. Saha Ray, S., Bera, R.K.: An approximate solution of a nonlinear fractional differential equation by Adomian decomposition method. Appl. Math. Comput. **167**, 561–571 (2005). https://doi.org/10.1016/j.amc.2004.07.020
22. Bekira, A., Guner, O.: Exact solutions of nonlinear fractional differential equations by $(\frac{G'}{G})$-expansion method. Chin. Phys. B **22**(11), 110202 (2013). https://doi.org/10.1088/1674-1056/22/11/110202
23. Zheng, B.: $(\frac{G'}{G})$-Expansion method for solving fractional partial differential equations in the theory of mathematical physics. Commun. Theor. Phys. **58**, 623–630 (2012). http://dx.doi.org/10.1088/0253-6102/58/5/02
24. Sirisubtawee, S., Koonprasert, S., Sungnul, S.: Some applications of the $(\frac{G'}{G}, \frac{1}{G})$-Expansion method for finding exact traveling wave solutions of nonlinear fractional evolution equations. Symmetry **11**(8), 952 (2019). https://doi.org/10.3390/sym11080952
25. Behera, S.: Dynamical solutions and quadratic resonance of nonlinear perturbed Schrödinger equation. Front. Appl. Math. Stat. **8**, 128 (2023). https://doi.org/10.3389/fams.2022.1086766
26. Behera, S., Aljahdaly, N.H., Virdi, J.P.S.: On the modified $(\frac{G'}{G^2})$-expansion method for finding some analytical solutions of the traveling waves. J. Ocean Eng. Sci. https://doi.org/10.1016/j.joes.2021.08.013
27. Odabasi, M.: Traveling wave solutions of conformable time fractional Zakharov-Kuznetsov and Zoomeron equations. Chinese J. Phys. (2019). https://doi.org/10.1016/j.cjph.2019.11.003
28. Elaheh Saberi, S.: Reza Hejazi, Ahmad Motamednezhad, Lie symmetry analysis, conservation laws and similarity reductions of Newell-Whitehead-Segel equation of fractional order. Journal of Geometry and Physics **135**, 116–128 (2019)
29. Graham, R.: Systematic derivation of a rotationally covariant extension of the 2-dimensional Newell-Whitehead-Segel equation. Fachbereich Physik, Universität-Gesamthochschule Essen, 5117 Essen, Germany
30. Prakash, A., Verma, V.: Numerical method for fractional model of newell-whitehead-segel equation. Front. Phys. **7**, 15 (2018). https://doi.org/10.3389/fphy.2019.00015
31. Ali, M., Alquran, M., Jaradat, I.: Asymptotic-sequentially solution style for the generalized Caputo time-fractional Newell-Whitehead-Segel system. Adv. Differ. Equs. **2019**, 70 (2019)

32. Wang, M.L., Li, X.Z., Zhang, J.L.: Phys. Lett. A **372**, 417 (2008)
33. Wazwaz, A.M.: A sine-cosine method for handling nonlinear wave equations. Math. Comput. Model. **40**, 499–508 (2004)
34. Alam, M.N., Akbar, M.A.: Some new exact traveling wave solutions to the simplified MCH equation. J. Ass. Arab Uni. Bas. App. Sci. **17**, 6–13 (2015)
35. Zhang, J., Jiang, F., Zhao, X.: Application to some non linear equations. Int. J. Comput. Math. **87**, 1716–1725 (2013)
36. Deng, Y.: Invariance of the Gibbs measure for the Benjamin-Ono equation. J. Eur. Math. Soc. **5**, 844–856 (2015)
37. Parkes, E.J., Vakhnenko, V.O.: Explicit solutions of the Camassa-Holm equation. Chaos Solitons Fractals **26**, 1309–1316 (2005)
38. Shiraishi, K., Hirenzaki, S.: Bogomol'nyi equations for vortices in Born-Infeld-Higgs systems. Int. J. Mod. Phys. A 2635–2647 (1991)
39. Whitham, G.B.: Linear and Nonlinear Waves. Wiley, New York, London (1927)
40. Claire, Y., Fenton, F.H.: Numerical solutions of reaction-diffusion equations: application to neural and cardiac models. Am. J. Phy. **84**, 626 (2016)
41. Yang, X.J., Gao, F., Srivastava, H.M.: A new computational approach for solving nonlinear local fractional PDEs. J. Comput. Appl. Math. (2017). https://doi.org/10.1016/j.cam.2017.10.007

Gel Electrophoresis of a Polarizable Charged Colloid with Hydrophobic Surface

Sirsendu Sekhar Barman and Somnath Bhattacharyya

Abstract We consider the electrophoresis of a charged colloid whose surface is assumed to be hydrophobic and whose particle is assumed to be polarizable. The current analysis is based on the numerical solution of the Navier-Stokes-Nernst-Planck-Poisson equations for the nonlinear electrokinetic equations, which are linked with the Laplace equation for the electric field inside the dielectric particle. The particle's hydrophobicity may affect its electric polarisation by making ions more convectively transportable. The nonlinear effects, such as double-layer polarization and relaxation, are also influenced by the hydrophobicity of the particle surface. A negatively charged colloid is created in the hydrogel medium and translated in the direction of the applied field by the strong electroosmotic flow caused by the hydrogel's immobile charges. By adjusting the gel's Debye length and volume charge density, the charged colloid can be trapped.

Keywords Gel electrophoresis · Hydrophobicity · Debye length · Permittivity ratio

1 Introduction

It has been discovered that electrophoresis in gel medium like polyacrylamide or agarose is a more effective method for separating and sorting macromolecules and nanoparticles (NPs) based on their size or charge [1]. Gel reduces large-scale convection, and thus allows for the execution of several runs on a single gel [2]. Additionally, the Joule heating impact is less than it would be with conventional electrophoresis since gel has a higher specific heat than water [3]. Hanauer et al. [4] experimentally

S. S. Barman (✉)
Department of Mathematics, Indian Institute of Information Technology Lucknow, Lucknow, India
e-mail: sirsendu012@gmail.com

S. Bhattacharyya
Department of Mathematics, Indian Institute of Technology Kharagpur, Kharagpur, India
e-mail: somnath@maths.iitkgp.ac.in

S. Bhattacharyya and H. S. Mahato (eds.), *Modeling, Analysis and Simulations of Multiscale Transport Phenomena*, Springer Proceedings in Mathematics & Statistics 491, https://doi.org/10.1007/978-981-96-3098-1_16

shown gel electrophoresis ability to separate gold and silver nanoparticles based on their size and structure. Deionized water with a bulk concentration of 10^{-4} mM is used as the supporting electrolyte in the experimental research by Matos et al. [5] and Hill [6] on the hydrogel medium with electrolyte saturation and electrokinetic transport of charged colloidal inclusions.

A uniformly distributed porous material that has been electrolyte-soaked in water can be used to model the hydrogel medium (Allison et al. [7]). The broad range Brinkman resistance and limited-range steric effects cause the gel electrophoresis to differ from the free-solution electrophoresis. The effective medium approach, in which the hydrogel is represented as a continuum and the hydrodynamics are controlled by the Brinkman equations, governs the long-range hydrodynamic interactions in hydrogel. Gel is characterized by its hydrodynamic screening length in the effective medium method. With an increase in agarose concentration, the screening length reduces. The friction between the gel skeleton and the migrating particle allows the short-range interactions to take place. For a dense gel in which the particle size is similar to the gel hydrodynamic pore size, the short-range effects are substantial. Here, the diluted gel will be taken into account in which the hydrodynamic interactions prevail and the specific gel structure is irrelevant. The shielding effect and electro-osmosis (EO) of the mobile counterions inside the gel medium need to be taken into consideration if the gel has a fixed charge density.

Several authors (Allison et al. [7], Hill [8], Hsu et al. [3]) treated the gel as a homogenous porous medium while analyzing the electrophoresis in an uncharged hydrogel. The Debye-Huckel approximation, a first-order perturbation analysis of the equilibrium Boltzmann distribution of ions under a weak-field assumption, or a low surface charge density of the particle were the foundations of these analyses. The experimental findings of Doane et al. [9] for Au nanoparticles with weak surface potentials (<8 mV) and a Debye length that is larger than the particle radius and hydrogel pore size serve as a foundation for the validation of these linear analyses. When the Debye length is in the order of the particle size, as is often the case for nanoparticles with a radius below 100 nm, the nonlinear effects resulting from the double layer polarisation (DLP) caused by the applied electric field may have a significant impact. The effect of convection on the ion distribution for a charged hydrogel is not negligible when the gel is very permeable, according to the nonlinear study of Bhattacharyya and De [10]. Hill [11] examined the electrophoresis of a bare nanoparticle in charge regulated hydrogels.

Due to the presence of the stationary charge density in hydrogels, the impact of the electric polarisation of a dielectric particle on electrophoresis in a charged hydrogel will be different from the corresponding free-solution electrophoresis. In contrast to O'Brien and White's [12] similar analysis of free-solution electrophoresis, Mohammadi and Hill [13] have shown that a particle's electrical response is dependent on its dielectric constant through a first-order linear perturbation analysis of the electrophoresis of an uncharged dielectric colloid in a charged hydrogel. Their study, however, agrees with O'Brien and White's [12] electrophoretic theory when the hydrogel is assumed to be uncharged. Results for the dielectric particle with a radius of 500 nm and a particle-to-electrolyte dielectric ratio of 0.02 were obtained by Mohammadi and Hill [13] under an applied electric field of 25V/cm. Based on a theoretical investigation Li et al. [14] came to the conclusion that the electrophoresis via a charged gel medium is substantially influenced by the hard core of the soft particle's dielectric constant or by large bare particles. The linear analysis demonstrates that for smaller ranges of particle size, the change in electrophoretic mobility induced by taking into account various values of the particle dielectric constant is low. Particularly when the applied electric field is sufficient to produce a potential drop across the particle in the order of the thermal potential and the Debye length is on the order of the particle size, for which the DLP effect is substantial, the linear analysis may not provide a precise result. A stronger applied electric field is commonly encountered in lab-on-a-chip devices. Under stronger applied electric fields, the dielectric polarisation becomes a significant phenomenon (Schnitzer and Yariv [15]). For a highly charged hydrogel with a longer screening time and a strong background EOF, the nonlinearity may become more evident.

As the dimensions of microfluidic devices shrink, the surface-to-volume ratio varies inversely on the characteristic length scale, making boundary conditions more crucial. The application of hydrophobic surfaces to increase transport in microfluidic devices has drawn the interest of researchers recently. The behaviour of hydrophobic/superhydrophobic surfaces while submerged in water has received a lot of attention [16]. Skin friction is drastically reduced by a long slip length in microscale flow, which can result in energy savings [17–19]. The ratio of the effective slip length to the electric double-layer (EDL) thickness determines how significantly the electro-osmotic enhancement is influenced by slip. The relationship between the slip length (of the fluid at the solid surface) and the Debye length regulates the velocity amplification factor adequately. However, Joly et al. [20] demonstrated that the surface charge density increases friction and, as a result, decreases the effectiveness of the hydrodynamic slip length. Joly et al. [20] proposed a model that explains the relationship between the solvent's Bjerrum length and surface charge density in terms of slip length. Several authors [21, 22] have explored the improvement of electrophoretic propulsion of charged colloids with a slipping surface both theoretically and empirically. Khair and Squires [23] carried out a comprehensive analysis of the electrophoresis of a colloid with a hydrophobic surface.

In this work we have considered the electrophoresis of a charged colloid in which the particle is considered to be polarizable with a hydrophobic surface. This present analysis is based on the numerical solution of the non-linear electrokinetic equations based on the Navier-Stokes-Nernst-Planck-Poisson equations coupled with the Laplace equation for the electric field within the dielectric particle. Knowing the effect of the dielectric rigid colloid's electric polarisation is one of the study's primary goals. No limitations on particle charge density, Debye length, or applied electric field are placed by the current investigation. Therefore, the effects of the double layer's relaxation and polarisation, the rigid colloid's solid polarisation, surface conduction, the surface's hydrophobicity, and their interactions with the background electro-osmotic flow are included into the formulation as it is. The governing equations have been quantitatively solved in their complete form. Regarding limiting scenarios like non-polarizable particles, hydrophobicity, and thin Debye layers at low charger densities of the hydrogel, our computed results are in good accord with the earlier studies.

2 Mathematical Formulation

We take into account the electrophoresis of a spherical, rigid particle of radius a with a uniform surface charge density of $\sigma\,(< 0)$ and electric permittivity of ϵ_p. The particle is submerged in a charged gel with permittivity ϵ_e, fixed charge density ρ_f, and viscosity μ that is responding to an applied electric field E_0 with an electrophoretic velocity U_E that is unknown in advance (Fig. 1).

The Brinkman equation governs the hydrodynamics of the gel, which is assumed to be a continuum media. The fixed immobile charge density of the homogeneous gel is calculated as $\rho_f = \mathcal{Z}en_f^*$, where n_f^* is the ionic number concentration of the immobile PEL ions, $\mathcal{Z}$ is the valence, and e is the elementary charge of the electron. The hydrodynamic screening length of the homogeneous gel is calculated as ℓ. The Nernst-Planck equation controls the ion transport in the hydrogel. The Poisson equation controls the electrical potential in the hydrogel. We scale the dimensional variables as follows: the length scale is the radius of the rigid particle a, the thermal potential $\phi_0 = k_B T/e$ is the potential scale, $U_0 = \epsilon_e \phi_0^2/a\mu$ is the velocity scale, $\tau = a/U_0$ is the time scale, $\epsilon_e \phi_0^2/a^2$ is the pressure scale, the bulk number concentration n_0 is the scale for the ionic concentration of the ith ionic species and $\Lambda = E_0 a/\phi_0$ is the scaled external electric field. Here k_B is the Boltzmann constant, T is the absolute temperature.

We take into account a set of spherical polar coordinates with the pole fixed at the sphere's centre and the initial line (z-axis) running parallel to the direction of the applied electric field. The momentum equation governed by the Darcy-Brinkman extended Stokes equations are

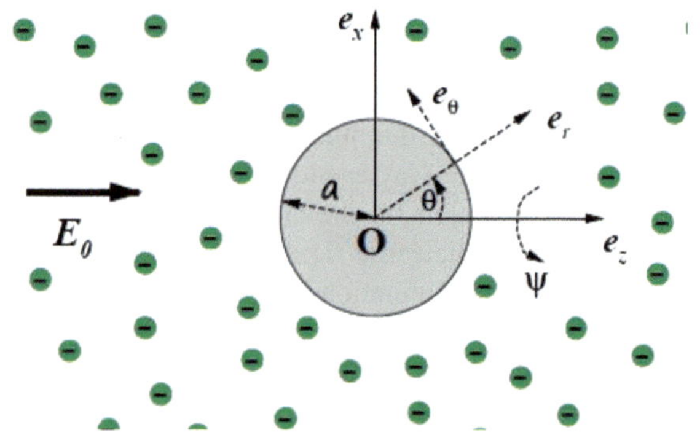

Fig. 1 Schematic diagram of the model geometry in Spherical co-ordinate system

$$Re\frac{\partial \mathbf{u}}{\partial t} + Re\,(\mathbf{u}.\nabla)\,\mathbf{u} - \nabla^2\mathbf{u} + \nabla p + \beta^2\mathbf{u} + \frac{(\kappa a)^2}{2}\rho_e\nabla\phi = 0 \tag{1}$$

$$\nabla.\mathbf{u} = 0 \tag{2}$$

where the velocity vector $textbfu = (u, v, 0)$ has radial and cross-radial components, respectively. Here, the variables are time (t), pressure (p), electric potential (phi), and scaled volumetric charge density ($\rho_e = \sum_i z_i n_i$) due to mobile electrolyte ions. The scaled ionic concentration of the ith ionic species with valence $z_i = \pm z$, $(i = 1, 2)$, is shown here as n_i. The inverse of the EDL thickness is given here by $\kappa = \sqrt{2zen_0/\epsilon_e\phi_0}$. The permeability of the hydrogel can be determined by the non-dimensional parameters $\beta = a/\ell$, respectively. $Re = U_0 a\rho/\mu$ is the Reynolds number.

The non-dimensional form of the Nernst-Planck equations governing the distribution of the ith ionic spacies is

$$Pe_i\frac{\partial n_i}{\partial t} + Pe_i\,(\mathbf{u}.\nabla n_i) = \nabla^2 n_i + \frac{z_i}{z}\nabla.\,(n_i\nabla\phi) \tag{3}$$

where the ionic Peclet number Pe_i is defined as $Pe_i = \epsilon_e\phi_0^2/\mu D_i (i = 1, 2)$. Here D_i is the diffusivity of the ith ion. Based on the Gauss law the electric potential in the gel is governed by the following Poisson equation

$$\nabla^2\phi = -\frac{(\kappa a)^2}{2}\rho_e - \beta^2 Q_f \tag{4}$$

where, Q_f measures the scaled fixed charge density of the hydrogel and are defined as $Q_f = \rho_f\ell^2/\epsilon_e\phi_0 = -n_f(\kappa\ell)^2/2$. In our present study we consider the constant charge properties of the hydrogel, which leads to the constant value of Q_f, respectively. The electric potential inside the particle is governed by

$$\nabla^2\phi_p = 0 \tag{5}$$

A slip boundary condition and no normal flux of ions are imposed on the surface of the solid core $(r = 1)$, i.e.,

$$\mathbf{u} = \lambda_s\frac{\partial \mathbf{u}}{\partial r}, \quad [\nabla n_i + (z_i/z)n_i\nabla\phi]\,.e_r = 0 \tag{6}$$

where, the unit vector along the radial direction is called e_r. The scaled hydrophobic parameter is λ_s. The following boundary condition results from the continuity in potential and displacement vector along the particle surface at $(r = 1)$.

$$\frac{\partial\phi}{\partial r} - \epsilon_r\frac{\partial\phi_p}{\partial r} = -\sigma, \quad \phi_p = \phi \tag{7}$$

$\epsilon_r = \epsilon_p/\epsilon_e$ is the particle to medium permittivity ratio. The corresponding variables go closer to their bulk values as they get farther from the particle ($r = R$, with $R >> 1$), and the fluid velocity in relation to the stationary particle gets closer to $-U_E$. The following are the far-field boundary conditions as a result.

$$\mathbf{u}.e_z = -U_E - Q_f\Lambda, \quad \phi = -\Lambda r \cos\theta, \quad n_1 = 1 + n_f, \quad n_2 = 1 \tag{8}$$

The unit vectors along the axial direction are represented here by e_z. The issue being discussed is axisymmetrical in nature. Along the axis of symmetry, which is defined as $\theta = 0$ and π, we apply the symmetry requirement. The balance between the particle's electric force and drag force yields the unknown electrophoretic velocity U_E. [24, 25].

3 Numerical Methods

The equations regulating fluid flow, along with the equations for ion transport and the Poisson equations for the electric field, are numerically solved in a coupled manner over the staggered grid configuration [26]. The discretized continuity equation is converted into a Poisson equation for pressure correction to establish the pressure connection between the continuity and momentum equations. This process is comparable to the SIMPLE method, which uses a cyclic guess-and-correct operation to solve the governing equations [26]. Through the use of the block elimination approach, the resulting linear algebraic equations are coupledly solved. A forward time marching approach is used to start abruptly from the initial equilibrium condition and continue until a steady state is reached.

4 Results and Discussions

We consider the thermal potential $\phi_0 = 0.02586$ V, $\mu = 10^{-3}$ Pas, ion diffusivity $D_1 = 1.33 \times 10^{-9}$ m^2/s and $D_2 = 2.03 \times 10^{-9}$ m^2/s [27]. Results were obtained by changing the Debye-Huckel parameter from 0.1 to 50 at a constant value for a. Scaled hydrophobicity λ_s ranges from 0 to 100 while scaled ϵ_r ranges from 0.01 to 100. We show a comparison of our numerical solutions (Fig. 2) for a hydrophobic particle in electrolyte media ($\beta \to 0$) with the pre-existing semi-analytical results of Gopmandal et al. [28] in order to assess the accuracy of our numerical method. Our results are in good agreement with the semi-analytical data for a thin Debye layer. Additionally, we've developed a strong comparison with the findings of a dielectric particle theory due to Li et al. [14].

4.1 Comparisons with Existing Results

When the equilibrium Boltzmann distribution of the ions is not sustained, nonlinear effects in electrophoresis arise. The distribution of ions and fluid flow are then coupled. The DLP effect is found to have a strong impact on the particle's electrokientics. Figure 2 shows that the semi-analytical model due to Gopmandal et al. [28] overestimate the mobility corresponding the numerical model when $\kappa a \sim O(1)$. The result is for liquid medium. Gopmandal et al. [28] gave an analytical expression for the mobility of a nonpolarizable ($\epsilon_r = 0$) particle with a constant surface charge density as a function of the slip length parameter under Debye-Huckel approximation in a weak electric field assumption. In their model, the polarisation and relaxation of the double layer as well as the EOF of counterions in the medium are neglected, and the Debye layer is considered to be coaxial. The nonlinear DLP and relaxation effects create a retardation to the electrophoresis and the impacts are significant for $\kappa a \sim O(1)$. However, when $\kappa a >> 1$, i.e., the bulk ionic concentration of the salt is high, a good agreement between the numerical model and the existing semi-analytical model [28] is achieved.

The analytical solution of Li et al. [14], which is based on the linear Debye-Hückel approximation under a weakly charged condition with a Boltzmann distribution of ions, also demonstrates the dependency of mobility on ϵ_r for uncharged or charged gel. The electrophoretic mobility can be reversed in a charged hydrogel by adjusting the ionic concentration of the electrolyte, as shown in Fig. 3b. Stout and Khair [29] demonstrated mobility reversal in the free-solution electrophoresis of charged dielectric colloids in a concentrated multivalent electrolyte. According to Stout and Khair [29], the charge inversion brought on by ion correlations results in mobility reversal. In the present situation, the mobile countercharges (positive ions) in the

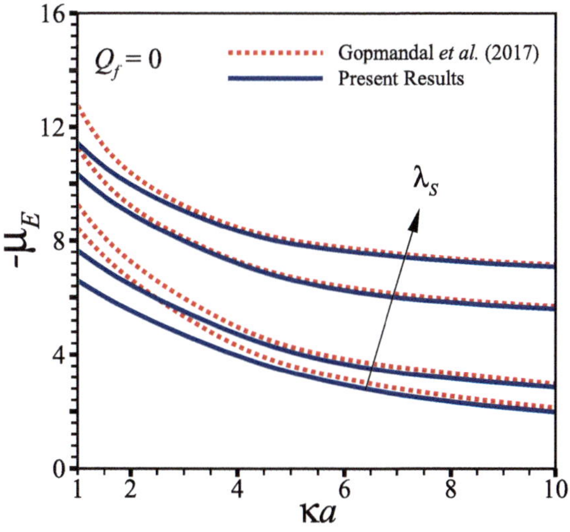

Fig. 2 Variation of electrophoretic mobility with κa for a hydrophobic particle ($\lambda_s = 0; 0.1; 1.0; 100.0$) at $\sigma = -5$ mCm^{-2} with the comparisons of Gopmandal et al. [28]

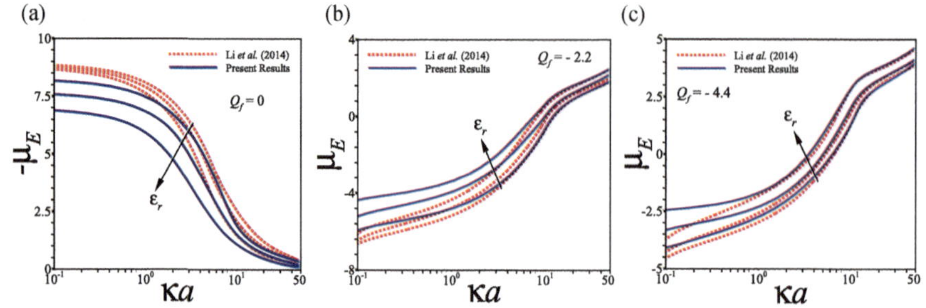

Fig. 3 Variation of electrophoretic mobility with κa and comparisons with Li et al. [14] at $\ell = 100\,\text{nm}$, $\sigma = -5\ \text{mCm}^{-2}$ for different $\epsilon_r(= 0; 1; 100)$ at **a** $\rho_f = 0\ (Q_f = 0)$; **b** $\rho_f = -4\text{kCm}^{-3}(Q_f = -2.2)$; **c** $\rho_f = -8\text{kCm}^{-3}(Q_f = -4.4)$. Dashed line represent the result of Li et al. [14]

negatively charged hydrogel produce an electro-osmotic flow that intensifies as the hydrogel charge density increases along the direction of the applied field. The particle's intrinsic mobility decreases with decreasing Debye layer thickness, and the particle is dragged by the background EOF. For instance, when the Debye length is increased from low to moderate values, we observe a reversal in particle translation. As the Debye length becomes thinner ($\kappa a \gg 1$), our computed solutions approach the result due to Li et al. [14].

4.2 *Effect of DLP Effect, Permittivity Ratio, Hydrophobicity and Fixed Hydrogel Charge Density*

In Fig. 4a–c, at various values of the Debye length, the impact of the nonlinear effects for low to higher values of hydrogel fixed charge density is depicted. Here ρ_f is varied from -10 to $-50\ \text{kC/m}^3$. The nonlinearity attenuates as the Debye length becomes thinner. Our computed results show a linear variation of U_E with ρ_f. Another finding from these results is that when the fixed charge density of the hydrogel increases, the effect of the colloid's solid polarisation on the electrophoresis increases. Here, ϵ_r is varied from 0 to 100. The findings in Fig. 4 demonstrate that gel electrophoresis is an effective method for separating dielectric particles according to their dielectric constant. As a result, sorting based on particle dielectric permittivity is efficient for electrolytes with low ionic concentrations. This implies that a hydrogel that has been saturated with a non-polar solvent that has a low dielectric permittivity, such as below 10, combined with surfactants might be a valuable tool for sorting particles according to their dielectric constant. The hydrodynamic drag that the particle experiences and, consequently, the particle electrophoretic velocity become sensitive to the particle's size when it is dragged by the background EOF caused by the countercharges in the charged gel, which is proportional to $-\rho_f E_0 \ell^2/\mu$. The background EOF is strong

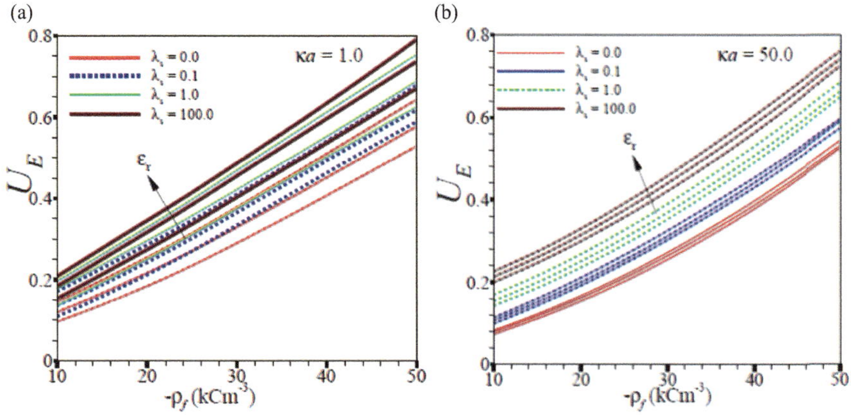

Fig. 4 Variation of elctrophoretic velocity with ρ_f for different $\lambda_s (= 0; 0.1; 1.0; 100.0)$ and $\epsilon_r (= 0; 1; 100)$ at $\ell = 100\,\text{nm}$, $\sigma = -5\ \text{mCm}^{-2}$ for **a** $\kappa a = 1$, **b** $\kappa a = 50$

when $\rho_f \ell^2$ becomes large. The electrophoretic velocity increases with the increase of λ_s and the enhancement of electrophoretic velocity makes a saturation for a thinner Debye length with the increase of ϵ_r.

5 Conclusions

For low to moderate ranges of Debye layer thickness, a quantitative evaluation of the effect of solid polarisation of a polarizable charged colloid in gel electrophoresis is made. The current analysis is based on the Nernst-Planck equation for ion transport, which accounts for the ion's electromigration, advective transport, and diffusion. The impact of dielectric polarisation in the context of gel electrophoresis has been investigated thoroughly. The DLP effect is quite prominent. There is a significant discrepency from the previous analytic results for thick Debye layer. The effect of slip ratio and dielectric polarization is quite prominent and it increases electrophoretic velocity. The electrophoretic velocity increases with the increase of fixed gel charge density and it increases with the increase of κa for charged gel.

The ion-ion steric interactions, which are significant for a highly charged surface, are not taken into account in the current study. Khair and Squires [30] provide an illustration of the impact of ions on the medium's ionic concentration or strongly charged surface on electrophoresis. Ion steric effect may not have a significant influence for this fixed value of surface charge density in the analysis because the surface potential decreases as the ionic concentration increases. The charge regulation model for the hydrogel charge density, which explains the association/dissociation reactions of the hydrogel charge groups, will be taken into consideration as a logical extension of the current investigation.

References

1. David Hames, B.: Gel Electrophoresis of Proteins: a Practical Approach, vol. 197. OUP Oxford (1998)
2. Li, F., Hill, R.J., Bare charged spheres in polyelectrolyte hydrogels: Nanoparticle gel electrophoresis. J. Colloid Interface Sci. **394**, 1–12 (2013)
3. Hsu, J.-P., Huang, C.-H., Tseng, S.: Gel electrophoresis of a charge-regulated, bi-functional particle. Electrophoresis **34**(5), 785–791 (2013)
4. Hanauer, M., Pierrat, S., Zins, I., Lotz, A., Sönnichsen, C.: Separation of nanoparticles by gel electrophoresis according to size and shape. Nano Lett. **7**(9), 2881–2885 (2007)
5. Matos, M.A., White, L.R., Tilton, R.D.: Electroosmotically enhanced mass transfer through polyacrylamide gels. J. Colloid Interface Sci. **300**(1), 429–436 (2006)
6. Hill, R.J.: Electric-field-enhanced transport in polyacrylamide hydrogel nanocomposites. J. Colloid Interface Sci. **316**(2), 635–644 (2007)
7. Allison, S.A., Xin, Y., Pei, H.: Electrophoresis of spheres with uniform zeta potential in a gel modeled as an effective medium. J. Colloid Interface Sci. **313**(1), 328–337 (2007)
8. Hill, R.J.: Electric-field-induced force on a charged spherical colloid embedded in an electrolyte-saturated brinkman medium. Phys. Fluids **18**(4), 043103 (2006)
9. Doane, T.L., Cheng, Y., Babar, A., Hill, R.J., Burda, C.: Electrophoretic mobilities of pegylated gold NPS. J. Am. Chem. Soc. **132**(44), 15624–15631 (2010)
10. Bhattacharyya, S., De, S.: Gel electrophoresis and size selectivity of charged colloidal particles in a charged hydrogel medium. Chem. Eng. Sci. **141**, 304–314 (2016)
11. Hill, R.J.: Hydrogel charge regulation and electrolyte ion-concentration perturbations in nanoparticle gel electrophoresis. Proc. R. Soc. A: Math. Phys. Eng. Sci. **471**(2184), 20150523 (2015)
12. O'Brien, R.W., White, L.R.: Electrophoretic mobility of a spherical colloidal particle. J. Chem. Soc. Faraday Trans. 2: Mol. Chem. Phys. **74**, 1607–1626 (1978)
13. Mohammadi, A., Hill, R.J.: Steady electrical and micro-rheological response functions for uncharged colloidal inclusions in polyelectrolyte hydrogels. Proc. R. Soc. A: Math. Phys. Eng. Sci. **466**(2113), 213–235 (2010)
14. Li, F., Allison, S.A., Hill, R.J.: Nanoparticle gel electrophoresis: soft spheres in polyelectrolyte hydrogels under the debye–hückel approximation. J. Colloid Interface Sci. **423**, 129–142 (2014)
15. Schnitzer, O., Yariv, E.: Dielectric-solid polarization at strong fields: breakdown of smoluchowski's electrophoresis formula. Phys. Fluids **24**(8), 082005 (2012)
16. Roach, P., Shirtcliffe, N.J., Newton, M.I.: Progess in superhydrophobic surface development. Soft Matter **4**(2), 224–240 (2008)
17. Stone, H.A., Stroock, A.D., Ajdari, A.: Engineering flows in small devices: microfluidics toward a lab-on-a-chip. Annu. Rev. Fluid Mech. **36**, 381–411 (2004)
18. Choi, C.-H., Ulmanella, U., Kim, J., Ho, C.-M., Kim, C.-J.: Effective slip and friction reduction in nanograted superhydrophobic microchannels. Phys. fluids **18**(8), 087105 (2006)
19. Ren, Y., Stein, D.: Slip-enhanced electrokinetic energy conversion in nanofluidic channels. Nanotechnology **19**(19), 195707 (2008)
20. Joly, L., Ybert, C., Trizac, E., Bocquet, L.: Liquid friction on charged surfaces: From hydrodynamic slippage to electrokinetics. J. Chem. Phys. **125**(20), 204716 (2006)
21. Bouzigues, C.I., Tabeling, P., Bocquet, L.: Nanofluidics in the debye layer at hydrophilic and hydrophobic surfaces. Phys. Rev. Lett. **101**(11), 114503 (2008)
22. Ajdari, A., Bocquet, L.: Giant amplification of interfacially driven transport by hydrodynamic slip: diffusio-osmosis and beyond. Phys. Rev. Lett. **96**(18), 186102 (2006)
23. Khair, A.S., Squires, T.M.: The influence of hydrodynamic slip on the electrophoretic mobility of a spherical colloidal particle. Phys. Fluids **21**(4), 042001 (2009)
24. Bhattacharyya, S., Gopmandal, P.P.: Migration of a charged sphere at an arbitrary velocity in an axial electric field. Colloids Surf. A: Physicochem. Eng. Aspects **390**(1–3), 86–94 (2011)

25. Barman, S.S., Bhattacharyya, S.: Electrokinetic transport of a non-conducting liquid droplet in a polyelectrolyte medium. Phys. Fluids **32**(1), 012011 (2020)
26. Fletcher, C.A.J.: Computational Techniques for Fluid Dynamics 2: Specific Techniques for Different Flow Categories. Springer Science and Business Media (2012)
27. Barman, S.S., Bhattacharyya, S., Gopmandal, P.P., Ohshima, H.: Impact of charged polarizable core on mobility of a soft particle embedded in a hydrogel medium. Colloid Polymer Sci. **298**(12), 1729–1739 (2020)
28. Gopmandal, P.P., Bhattacharyya, S., Ohshima, H.: Effect of hydrophobic core on the electrophoresis of a diffuse soft particle. Proc. R. Society A: Math. Phys. Eng. Sci. **473**(2199), 20160942 (2017)
29. Stout, R.F., Khair, A.S.: A continuum approach to predicting electrophoretic mobility reversals. J. Fluid Mech. 752 (2014)
30. Khair, A.S., Squires, T.M.: Ion steric effects on electrophoresis of a colloidal particle. J. Fluid Mech. **640**, 343–356 (2009)

Photon Echo on DEOXYRIBONUCLEIC ACID Molecules: A Simple Theoretical Study

Subhamoy Singha Roy and Soumili Ghosh

Abstract Here, we have explored how DEOXYRIBONUCLEIC ACID molecules may function as spin systems when spins are positioned on the union set up as a sequence that is antiferromagnetic. As it spins tell of within the Lie algebra of the connecting form may be identified from the spin structure-related Chern-Simons theory. In addition to being one-to-one operations and not two-part processes, it is shown that difficulties between two DEOXYRIBONUCLEIC ACID molecules cause one spin-echo to signal the transformation of the Berry phase, which can be retrieved using the entangled photon echo method. It is found that this and the thermodynamic entropy coincide rather well as the charge of encircling an inflexible polymer thread that is protected by a thin tube or entropy of entanglement.

Keywords Chern-Simons theory · Spin-echo · Berry phase · Thermodynamic entropy · Entanglement entropy

1 Introduction

It is well known that quantum physics demonstrates an extremely irregular association among two actually separated components between the entire arrangement. Right-handed (left-handed) helical sense is possessed by B-DEOXYRIBONUCLEIC ACID (Z-DEOXYRIBONUCLEIC ACID) due to the twisting of two poly-nucleotide chains around the same axis and two alpha-helix double helix strands. We previously knew that there is supercoiled DEOXYRIBONUCLEIC ACID since it was discovered in vivo chromosomal DEOXYRIBONUCLEIC ACID molecules that had topological domains that allow for supercoiling [1–3]. Prokaryotes, or cells without nuclear membranes, have DEOXYRIBONUCLEIC ACID molecules that frequently embrace the interwound structures known as "plectonemic" supercoils. Chromosome

S. S. Roy (✉) · S. Ghosh
Department of Physics, JIS College of Engineering (Autonomous), West Bengal University of Technology, Kalyani, Nadia, India
e-mail: subhomay.singharoy@jisacolege.ac.in

S. Bhattacharyya and H. S. Mahato (eds.), *Modeling, Analysis and Simulations of Multiscale Transport Phenomena*, Springer Proceedings in Mathematics & Statistics 491, https://doi.org/10.1007/978-981-96-3098-1_17

DEOXYRIBONUCLEIC ACID molecules are recognized as arranged keen on topologically self-determining circles in eukaryotes, which include organelles through distinct inside membranes and cells with nuclei [2–5]. The qualitative mechanism of supercoiled DEOXYRIBONUCLEIC ACID has been examined by a number of writers [6].

According to Boles et al. tests [7], temperature shift controls the supercoil's shape. DEOXYRIBONUCLEIC ACID-DEOXYRIBONUCLEIC ACID attraction has been shown in experiments by BeDeoxyribonucleic acid r et al. [8] to be resistant to fluctuation entropy. Since a DEOXYRIBONUCLEIC ACID molecule may be thought of as a chain of spin systems, we will be studying several multivariate mechanical features of DEOXYRIBONUCLEIC ACID molecules in this note. Should a spin through a set direction is introduced within the coil's axis, observably two coils have opposing intentions within the spin when two polynucleotide manacles are twisted concerning the similar axis in a DEOXYRIBONUCLEIC ACID molecule with a certain helical sense. Every time a strand turns, two strands travel against the axis, consequently, two neighboring coils spin directions should be facing apart. This means that a DEOXYRIBONUCLEIC ACID molecule may appear to be a spin sequence that is antiferromagnetic positioned along the supercoil's vertical axis. From a study of this spin system, we will look at the topological features, elastic properties, and thermodynamic properties of DEOXYRIBONUCLEIC ACID molecules. We will also look at the intricate network of 2 numbers DEOXYRIBONUCLEIC ACID molecules placing spin-echo to one another, which creates a transmutation of Berry phase that can be precisely calculated as entanglement. We will formulate the entangled state produced creating 4-wave amalgamation technique known as the photon-echo approach, from end to end pulse individual estranged in time, subsequent to spin-echo generates the voltage between entangled spins and entangled opposite spins. Assuming that a spin of a particular orientation is inserted on the axis (the vertical line discussed) of the coil in order for two neighboring coils have contrasting directions of the rotating, summarize the efforts on two poly-nucleotide chains that are twisted about the same vertical line having a precise spiral intelligence inside a DEOXYRIBONUCLEIC ACID chain.

2 Theoretical Background

The reason for this is that in every turn, two strands travel in different information on the axis downward, hence the spin direction assigned to the 2-neighboring coils ought to really be opposed to each other. Consequently, when the spin is thought to be positioned on the supercoil's axis, a DEOXYRIBONUCLEIC ACID molecule can be seen as a lengthy a arrangement of antiferromagnetic spin manacles. As soon as the spin is positioned on the axis at the location in space x, a element vector representing the digression $\vec{t} = \left[\partial_s(\vec{r})\right]$ wherever $\left[\vec{r}(s)\right]$ is a breathing space curvature defined parameters by the curve length s may be linked to a vector of turn. Bosonic or fermionic oscillators can be used to produce in the Lie algebra of the $(SU.(2))$ depiction, a spin vector.

The notation for the spin vector $\left\{\vec{S}.(x)\right\}$ is

$$\vec{S}((x)) = \left[\left\{\psi_{\alpha}^{\dagger}(x) \times \left(\vec{\sigma}_{\alpha\beta}\right) \times \psi_{\beta}(x)\right\}\right] \tag{1}$$

wherever $\left[\vec{\sigma}_{(\alpha\beta)}\right]$ is the Pauli matrix vector along with $\left\{\left[\psi^{\dagger}.(\psi)\right]\right\}$ is the fermionic oscillator function. In order to generate a unit vector, $\vec{n}$,

$$\vec{n} = \left\{(\psi_1^* \psi_2^*)\vec{\sigma}\begin{pmatrix}\psi_1\\ \psi_2\end{pmatrix}\right\} \tag{2}$$

with

$$\psi_1 = \left(\cos\left({}^{\theta}/_{2}\right)\right)e^{{}^{i\varphi}/_{2}} \tag{3.a}$$

$$\psi_2 = \left(\sin\left({}^{\theta}/_{2}\right)\right)e^{{}^{-i\varphi}/_{2}} \tag{3.b}$$

at the present examine the emergence among the Berry phase within the entanglement involving two quantized 1/2 fragments with the same spin. 2-spin 1/2 DEOXYRIBONUCLEIC ACID compounds in their antisymmetric Bell State are

$$|\psi_2\rangle = (\cos\theta|\psi_1\rangle) = \left(\frac{1}{2\pi}\right)(\Upsilon\uparrow - \Upsilon\downarrow)|\psi_1\rangle \tag{3.c}$$

Based on variations in the Berry phase factor.

Meant for 2-particles, both A and B, that are symmetric at coordinates both x and y, the most generic antisymmetric Bell state is

$$\begin{aligned}|\psi_2(t)\rangle &= \{(\alpha|)\uparrow(t)\rangle|\downarrow(t)\rangle\} \\ &-\{\beta|\uparrow(t)\rangle|\downarrow(t)\rangle\}\end{aligned} \tag{3.d}$$

particular two complex coefficients, α and β.

When one fermion rotates a single DEOXYRIBONUCLEIC ACID molecule over a time interval of τ, the spinor returns to its initial state, gaining only the Berry phase and losing the dynamical phase. The new entangle state form that we have is as

$$|\psi_2(t=\tau)\rangle = \left\{(\alpha)|\uparrow(t)\rangle\middle|\downarrow(t)\rangle - e^{2i\Upsilon\uparrow}|\uparrow(t)\rangle\beta|\downarrow(t)\rangle\right\} \tag{3.e}$$

The Berry phase and the dynamic phase are eliminated in the "spin-echo" approach when we examine $(\theta = \pi)$.

This facilitates our writing

$$\vec{S}(x) = \left\{\left(\sqrt{3}/2\right)\psi_{\alpha}^{\dagger}(x)\vec{\sigma}_{\alpha\beta}\psi_{\beta}\right\} \tag{4}$$

Using the vector unit $\vec{n}$ provided by Eq. (2), We are able to generate a element vector n_{μ} through $(\mu) = 0, 1, 2, 3$ in $3 + 1$ dimensions.

$$n_{\mu} = \left(1/\sqrt{2}\right)\left\{(\psi_1^* \psi_2^*)\sigma_{\mu}\begin{pmatrix}\psi_1 \\ \psi_2\end{pmatrix}\right\} \tag{5}$$

with $\vec{\sigma}$ representing the Pauli matrices and $(\sigma_0 = \mathrm{I})$, I serving as the identity matrix. Next, we build the topological current

$$J_{\mu} = \left\{(1/12\pi^2)\varepsilon_{(\mu\nu\lambda\sigma)}\varepsilon_{(abcd)}n_a\partial_{\nu}n_b\partial_{\lambda}n_c\partial_{\sigma}n_d\right\} \tag{6}$$

where $(\mu, \nu, \lambda, \sigma)$ stands for space–time indices and $[a, b, c, d]$ stands for [0, 1, 2, 3]. The expression [11] can be used to express the current J_{μ}.

$$J_{\mu} = \left[(1/24\pi^2)\left\{\varepsilon_{(\mu\nu\lambda\sigma)}Tr(g^{-1}\partial_{\nu}g) \times (g^{-1}\partial_{\lambda}g) \times (g^{-1}\partial_{\sigma}g)\right\}\right] \tag{7}$$

with $g = (n_0 \times \mathrm{I} + i \times \vec{\mathrm{n}}.\vec{\sigma})$, who is a member of group $\{SU(2)\}$. The field force $\left(F_{\mu\nu}\right)$ of a gauge budding A_{μ} inside Euclidean 4-dimension in gap-time would vanish at all places atop the border S^3 of a confident volume V^4, within which $\left(F_{\mu\nu} \neq 0\right)$ the gauge budding incline to a unadulterated gauge in the direction of the border line, if we were to demand this immediately.

$$A_{\mu} = \left(g^{-1} \times (\partial_{\mu}g)\right) \tag{8}$$

with $(g\varepsilon) \times \{SU(2)\}$.

The topological current generated by (7) may now be expressed as [10]

$$J_{\mu} = (1/16\pi^2)\varepsilon^{\mu\nu\lambda\sigma}\left[Tr\left\{(A_{\nu} \times F_{\lambda\sigma}) + \left(\frac{2}{3}\right)A_{\nu} \times A_{\lambda} \times A_{\sigma}\right\}\right] \tag{9}$$

with A_{μ} provided by Eq. (8).It is noticed that we may associate spin with this current J_{μ} given by (2), the unit vector $\vec{n}$ serves as the foundation for the spin vector, which is contained in the current J_{μ} as is obvious from Eq. (6). Indeed, the topological Lagrangian may be seen regarding the affine gauge fields within $\{SU(2)\}$,

$$L = -\left\{(1 \div 4)\left(Tr\varepsilon^{\mu\nu\alpha\beta}\right)\left(F_{\mu\nu} \times F_{\alpha\beta}\right)\right\} \tag{10.a}$$

Now, the gauge connection in this equation related to the Lagrangian

$$\left[L_{eff}^{\uparrow}\right] = \left[-\{(i \div 2)(\dot{\xi} - \dot{\varphi} \times Cos\theta)\}\right] \tag{10.b}$$

some modification within θ, ϕ, ξ causes a gauge conversion, which in turn gives birth to the Berry phase.

The required geometric phase of the sole quantized spinor is now available,

$$\Upsilon^{\uparrow} = i \times \int \left(\left(L_{eff}^{\uparrow}\right)dt\right) = i \times \oint \left\{A^{\uparrow}(\lambda) \times d\lambda\right\}$$
$$= (0.5) \times (\oint d\xi - (\cos\theta) \times \oint d\varphi) = \{\pi \times (1 - \cos\theta)\} \qquad (10.c)$$

The topological current is created as a result [12].

$$\vec{J_\mu} = \left(\varepsilon^{\mu\nu\lambda\sigma}\vec{a}_\nu \times \vec{f}_{\lambda\sigma}\right) = \left(\varepsilon^{\mu\nu\lambda\sigma}\partial_\nu\vec{f}_{\lambda\sigma}\right) \qquad (11)$$

where the equivalent field strength $F_{\mu\nu}$ and the $\{SU(2)\}$ gauge field A_μ are taken as

$$[A_\mu] = \left(\left(\vec{a_\mu}\right).\left(\vec{\sigma}\right)\right) \text{ and } [F_{\mu\nu}] = \left(\left(\vec{f_{\mu\nu}}\right).\left(\vec{\sigma}\right)\right) \qquad (12)$$

$\vec{\sigma}$ is the Pauli matrices' vector. As such suggests the fact thetopological current $\vec{J_\mu}$, which is provided by Eq. (11) may be represented as the spin vector $\vec{S}(x)$. When gauge fields are positioned on links, a spin system on a lattice may be understood in terms of this current as if currents are situated on the vertices [9]. This makes it easier for us to think in relation to spin arrangement connected to a DEOXYRIBONUCLEIC ACID superposition of coils in stipulations respect to the Chern-Simons topology, which we shall cover in the following section. Currently, the only quantized spinor's required geometrical phase is found in Eq. 10.c.

This demonstrates that the Berry Phase for quantized spinor is a solid angle expressed concerning the quantization axis. Intended for ($\theta = 0$) the smallest amount of $\Upsilon^{\uparrow}$ is Zero and ($\theta = 180°$) upper limit.

We have a spin-up case

$$\Theta^{\uparrow}(\lambda) = \{(1/2)(1 - \cos\theta)\} \qquad (13)$$

This results in the quantized spinor's noise-dependent Berry Connection,

$$\Theta^{\uparrow}(\lambda) = \{(1/2)(1 - \cos\theta + \sin\theta\delta\theta)\} \qquad (14)$$

The outcome is now a modified Berry phase

$$\Pi^{\uparrow} = \{\pi(1 - \cos\theta + \sin\theta\delta\theta)\} \qquad (15)$$

And the same goes for down spinor

$$\Pi^{\downarrow} = \{\pi(1 - \cos\theta - \sin\theta\delta\theta)\} \quad (16)$$

Regarding the quantized techniques for spin up and spin down in that sequence, we take $\Pi^{\uparrow}$,$\Pi^{\downarrow}$ to be the noise-induced Berry phase [13–15]. As we can see from Eq. (3.c) for the entangled condition of 2-indistinguishable spinors, the development inside the state at a precise immediate is dependent regarding the division of $\Upsilon^{\uparrow}$ and $\Upsilon^{\downarrow}$, implying a twofold spike in noise. Based on Eq. (3.e), it can be observed that noise will have a less impact on the entangled state that results from "spin-echo." After spin-echo generates entangled-spin voltage (ESV), the entangled state is now created.

$$\Im_{ESV} = \left(\Upsilon^{\uparrow} - \Upsilon^{\downarrow}\right) \quad (17)$$

and entangled-inverse spin voltage (EISV)

$$\Im_{EISV} = \left(\Upsilon^{\downarrow} - \Upsilon^{\uparrow}\right) \quad (18)$$

$\Upsilon^{\uparrow}$ and $\Upsilon^{\downarrow}$ and are two indistinguishable spinors that are intertwined in an up and down condition.

Currently, spinor produces photon-echo through four-wave entangled-inverse-spin-voltage (EISV) and entangled-spin-voltage (ESV).

$$(\Im_{ESV} + \Im_{EISV}) = \left(\Upsilon^{\downarrow} - \Upsilon^{\uparrow} + \Upsilon^{\uparrow} - \Upsilon^{\downarrow}\right) = 0 \quad (19)$$

$$\begin{aligned} &\Rightarrow (\Im_{ESV} + \Im_{EISV}) = 0 \\ &\Rightarrow \Im_{ESV} = -\Im_{EISV} \end{aligned} \quad (20)$$

3 Discussion

Based on the quantum structure of DEOXYRIBONUCLEIC ACID molecules regarding an anti-ferromagnetic spin sequence, the current study reveals that the entanglement-entropy(EE) that causes the entropic impending linked through the open energy for each length analogous to the entropy rate of impounding a hard polymer surrounded by a thin cylinder is the result of it. At zero temperature, the intrinsic entropic impending resists the supplely induced implosion of a supercoil, and the entanglement entropy serves as an effective representation of the thermodynamic entropy. An important conclusion is that the entanglement entropy quantifies the system's overall information richness. In fact, entropy indicates the quantity of information that is accessible for interpretation, and entanglement may be utilized to interpret information when a DEOXYRIBONUCLEIC ACID is thought of like a one-way depository of sequential in turn. Given this, the amount of genetic information that may be deciphered can be determined by taking a measure of the entanglement entropy. An attractive replacement to a particular DEOXYRIBONUCLEIC ACID molecular arrangement represents a photon-echo quantum reminiscence [16, 17]. The quantum condition transit between a photon and an entity of unevenly stretched molecular soaks causes the configuration to become increasingly complicated [18, 19]. It may also be claimed that the primary cause of the quantization variation that the interaction of Quantum Hall particles in the area between the plateau and non-plateau is possible is noisy.

References

1. Worcel, A., Burgi, E.: J. Mol. Biol. **71**, 127 (1972)
2. Benyajati, C., Worcel, A.: Cell **9**, 393 (1976)
3. Jackson, D.A., Dickinson, P., Cook, P.R.: EMBO J. **9**, 567 (1990)
4. Wolffe, A.P.: Chromatin. Academic Press, New York (1993)
5. Freeman, L.A., Garrard, W.T.: Crit. Rev. Euk. Gene Exp. **2**, 165 (1992)
6. Marko, J.F., Siggia, F.D.: Phys. Rev. E **52**, 2912 (1990)
7. Boles, T.C., White, J.H., Cozzarelli, N.R.: J. Mol. Biol. **213**, 931(1990)
8. BeDEOXYRIBONUCLEIC ACID, Furrer, P., Stasrak, A., Dubochet, J., Engelman, C.H., Bates, A.D.: J. Mol. Biol. **235**, 825 (1994)
9. Goswami, G., Bandyopadhyay, P.: J. Math. Phys. **34**, 749 (1995)
10. Bandyopadhyay, P.: Proc. Roy. Soc. Lond. A **466**, 2917 (2010)
11. Abanov, A.I., Wiegmann, P.B.: Nucl. Phys. B **570**, 685 (2000)
12. Carmeli, M., Malin, S.: Ann. Phys. **103**, 208 (1977)
13. Roy, S.S.: Theor. Phys. **2**(3), 141 (2017)
14. Roy, S.S., Bandyopadhyay, P.: Phys. Lett. A **382**, 1973 (2018)
15. Roy, S.S., Bandyopadhyay, P.: Phys. Lett. A **337**, 2884 (2013)
16. Cirac, I., et al.: Phys. Lett. **78**, 3221 (1997)
17. Swigon, D.: In: Benham, C.J., et al. (eds.) Mathematics of DEOXYRIBONUCLEIC ACID Structure, Function and Interactions. Springer Science Business Media (2009)
18. Roy, S.S., Bandyopadhyay, P.: Euro. Phys. Lett. **109**(4), 48002 (2015)
19. Anderson, P.W.: Science **235**, 1196 (1987)

Numerical Simulation of a Simplified Stratification Model of Suspended Sediment Concentration in an Open-Channel Turbulent Flow

Sourav Hossain, Koeli Ghoshal, and Anirban Dhar

Abstract This study predicts the dynamics of suspended sediment particles in open channels by considering a two-dimensional unsteady sediment transport model with turbulence-suppressing density stratification. The governing equation is a partial differential equation, which has been solved using OpenFoam, an open-source finite volume framework. The calculated FVM-based OpenFoam solution was compared with the FDM-based solution for validation and found to be in good agreement. It has been observed that the suspended concentration profiles are significantly affected by density stratification and should be accounted for in any theory of suspended sediment transport problem. Different contour and 3D plots have been depicted and physically illustrated for a better understanding of the sediment transport process.

Keywords Stratification · Suspended sediment · OpenFoam

1 Introduction

The transport of suspended sediment has been studied by engineers and hydrologists for centuries. Sediment in suspension is different from sediment transported by bed-load in that the sediment can be surrounded by the fluid for a considerable amount of time without frequent contact with the bed. Those models are based on either one-dimensional steady [5, 10, 15] or one-dimensional unsteady [9] formulations. In order to investigate the suspended concentration

S. Hossain (✉) · K. Ghoshal
Department of Mathematics, IIT Kharagpur, Kharagpur, India
e-mail: shossain.iitkgp@gmail.com

K. Ghoshal
e-mail: koeli@maths.iitkgp.ac.in

A. Dhar
Department of Civil Engineering, IIT Kharagpur, Kharagpur, India
e-mail: anirban@civil.iitkgp.ac.in

S. Bhattacharyya and H. S. Mahato (eds.), *Modeling, Analysis and Simulations of Multiscale Transport Phenomena*, Springer Proceedings in Mathematics & Statistics 491, https://doi.org/10.1007/978-981-96-3098-1_18

distribution, the unsteady two-dimensional non-equilibrium transport model has been considered in this work incorporating the stratification effect in a simplified form. OpenFOAM [7], an open-source framework, provides a configuration for solving differential equations using Finite Volume Method (FVM). The present model equations which is a PDE, has been solved using OpenFOAM [7]. The same governing equation has also been solved using Finite Difference Method (FDM) to validate the numerical solution. A comprehensive discussion of the obtained results is presented both graphically and theoretically, resulting in important conclusions. As suspended sediments play a key role in erosion, transport and deposition, the hydrological research community pays a great deal of attention to that process. Because sediment-water mixed flow is turbulent and unsteady, it is very challeging to predict the distribution of suspended sediment. Turbulence has a fluctuating behavior, which complicates the process and makes it difficult to build a model that can be applied under any hydraulic condition. Therefore, there is no end to the exploration of turbulent flow. This study attempts to predict particle concentrations in suspension during non-equilibrium sediment transport.

There have been several studies on the non-equilibrium transport of suspended sediment particles. For example, Hjelmfelt and Lenau [6], Liu [11] and Kumbhakar et al. [8] studied the steady two-dimensional model; whereas Liu [11], Mohan et al. [16] and Ghoshal et al. [4] studied the unsteady one-dimensional model for concentration distribution. Partial differential equations (PDEs) commonly arise to model non-equilibrium sediment transport problems. The stratification phenomenon, which is influenced by the presence of sediment in the flow, is a significant phenomenon occurring in the suspended sediment transport problems. The fluid is separated into two zones: 'a heavy fluid zone' near the bed with a high concentration of sediment and a 'light fluid zone' with a low concentration of sediment far from the bed [2]. As a result, the concentration distribution changes and the entire region behaves like a region of stably stratified flow. Several models are available in literature including this effect.

2 Problem Formulation

The motion of suspended sediment in sediment-laden flows is governed by an advection-diffusion equation. Based on the assumption that the flow is uniform along the stream-wise direction and also independent of time, the transport equation for suspended sediment particles in a wide open-channel can be expressed as the following:

$$\frac{\partial \tilde{C}}{\partial \tilde{t}} + \tilde{u}_f(\tilde{z})\frac{\partial \tilde{C}}{\partial \tilde{x}} - \tilde{w}_s\frac{\partial \tilde{C}}{\partial \tilde{z}} = \frac{\partial}{\partial \tilde{z}}\left(\tilde{\epsilon}_s\frac{\partial \tilde{C}}{\partial \tilde{z}}\right) \tag{1}$$

where $\tilde{t}$ denotes time, $\tilde{C}$ represents time-averaged sediment concentration; $\tilde{x}$ and $\tilde{z}$ are the stream-wise and vertical directions, respectively; $\tilde{u}_f$ is time-averaged velocity component in $\tilde{x}$ -direction; $\tilde{\epsilon}_s$ is the sediment diffusivity; $\tilde{w}_s$ is the settling velocity of the sediment particles.

2.1 Different Closures

To obtain the concentration profiles from Eq. (1), one has to introduce the appropriate expressions for stream-wise velocity distribution $\tilde{u}_f(\tilde{z})$, sediment diffusivity $\tilde{\epsilon}_s$ and sediment settling velocity $\tilde{w}_s$. In sediment-laden flows, the majority of the sediment particles are conveyed in the inner layer of the flow towards the bed; whereas the outer layer has a lower sediment concentration [1]. The density of a sediment-water mixture reduces as elevation from the bed increases, resulting in stratification. In response to this stratification, turbulent momentum transfer is modified, thereby influencing velocity distribution [18]. According to Wright and Parker [18], the vertical concentration gradient of sediment impacts momentum flux via a change in turbulent diffusivity. Therefore, they suggested a modified equation for turbulent diffusivity $\tilde{\epsilon}_t$ that incorporates the flux Richardson number. They also demonstrated that the usual vertical variation in Richardson number in sand-bed rivers is minimal enough to be deemed constant. Considering a constant Richardson number throughout the flow depth, turbulent diffusivity can be calculated as follows [18]:

$$\tilde{\epsilon}_t = \alpha \kappa \tilde{u}_* \tilde{z}\left(1 - \frac{\tilde{z}}{\tilde{h}}\right) \tag{2}$$

where $\tilde{u}_*$ is the fluid shear velocity, $\kappa = 0.41$ is the von-Karman constant and $\tilde{h}$ denotes the flow depth. The parameter α is the depth average correction to the eddy viscosity profile that accounts for density stratification effects. In general, α needs to be computed empirically. The current study uses the expression for α suggested by Wright and Parker [18] since it has been validated for a wide range of data for various sand-bed rivers. According to Wright and Parker [18], the mathematical expression for α is as follows [18]

$$\alpha = \begin{cases} 1 - 0.06\left(\frac{\tilde{C}_{5t}}{S_0}\right)^{0.77} & \text{when } \frac{\tilde{C}_{5t}}{S_0} \leq 10 \\ 0.67 - 0.0025\left(\frac{\tilde{C}_{5t}}{S_0}\right) & \text{when } \frac{\tilde{C}_{5t}}{S_0} > 10 \end{cases}$$

where S_0 is the slope of the bed and and $\tilde{C}_{5t}$ is the total concentration at 5% of the flow depth ($\frac{\tilde{z}}{\tilde{h}} = 0.05$). Accordingly, $\alpha = 1$ corresponds to clear water eddy diffusivity distribution ($\tilde{\epsilon}_{t0}$) whereas a value of $\alpha < 1$ parameterizes density stratification effect on the redistribution of energy by turbulent eddies. The turbulent and sediment diffusivities are connected as $\tilde{\epsilon}_s = \sigma \tilde{\epsilon}_t$ in which σ is a constant of proportionality, known as inverse of Schmidt number. For the simplicity, $\sigma = 1$ is used in this work. Several studies used the value of σ as unity [11, 12]. The stream-wise velocity distribution

$\tilde{u}_f(\tilde{z})$ considered for this study is as follows [18]

$$\frac{\tilde{u}_f(\tilde{z})}{\tilde{u}_*} = \frac{1}{\alpha\kappa} \ln \frac{\tilde{z}}{\tilde{z}_0} \tag{3}$$

where $\tilde{z}_0$ is the start elevation of the log-law. Equation (3) is characterized by the stratification corrected log-law [18]. Apart from the sediment diffusivity and stream-wise velocity distribution, the settling velocity of suspended sediment particles is determined by the expression of Song et al. [19], which is

$$\tilde{w}_s = \frac{\nu_f}{D_p} D_{p_*}^3 \left(38.1 + 0.93 D_{p_*}^{12/7}\right)^{-7/8} \tag{4}$$

where D_p is the particle diameter in dimensional form whereas $D_{p_*}(= D_p[(s-1)g/\nu_f^2]^{1/3})$ is the non-dimensional particle diameter in which $g(= 9.81\, m/s^2)$ is the acceleration due to gravity, $\nu_f(= 10^{-6}\, m^2/s)$ is kinematic viscosity of fluid; $s(= 2.65)$ is the relative density of sediment particles.

2.2 Non-dimensional Equations

We introduce here the following non-dimensional variables:

$$x = \frac{\tilde{x}}{\tilde{h}},\ z = \frac{\tilde{z}}{\tilde{h}},\ a = \frac{\tilde{a}}{\tilde{h}},\ t = \frac{\tilde{t}\tilde{u}_*}{\tilde{h}},\ w_s = \frac{\tilde{w}_s}{\tilde{u}_*},\ u_f = \frac{\tilde{u}_f}{\tilde{u}_*},\ C = \frac{\tilde{C}}{\tilde{C}_{\tilde{a}}},\ \epsilon_t = \frac{\tilde{\epsilon}_s}{\sigma\tilde{u}_*\tilde{h}}$$

where $\tilde{C}_{\tilde{a}}$ is the reference concentration at the reference height $\tilde{a}$. Using the above dimensionless variables, one can easily get the following non-dimensional form of Eq. (1) as

$$\frac{\partial C}{\partial t} + u(z)\frac{\partial C}{\partial x} - w_s\frac{\partial C}{\partial z} = \frac{\partial}{\partial z}\left[\sigma\epsilon_t\frac{\partial C}{\partial z}\right] \tag{5}$$

The transformed expressions for $u(z)$ and ϵ_t are as respectively follows:

$$\epsilon_t(z) = \alpha\kappa z(1-z) \tag{6}$$

and

$$u(z) = \frac{u_f(z)}{\overline{u}_f(z)} = \frac{1-z_0}{z_0 - \ln z_0 - 1} \ln \frac{z}{z_0} \quad \text{for } z_0 \le z \le 1 \tag{7}$$

where $\overline{u}_f(z)$ is depth-averaged velocity and $z_0 = 0.001$ [11]. Equation (5) is a two-dimensional initial boundary value problem (IBVP). To solve Eq. (5), we consider the following initial and boundary conditions

$$C(t=0, x, z) = C_{t_0} \tag{8}$$

$$C(t, x = 0, z) = C_{x_0} \tag{9}$$

$$C(t, x, z = a) = 1 \tag{10}$$

$$C(t, x, z = 1) = 0 \tag{11}$$

where C_{t_0} and C_{x_0} can have values of zero or unity. Equation (5), along with the above initial and boundary conditions, represents the two-dimensional unsteady distribution of suspended sediment in an open channel incorporating the effect of density stratification.

3 Numerical Methodology

3.1 Discretization Technique

By using Eqs. (6) and (7) in Eq. (5), the suspended concentration becomes

$$\frac{\partial C}{\partial t} + \frac{1 - z_0}{z_0 - \ln z_0 - 1} \ln \frac{z}{z_0} \frac{\partial C}{\partial x} - w_s \frac{\partial C}{\partial z} = \alpha\kappa \frac{\partial}{\partial z}\left[z\left(1 - z\right)\frac{\partial C}{\partial z}\right] \tag{12}$$

This work uses the standard FVM for the spatial domain and the fully implicit Euler time marching scheme for the temporal domain. Invoking a backward Euler method (fully implicit), the semi-discretized form of the Eq. (12) can be expressed as

$$\frac{C^{n+1} - C^n}{\Delta t} + \frac{1 - z_0}{z_0 - \ln z_0 - 1} \ln \frac{z}{z_0} \left.\frac{\partial C}{\partial x}\right|^{n+1} - w_s \left.\frac{\partial C}{\partial z}\right|^{n+1} = \alpha\kappa(1 - 2z) \left.\frac{\partial C}{\partial z}\right|^{n+1} + \alpha\kappa\left(z - z^2\right) \left.\frac{\partial^2 C}{\partial z^2}\right|^{n+1} \tag{13}$$

The superscripts n and $n + 1$ represent the old and present time levels $t = t^n$ and $t = t^{n+1}$ respectively. C^{n+1} is the value of C at present time level $n + 1$ and is assumed to be unknown whereas C^n is the old time level value of C at time level n which is known. $\Delta t (= t^{n+1} - t^n)$ denotes the time step.

3.2 Implementation in OpenFOAM

This study uses a FVM-based open-source framework OpenFoam [7] for the numerical computation of the present problem. OpenFoam has recently gained a lot of popularity across a variety of research fields, particularly in fluid dynamics. It contains all of the infrastructure required to discretize the governing equations in the spatial and temporal domains and to solve the system of linear algebraic equations that results from the discretization process. The numerical schemes for the various derivative terms present in Eq. (13) are listed in Table 1. A Pre-conditioned Conjugate

Table 1 Numerical scheme and its order of accuracy for different differential terms

Term	Scheme	Order of accuracy	Interpolation scheme
Temporal	Backward Euler	1st order	
Gradient	Gauss	2nd order	Linear
Laplacian	Gauss	3rd order	Linear corrected

Gradient (PCG) solver with a diagonal incomplete Cholesky (DIC) preconditioner has been used to solve the set of discrete linear algebraic equations derived from Eq. (13). In comparison to other iterative matrix solvers, the above solver exhibits faster convergence and is less expensive to compute.

4 Results and Discussion

4.1 *Validation of the Present FVM-Based OpenFoam Solution with FDM-Based Numerical Solution*

To validate the present numerical method, the performance of the OpenFOAM solution was compared with the FDM-based solution. In order to do so, the same governing equation has been discretized through FDM. A second-order finite difference approach was used to approximate the gradient and Laplacian terms. First order backward Euler method approximates the time derivative implicitly. For obtaining FDM-based numerical solutions, the resulted discretized linear algebraic equations have been solved using Gauss-Seidal method. In Fig. 1, the FVM-based OpenFoam solution is compared with the FDM-based solution. Figure 1a shows the vertical concentration profile at $t = 15$ and $x = 5$, while Fig. 1b depicts the variation of depth-averaged concentration along the stream-wise direction at $t = 15$. Each of the required parameters is mentioned in the figures. In both the cases, OpenFoam solution agrees well with the FDM-based solution.

4.2 *Impact of Stratification on Suspended Concentration Profiles*

There are many studies ([3], [13], [14]) that have developed parameterizations of density stratification due to the computational convenience of representing multiple physical processes via a limited number of parameters. To parameterize stratification, the depth-averaged coefficient α (mentioned in Eq. 2) has been used in this work. A value of $\alpha = 1$ corresponds to clear eddy viscosity, which means that stratification effect is left out. However, $\alpha < 1$ parameterizes density stratification effect.

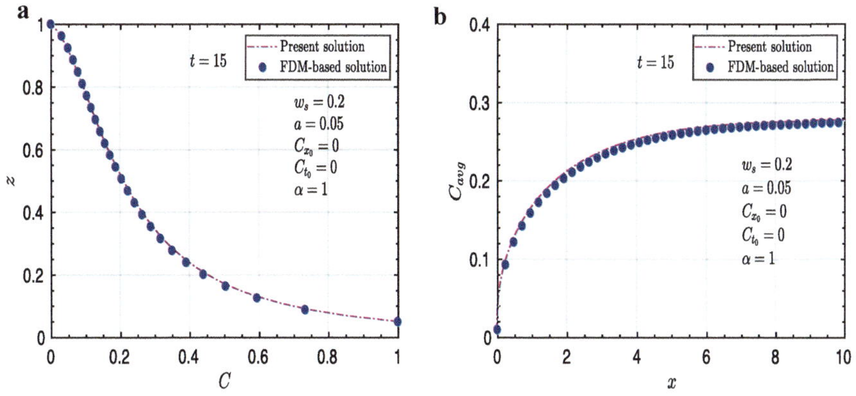

Fig. 1 Comparison between the FVM-based OpenFoam solution and FDM-based solution: **a** (the left one) vertical concentration profile at $x = 5$ and $t = 15$ and **b** (the right one) depth-averaged concentration at $t = 15$

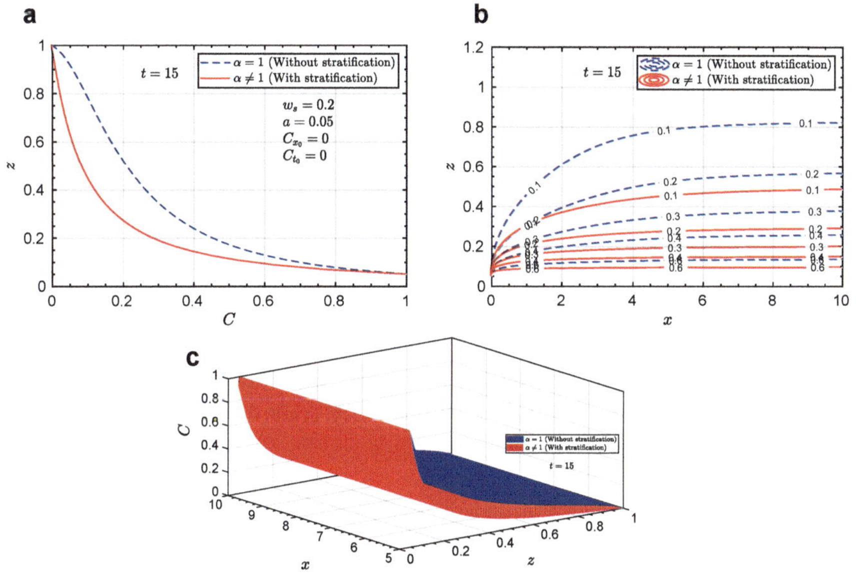

Fig. 2 Effect of stratification of sediment concentration: **a** (the left one) vertical concentration profile at $x = 5$ and $t = 15$, **b** (the right one) contour line plot of concentration at $t = 15$ and **c** (the bottom one) 3D surface plot of concentration at $t = 15$

Figures 2a, b illustrate the effect of stratification on vertical concentration profiles and concentration contour lines respectively, while Fig. 2c shows how the 3D surface plot of sediment concentration changes when stratification is considered. At

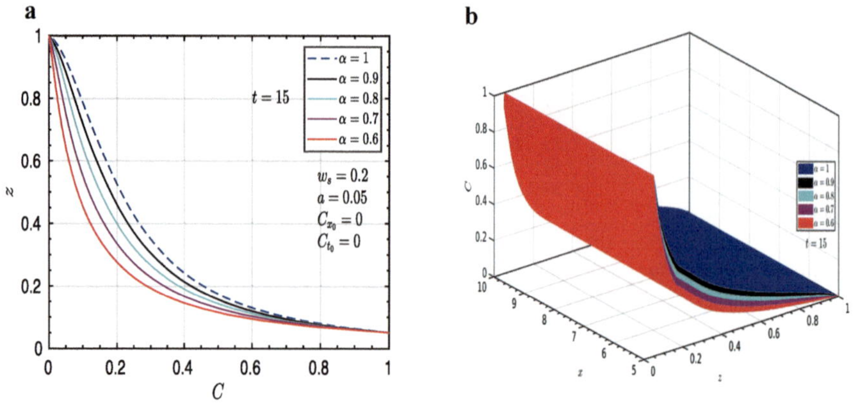

Fig. 3 Sensitivity of α—values on sediment concentration: **a** (the left one) vertical concentration profile at $x = 5$ and $t = 15$ and **b** (the right one) 3D surface plot of concentration at $t = 15$, for five different values of α

time $t = 15$, Fig. 2a is plotted at a specific downstream location $x = 5$. Figures 2b, c are plotted at the same time $t = 15$. All the required parameters are listed inside the figure. As shown in Fig. 2, stratification reduces sediment concentration in comparison to neutral conditions. The reason can be explained as follows. As sediment is suspended in the fluid, it increases the effective density of the fluid, and near the bed, there is a vertical gradient of density that produces a negative buoyancy ([17]). This results in increased dissipation of turbulent eddies, causing eddy viscosity to decrease. Consequently, the concentration profile decreases in magnitude. Figure 3 describes the sensitivity of α-values to sediment concentration magnitudes. Vertical concentration profiles at $t = 15$ and $x = 5$ and 3D surface plots at $t = 15$ were plotted using five different values of α (< 1) in Fig. 3(a) and 3(b) respectively. It is clear from the figures that the magnitude of concentration decreases with decreasing α.

4.3 Temporal and Horizontal Evolution Suspended Concentration Profiles

The transient suspended sediment concentration profiles are plotted in Fig. 4a for different values of t at a particular downstream location $x = 5$. On the other hand, the stream-wise variation of concentration profiles is depicted in Fig. 4b at a particular time $t = 15$. In the figure, all the necessary parameters are listed. Based on the figures, it can be seen that the profile is similar to a Rouse-type concentration profile at large time periods and far fields.

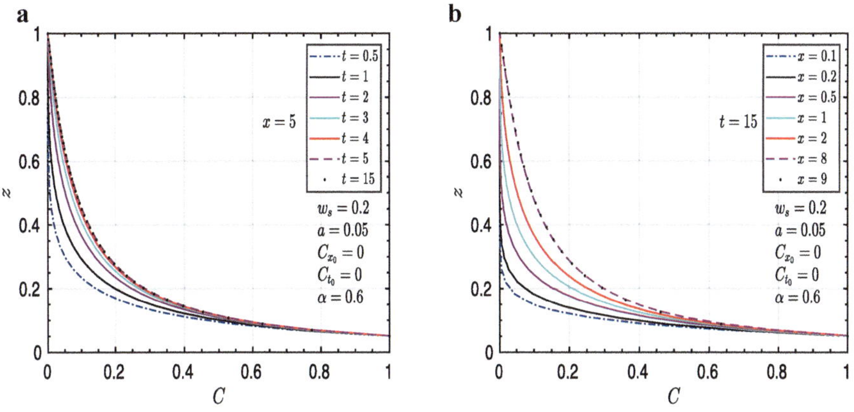

Fig. 4 Temporal and stream-wise variation of sediment concentration: **a** (the left one) time advancement vertical concentration profile at $x = 5$ and **b** (the right one) stream-wise advancement of vertical concentration profile at $t = 15$

5 Concluding Remarks

In the present work, the distribution of suspended sediment in an unsteady, two-dimensional turbulent flow through an open channel is investigated. The presence of sediments modifies the characteristics of the flow by creating a vertical density-gradient of sediment. Density stratification is a process that results in the formation of distinct, separate horizontal layers. This phenomenon is taken into account in the mathematical modelling. The PDE arising from the problem formulation has been solved with OpenFOAM, an open source FVM-based computational fluid dynamics library. Good matching has been seen when the solutions have been tested against FDM-based solutions. It is observed that stratification results in a decrease in the magnitude of concentration distribution. Different 3D surface plots and contour plots have been depicted to analyze the effects of stratification on concentration distribution. To visualise the pattern of concentration profiles over time and horizontal distance, the horizontal and temporal variation of concentration distribution has been plotted. The concentration profile exhibits similarity to the Rouse profile at large time and far from the downstream.

References

1. Coleman, N.L.: Effects of suspended sediment on the open-channel velocity distribution. Water Res. Res. **22**(10), 1377–1384 (1986)
2. Dey, S.: Fluvial Hydrodynamics. Springer (2014)

3. Gelfenbaum, G., Smith, J.D.: Experimental evaluation of a generalized suspended-sediment transport theory (1986)
4. Ghoshal, K., Jain, P., Absi, R.: Nonlinear partial differential equation for unsteady vertical distribution of suspended sediments in open channel flows: Effects of hindered settling and concentration-dependent mixing length. J. Eng. Mech. **148**(1), 04021123 (2022)
5. Herrmann, M.J., Madsen, O.S.: Effect of stratification due to suspended sand on velocity and concentration distribution in unidirectional flows. J. Geophys. Res.: Oceans **112**(C2) (2007)
6. Hjelmfelt, A.T., Lenau, C.W.: Nonequilibrium transport of suspended sediment. J. Hydraul. Div. **96**(7), 1567–1586 (1970)
7. Jasak, H.: Error analysis and estimation for the finite volume method with applications to fluid flows (1996)
8. Kumbhakar, M., Mohan, S., Ghoshal, K., Kumar, J., Singh, V.P.: Semianalytical solution for nonequilibrium suspended sediment transport in open channels with concentration-dependent settling velocity. J. Hydrol. Eng. **27**(2), 04021048 (2022)
9. Kundu, S.: Study of unsteady nonequilibrium stratified suspended sediment distribution in open-channel turbulent flows using shifted chebyshev polynomials. ISH J. Hydraul. Eng. **28**(1), 42–52 (2022)
10. Kundu, S., Ghoshal, K.: Effects of secondary current and stratification on suspension concentration in an open channel flow. Environ. Fluid Mech. **14**(6), 1357–1380 (2014)
11. Liu, X.: Analytical solutions for steady two-dimensional suspended sediment transport in channels with arbitrary advection velocity and eddy diffusivity distributions. J. Hydraul. Res. **54**(4), 389–398 (2016)
12. Liu, X., Nayamatullah, M.: Semianalytical solutions for one-dimensional unsteady nonequilibrium suspended sediment transport in channels with arbitrary eddy viscosity distributions and realistic boundary conditions. J. Hydraul. Eng. **140**(5), 04014011 (2014)
13. McLean, S.: Depth-integrated suspended-load calculations. J. Hydraul. Eng. **117**(11), 1440–1458 (1991)
14. McLean, S.: On the calculation of suspended load for noncohesive sediments. J. Geophys. Res.: Oceans **97**(C4), 5759–5770 (1992)
15. Mohan, S., Kumbhakar, M., Ghoshal, K., Kumar, J.: Semianalytical solution for simultaneous distribution of fluid velocity and sediment concentration in open-channel flow. J. Eng. Mech. **145**(11), 04019090 (2019)
16. Mohan, S., Kumbhakar, M., Ghoshal, K., Kumar, J.: Semi-analytical solution for one-dimensional unsteady sediment transport model in open channel with concentration-dependent settling velocity. Physica Scripta **95**(5), 055204 (2020)
17. Moodie, A.J., Nittrouer, J.A., Ma, H., Carlson, B.N., Wang, Y., Lamb, M.P., Parker, G.: Suspended sediment-induced stratification inferred from concentration and velocity profile measurements in the lower yellow river, china. Water Res. Res. **58**(5), e2020WR027192 (2022)
18. Wright, S., Parker, G.: Flow resistance and suspended load in sand-bed rivers: simplified stratification model. J. Hydraul. Eng. **130**(8), 796–805 (2004)
19. Zhiyao, S., Tingting, W., Fumin, X., Ruijie, L.: A simple formula for predicting settling velocity of sediment particles. Water Sci. Eng. **1**(1), 37–43 (2008)

Effects of Hydrodynamic Phenomena on Two-Dimensional Distribution of Suspended Sediment Concentration in an Open Channel Flow

Sumit Sen, Sourav Hossain, and Koeli Ghoshal

Abstract The present study develops a steady two-dimensional model for the distribution of suspended sediment concentration in an open channel turbulent flow carrying sediments. The theoretical model takes into account the effect of hindered settling through the exponent of reduction of settling velocity. Unlike most others, this study looks into the effect of secondary current on concentration distributions in both narrow and wide open channels. Apart from this, the model considers the influence of fluid-induced lift force and the force exerted owing to the gradient of Reynolds normal stress. The inclusion of these hydrodynamic phenomena makes the governing equation highly non-linear. Thus, the governing equation, along with two generalised boundary conditions and an initial condition, has been numerically solved. The obtained results show that the hindered settling increases the magnitude of the concentration profile. On the other hand, the sediment concentration decreases due to the presence of the secondary current; whereas it increases because of the exerted force. Finally, the model has been validated with a wide range of experimental data at a far-field position.

Keywords 2D cocentration distribution · Hindered settling · Secondary current

1 Introduction

Sediment transport in open channels is an important topic of research due to its wide range of applications in several fields like sedimentary geology, civil engineering, hydraulic engineering etc. Transport of sediment in fluvial streams is classified into three categories—bed load, suspended load and wash load. The study of suspended load/transport is still a topic of ongoing research interest because the irregular behavior of turbulence makes it impossible for any model to be claimed to be universal.

S. Sen · S. Hossain · K. Ghoshal (✉)
Department of Mathematics, IIT Kharagpur, Kharagpur, India
e-mail: koeli@maths.iitkgp.ac.in

S. Bhattacharyya and H. S. Mahato (eds.), *Modeling, Analysis and Simulations of Multiscale Transport Phenomena*, Springer Proceedings in Mathematics & Statistics 491, https://doi.org/10.1007/978-981-96-3098-1_19

To compute the suspended load, knowledge of suspended sediment concentration distribution is important. Rouse [20] was the pioneer in the studies of sediment concentration in suspension. He proposed an equation for the vertical concentration distribution by balancing the upward diffusion force to the downward gravitational force. After that, several models have been developed for the vertical concentration distribution using different approaches/theories. However, in comparison to the steady one-dimensional (1D) models, few steady two-dimensional (2D) or unsteady 1D models have been developed for the distribution of concentration. Details can be seen from Kumbhakar et al. [9]. Since the steady 2D distribution of concentration is the subject of the current study, steady 2D models will receive the majority of our attention from now on. Most of the studies that are present in the literature on steady 2D models deal with linear PDE, as they didn't include any hydrodynamic phenomena. Also, except Liu [11], none of the models present in the literature considered the generalized boundary condition at the bottom. Jing et al. [8] and Kumbhakar et al. [9] considered the hindered settling effect on their non-equilibrium transport model for steady 2D distribution of concentration. Upon consideration of concentration dependent hindered settling expression, their governing equation became nonlinear PDE, which had been solved numerically by Jing et al. [8] and semi-analytically by Kumbhakar et al. [9].

Secondary currents in open channel flows are ubiquitous irrespective of the channel geometry and play an important role in sediment transport. According to Prandtl [18], there are two different kinds of secondary currents presence in fluid flow. The present study considers secondary current of Prandtl's second kind, which are generated due to turbulence. Previous studies concerning secondary currents mainly focused on their effect on the flow velocity profile rather than the concentration profile. There are very few works in the literature that investigate the concentration distribution while accounting for the effect of secondary current in the model. Yang [23] incorporated the effect of secondary current into their model and studied both velocity and concentration distribution. But the study didn't show the effect of secondary current on concentration distribution explicitly. Although Kundu and Ghoshal [10] and Pal and Ghoshal [17] studied the concentration distribution considering the effect, their studies were limited to steady vertical distribution, and neither of the model took into account the generalized boundary condition at the bottom and the free surface. The fluid induced lift force and the force induced by the gradient of Reynolds shear stress are two other important hydrodynamic phenomena that are frequently overlooked in concentration studies. Fu et al. [5] showed the importance of these two forces on concentration distribution and later, Pal and Ghoshal [17] used these effects in their study of steady vertical concentration distribution.

The aforementioned literature review reveals that no studies have been done on steady 2D distribution of concentration considering all the above described hydrodynamic phenomena. Therefore, the objectives of the present study are to—(1) develop a more generalized non-equilibrium transport model for the steady 2D distribution of sediment concentration with the generalized boundary condition by considering the hindered settling effect, secondary current, lift force and Reynolds normal stress gradient, (2) analyze the effects of the hydrodynamic phenomena on the concentration distribution and (3) verify the proposed model with the experimental data under certain restrictions.

2 Problem Formulation

2.1 Governing Equation

This study considers the motion of suspended sediment in incompressible turbulent flow through open channels. The flow is assumed to be uniform in the longitudinal direction and independent of time and the sediment concentration varies along stream-wise and vertical directions. It is also assumed that the flow is sediment free initially. It enters into an erodible sediment bed and start interacting with sediment particles. A schematic diagram of the flow in the xz plane is presented in Fig. 1. Governing equation for non-equilibrium suspended sediment is the advection-diffusion equation that can be expressed as

$$u_s \frac{\partial c}{\partial x} + \frac{\partial (w_s c)}{\partial z} = \frac{\partial}{\partial x}\left\{(\varepsilon_m + \varepsilon_{sx}) \frac{\partial c}{\partial x}\right\} + \frac{\partial}{\partial z}\left\{(\varepsilon_m + \varepsilon_{sz}) \frac{\partial c}{\partial z}\right\} \tag{1}$$

where c is the volumetric sediment concentration, x and z represent the longitudinal and vertical directions in the cartesian co-ordinate system, u_s and w_s are the time-averaged mean velocity components of sediments in x and z directions, respectively. Also, ε_m is the molecular diffusion coefficient and ε_{sx} and ε_{sz} are the sediment diffusion coefficients in x and z directions, respectively. In turbulent flow, turbulent diffusion coefficients are predominant and the effect of molecular diffusion coefficient

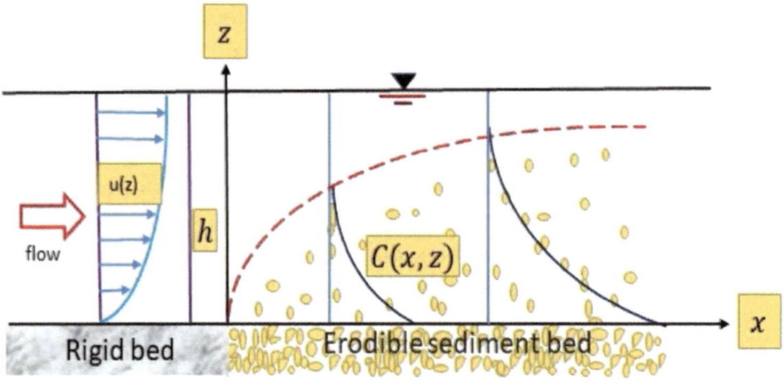

Fig. 1 Schematic pattern of sediment transport and flow velocity over an erodible bed

is negligible ($\varepsilon_m \approx 0$) in comparison to turbulent diffusion coefficients. Diffusion components in the longitudinal direction can be neglected as advection dominates in most cases [11]. Also, the vertical velocity of sediment particles w_s in sediment-mixed fluid is composed of the downward settling velocity ω_s of sediment particles and the mean velocity w_f of the fluid. Then Eq. (1) becomes

$$u_s \frac{\partial c}{\partial x} + \frac{\partial}{\partial z}\left\{\left(w_f - \omega_s\right)c\right\} = \frac{\partial}{\partial z}\left(\varepsilon_s \frac{\partial c}{\partial z}\right) \quad (2)$$

where ε_s is written in place of ε_{sz}. In sediment-mixed flow, it is reported by many researchers that flow around surrounding settling sediment particles exhibit a larger drag on the particles in comparison to clear water flow, which in turn reduces the settling velocity of the suspended sediments. This phenomena is known as hindered settling effect, which can be mathematically expressed as [19]

$$\omega_s = \omega_0(1 - c)^{n_H} \quad (3)$$

where ω_s and ω_0 are the settling velocities of sediment particles in sediment-laden fluid and clear fluid, respectively; n_H is the exponent of reduction of settling velocity. The exponent n_H depends on particle Reynolds number and its value varies between 2.3 and 4.9. Here, an average value $n_H = 4$ has been considered for mathematical convenience.

2.2 Secondary Current

The existence of secondary current in laboratory and field measurement was reported a century ago. Irrespective of the channel's geometry, open channel flows are three-dimensional owing to the presence of secondary current . It is well known that in narrow channels, maximum velocity occurs below the free surface due to the presence of the secondary currents, which are generated due to the side wall effect. Though weak, many researchers reported the existence of secondary current in wide open channel flows also. The turbulence induced secondary currents in wide channels are generated due to various reasons such as asymmetry of the channel boundaries, free surface effects, variations in bed conditions and instabilities in turbulent flows [22]. Secondary flows or currents are composed of lateral and vertical fluid flow components, whereas the flow parallel to the stream-wise direction is known as primary flow. As we are interested only in the sediment transport phenomena at the central region of the open channel, the lateral variations are neglected in Eq. (1). Therefore, the effect of secondary current has been considered through the vertical component of the velocity w_f of the fluid. Also, researchers [5, 23] pointed out the importance of the non-zero value of w_f in sediment-mixed fluid and suggested that it depends on the concentration of suspended sediment particles. Analysing the experimental data of Muste and Patel [14], Yang [23] concluded that the presence of sediment particles modify the Reynolds shear stress distribution. On the other hand, from the momentum equation for steady, uniform sediment-laden turbulent flow, one can derive Reynolds shear stress as a function of momentum flux uw_f in an open channel as

$$-\frac{\overline{u'_f w'_f}}{u_*^2} = \left(1 - \frac{z}{h}\right) + \frac{u_f w_f}{u_*^2} \tag{4}$$

in which u_f is the mean velocity of the fluid in the stream-wise direction, u_* is the shear velocity and h is the depth of the channel. Following Eq. (4), Yang [23] inferred that the Reynolds shear stress could be affected by the momentum flux $u_f w_f$ or by the non-zero w_f. In other words, Reynolds shear stress increases due to the presence of sediments in the flow. This implies that the magnitude of downward velocity w_f is reduced or an upward velocity is induced by the presence of particles that compensates the downward w_f caused by the secondary currents. Hence, the vertical velocity w_f of the fluid can be expressed as [23]

$$w_f = w_1 + w_2 \tag{5}$$

where w_1 denotes the vertical velocity of sediment free fluid caused by the secondary currents and w_2 denotes the additional upward velocity of fluid caused by the presence of suspended sediments.

Due to the irregularities of the secondary currents, reliable expression of w_1 is lacking in the literature. For clear water flows, Yang et al. [24] assumed that uw_1/u_*^2 follows a linear relation, after noticing that u_f is negligible near a solid bed and approaches its maximum in the upper layer. Following [24], the expression is given as

$$\frac{u_f w_1}{u_*^2} = -\frac{\alpha z}{h} \tag{6}$$

where α is a positive coefficient, which depends on the aspect ratio of the channel and can be determined from the expression of Yang et al. [24] as

$$\alpha = 1.3exp(-Ar/2) = 1.3exp(-b/h) \tag{7}$$

where Ar is the aspect ratio of the channel and b is the half width of the channel.

For the other component of fluid velocity w_2, Chien and Wan [2] analysed the settlement of sediment in river flows and found that sediment particles induce an fluid velocity w_2 that can be expressed as

$$w_2 = \beta_1 \frac{c\omega_s}{1 - c} \tag{8}$$

where β_1 is a factor, whose value has been considered as 1 [2]. However, they considered w_2 simply as a function of sediment concentration and settling velocity and didn't include the effect of particle-fluid interactions. Later on, Fu et al. [5]

proposed an expression for the upward wall normal velocity w_2 considering the interaction effect of particle-fluid through the lift force effect and sediment stress gradient. Thus, the present study considers the expression of Fu et al. [5] for w_2, which is given by

$$w_2 = \tau_p \left(F_L + F_S\right) \tag{9}$$

where τ_p is the particle relaxation time, F_L is the fluid induced lift force and F_S is a force produced by the sediment stress gradient in the vertical direction. The expression of particle relaxation time or particle time scale τ_p has been considered from the study of Greimann et al. [6] and can be expressed as

$$\tau_p = \frac{\omega_0 \Phi^{\nu}_{f} \rho_s}{g(\rho_s - \rho_f)} = \frac{\omega_0 (1-c)^{\nu}}{g\left(1 - \rho_f/\rho_s\right)} \tag{10}$$

where ω_0 is the settling velocity of sediment particles in clear fluid, Φ_f is the volumetric concentration of fluid phase, ρ_s and ρ_f are the mass densities of sediment and fluid, respectively, g is the gravitational acceleration and ν is the exponent which accounts for the effects of particle crowding. According to Greimann et al. [6], the value of the exponent ν can be considered as 1, provided the flow is not dense, i.e., if $\Phi_s < 0.1$, in which Φ_s is the volumetric concentration of solid phase. Using Eqs. (4), (5), (6), (8) and (9) in Eq. (2), one gets

$$u_s \frac{\partial c}{\partial x} + \frac{\partial}{\partial z}\left[\left\{-\frac{\alpha z u_*^2}{h u_f} + \frac{\omega_0 (1-c)^{\nu}}{g\left(1 - \rho_f/\rho_s\right)}(F_L + F_s) - \omega_0 (1-c)^{n_H}\right\} c\right] = \frac{\partial}{\partial z}\left(\varepsilon_s \frac{\partial c}{\partial z}\right) \tag{11}$$

To obtain the steady two-dimensional distribution of concentration from Eq. (11), one needs to find appropriate expressions of u_s, u_f, F_L, F_s and ε_s.

2.3 Fluid Induced Lift Force

Lift is typically thought to act upward from the horizontal plane in opposition to gravity. However, according to researchers, it is more practical to take it as a force acting perpendicular to the streamwise direction. The fluid induced lift force plays an important role in the transport of sediment, specially in the near wall regions [21]. McLaughlin [13] extended the Saffman's formula for lift force by removing the limitations and including the effects of the presence of wall as

$$F_L = \frac{27}{2\pi^2} \frac{\rho_f \nu_f}{\rho_s d} \sqrt{\frac{1}{\nu_f} \frac{\partial u_f}{\partial z}}\, u_l J \tag{12}$$

where d is the diameter of the particle, ν_f is the kinematic viscosity, u_l ($\equiv u_s - u_f$) is the stream-wise velocity lag between sediment particle and fluid and J is a parameter, which depends on vertical gradient of fluid velocity $\partial u_f/\partial z$, velocity lag u_l, and vertical distance of particle from the sediment bed z. According to McLaughlin [13], the maximum value of J is 2.255.

2.4 Exerted Force Due to the Gradient of Reynolds Normal Stress

The force produced by the vertical gradient of Reynolds normal stress on sediments in suspension is given by

$$F_S = -\frac{\partial}{\partial z}\left(\overline{w_s'^2}\right) \tag{13}$$

where w_s' is the fluctuation part of the vertical velocity component w_s and $\sqrt{\overline{w_s'^2}}$ is known as turbulence intensity of sediment particles along the vertical. Greimann and Holly [7] proposed a relation between vertical turbulence intensity of sediment w_s and that of the fluid w_f using an empirical parameter D_s as

$$\overline{w_s'^2} = D_s \overline{w_f'^2} \tag{14}$$

where D_s represents the damping of particle turbulence relative to the fluid. They suggested that it was sufficient to consider $D_s = 1.2$ for most of the fluids. Nezu [15] proposed universal models for predicting turbulence intensities of fluid in the vertical and horizontal directions in clear fluid considering that the turbulent energy production is approximately in balance with the viscous dissipation. The expression for vertical turbulence intensity is as follows [15]

$$\frac{\sqrt{\overline{w_f'^2}}}{u_*} = D_w exp\left(-\lambda_w z/h\right) \tag{15}$$

where D_w and λ_w are empirical constants, whose values are taken as 1.23 and 0.67, respectively [16].

Considering that there is no stream-wise velocity lag between fluid and sediment particle (i.e., $u_s = u_f$) and using Eqs. (12)–(15), the non-dimensional expression of Eq. (11) yields

$$\begin{aligned} u^+\frac{\partial c^+}{\partial x^+} + \frac{\partial}{\partial z^+}\Bigg[\Bigg\{-\frac{\alpha z^+}{u^+} + \frac{2.43u_*^2\omega_0^+(1-c_a c^+)^\nu}{gh\left(1-\rho_f/\rho_s\right)} exp(-1.34z^+) \\ - \omega_0^+(1-c_a c^+)^{n_H}\Bigg\}c^+\Bigg] = \frac{\partial}{\partial z^+}\left(\varepsilon_t^+\frac{\partial c^+}{\partial z^+}\right) \end{aligned} \tag{16}$$

where the dimensionless variables are

$$u^+ = \frac{u_f}{u_*},\ c^+ = \frac{c}{c_a},\ x^+ = \frac{x}{h},\ z^+ = \frac{z}{h},\ \omega_0^+ = \frac{\omega_0}{u_*},\ \text{and } \varepsilon_t^+ = \frac{\gamma \varepsilon_t}{u_* h} \tag{17}$$

in which c_a is the reference concentration at the reference height $z = a$ and ε_t is the eddy viscosity or turbulent diffusivity and $\gamma = \varepsilon_s/\varepsilon_t$ is the inverse Schmidt number. Expressions of u^+ and ε_s^+ are needed to assess the two-dimensional distribution of concentration from Eq. (16). For that purpose, logarithmic velocity profile and parabolic profile of eddy viscosity has been considered as

$$u = \frac{u_*}{\kappa} \ln \frac{z}{h} \tag{18}$$

and

$$\varepsilon_t = \kappa u_* z(1 - z/h) \tag{19}$$

3 Initial and Boundary Conditions

At the inlet $x^+ = 0$, concentration can be considered as arbitrary following Liu [11]. In this study, we are assuming that the inlet is sediment free as

$$c^+(x^+ = 0,\ z^+) = 0 \tag{20}$$

At the free surface, zero sediment flux is specified and at the bottom, the most generalized boundary condition suggested by Cheng [1] has been considered. Therefore the non-dimensional form of the boundary conditions are expressed as

$$\varepsilon_s^+ \frac{\partial c^+}{\partial z^+} + \omega^+ c^+ = 0 \text{ at } z^+ = 1 \tag{21}$$

and

$$\varepsilon_s^+ \frac{\partial c^+}{\partial z^+} + (\omega^+ - B)c^+ = -BC^* \text{ at } z^+ = \frac{a}{h} \tag{22}$$

where B and C^* are the non-dimensional depth independent parameter. The governing equation (16) is a highly non-linear PDE with initial condition Eq. (20) and non-homogeneous Robin type boundary conditions Eqs. (21) and (22), analytical solution of which may not be possible. Thus, the present model has been solved numerically. In the next section, solution methodology has been discussed.

4 Solution Methodology

Discretizing the governing PDE using finite difference scheme, one gets

$$\frac{dc_i^+}{dx} = \frac{1}{u^+}\left[\delta\left(\varepsilon_t^+\delta c^+\right)\right]_i - \frac{1}{u^+}\left[\delta\left\{\left(-\frac{\alpha z^+}{u^+} + \frac{2.43u_*^2\omega_0^+(1-c_a c^+)^{\nu}}{gh\left(1-\rho_f/\rho_s\right)}\, exp(-1.34z^+) - \omega_0^+(1-c_a c^+)^{n_H}\right)c^+\right\}\right]_i \tag{23}$$

where $i = 2, 3, ..., N$, in which $N - 1$ is the total number of internal grid points and $\Delta z^+ = (1 - A)/N$ is the distance between consecutive grid points. Vertical grid points are defined as $z_i^+ = A + (i - 1)\Delta z^+$ and δc^+ and $\delta^2 c^+$ are defined as

$$(\delta c^+)_i = \left(\frac{\partial c^+}{\partial z^+}\right)_i = \frac{c_{i+1}^+ - c_{i-1}^+}{2\Delta z^+}, \tag{24}$$

$$(\delta^2 c^+)_i = \left(\frac{\partial^2 c}{\partial z^{+2}}\right)_i = \frac{c_{i+1}^+ - 2c_i^+ + c_{i-1}^+}{(\Delta z^+)^2}, \tag{25}$$

Grid points $i = 1$ and $N + 1$ corresponds to the reference height $z^+ = A$ and free surface $z^+ = 1$, respectively. Three points forward difference and three points backward difference scheme have been used to discretize the boundary conditions. The discretized boundary conditions can be expressed as

$$\varepsilon_t^+(1)\left(\frac{-3c_1^+ + 4c_2^+ - c_3^+}{2\Delta z^+}\right) + (\omega^+ - B)c_1^+ = -BC^* \tag{26}$$

$$\varepsilon_t^+(N+1)\left(\frac{3c_{N+1}^+ - 4c_N^+ + c_{N-1}^+}{2\Delta z^+}\right) + \omega^+ c_{N+1}^+ = 0 \tag{27}$$

Finally the governing equation with the inlet concentration and boundary conditions has been solved numerically.

5 Results and Discussions

5.1 Effect of Hydrodynamic Mechanisms on Concentration Distribution

Fig. 2 shows the hindered settling effect on the distribution of suspended sediment concentration through the exponent of reduction n_H. The effect of hindered settling on the vertical distribution of concentration at a far field position, i.e., $x^+ \to \infty$ has been shown in Fig. 2a. On the other side, Fig. 2b depicts the effect of hindered

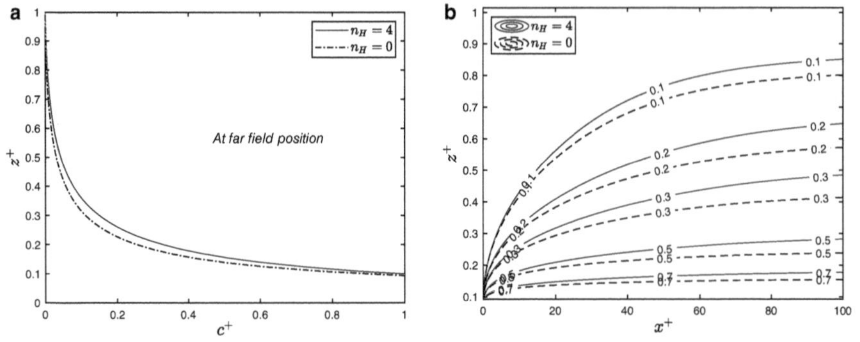

Fig. 2 Effect of hindered settling on **a** vertical concentration profile at a far field position and **b** concentration contour lines

settling on the 2-D suspended concentration contour. In both the figures, values of the parameters are the same and are $A = 0.093$, $\omega^+ = 0.2$, $C_a = 0.057$, $B = 2$ and $C_* = 1$. From the figure, one can observe that the hindered settling effect increases the magnitude of the concentration profiles and the effect is prominent in the main region of the channel. The reason behind the increase in magnitude of concentration in the main suspension region is the presence of surrounding sediment particles, which reduces the fall velocity of the suspended sediments. Near the free surface, concentration is very low and thus the effect is negligible. Though near the bed, concentration is very high, but sediment particles are very close to the bed and no significant difference is observed. Also, from the contour plot, it can be observed that as x^+ increases, concentration distribution becomes more uniform. The effect of secondary current on the vertical concentration profiles has been shown graphically in Fig. 3 at $x^+ = 5$. In Fig. 3, one curve is plotted neglecting the effect of secondary current, i.e., $w_1 = 0$ in the model and two other curves, in which secondary current is considered for different values of aspect ratio Ar. It can be observed from the figure that magnitude of the concentration profiles decreases with the presence of secondary currents, irrespective of the channel. Also, from the figure it can be noticed that with the decrease of the aspect ratio, i.e., with the increase of the value α, the magnitude of concentration decreases. Therefore, from Fig. 3, one can conclude that, secondary current affects the concentration more in narrow-dip channel than in wide channel.

Figure 4 shows the effects of exerted force due to the gradient of Reynolds normal stress at a vertical non-dimensional height $z^+ = 0.6$. It has been observed that due to the presence of this effect, magnitude of the concentration profile increases and as the distance from the channel entrance increases, effect on the concentration profile also increases. It is worth mentioning here that the effect F_s is very less compared to the effect of secondary current and hindered settling effect on concentration distribution. Still, the parameter is very important for the mathematical modelling of suspended sediment concentration due to its relation with the turbulence intensity.

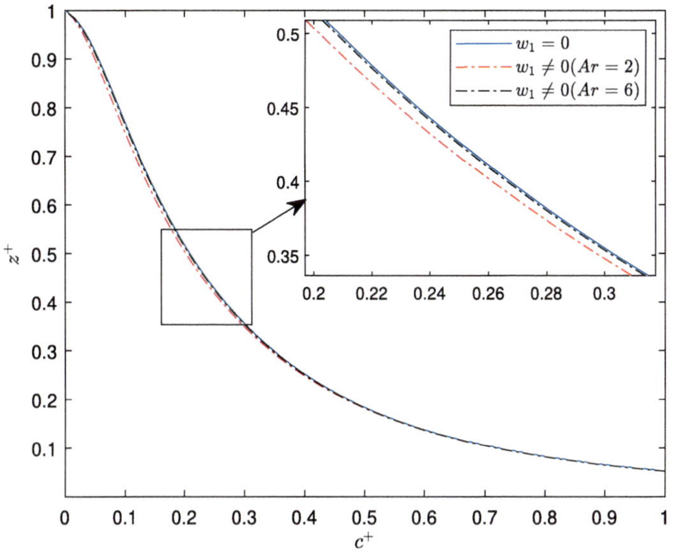

Fig. 3 Effect of secondary current on vertical concentration profile at $x^+ = 5$

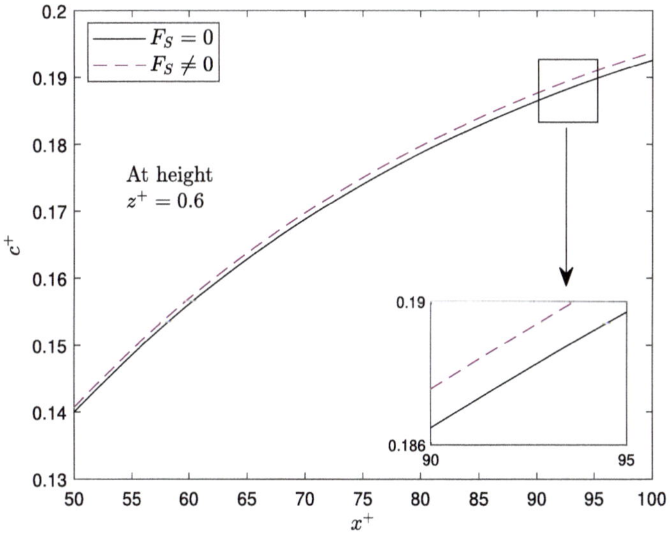

Fig. 4 Effect of gradient of Reynolds normal stress on concentration distribution at $z^+ = 0.6$

5.2 Comparison of the Present Model with Experimental Data

To the best of the authors' knowledge, no experimental data has been available in the literature for steady two dimensional distribution of suspended concentrations. However, for the vertical distribution of concentration many data are available. It has been observed that as x^+ tends to infinity, a steady 2D model of concentration behaves like a steady uniform 1D distribution of concentration [9]. Thus, the solution of the proposed model can be compared with the experimental data for the vertical distribution of concentration. For that purpose, experimental data of Coleman [3], Einstein and Chien [4] and Lyn [12] have been considered. To validate the proposed model, runs 7 and 24 of Coleman [3], runs S-4 and S-12 of Einstein and Chien [4] and runs 1565EQ and 1965EQ of Lyn [12] and have been considered. Details of the flow characteristics of these runs have been listed in Table 1. Figures 5, 6 and 7 show the comparison of the proposed model with the experimental data of Coleman [3], Einstein and Chien [4] and Lyn [12], respectively. The test cases contained different characteristics of the channel (such as narrow and wide) and different sizes of the

Table 1 Flow characteristic of the considered experimental runs

Experiments	Runs	Diameter D (mm)	Height h (cm)	Shear velocity u_* (cm/s)	Reference height A	Reference concentration $c_a(\times 100)$
Coleman [3]	7	0.105	17.10	4.10	0.035	0.620
	24	0.210	16.90	4.10	0.036	0.340
Einstein and Chien [4]	S-4	1.300	11.50	14.28	0.070	7.453
	S-12	0.274	11.90	12.51	0.041	23.434
Lyn [12]	1565EQ	0.190	6.45	3.58	0.093	9.640
	1965EQ	0.190	6.51	3.75	0.072	6.220

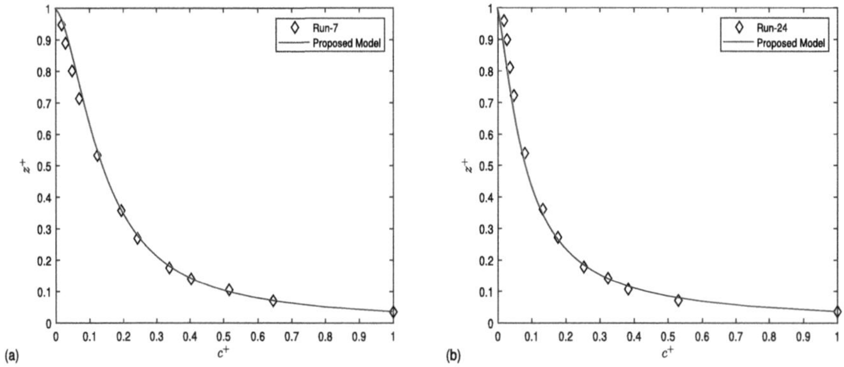

Fig. 5 Comparison of the present model with the experimental data of Coleman [3]—**a** Run-7 and **b** Run-24

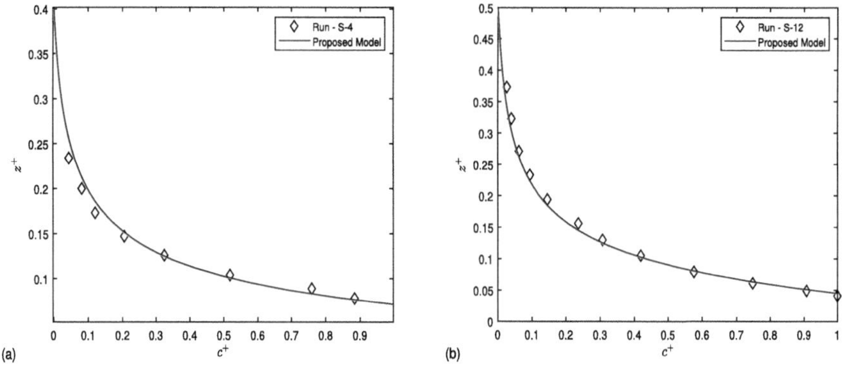

Fig. 6 Comparison of the present model with the experimental data of Einstein and Chien [4]—**a** Run-S-4 and **b** Run-S-12

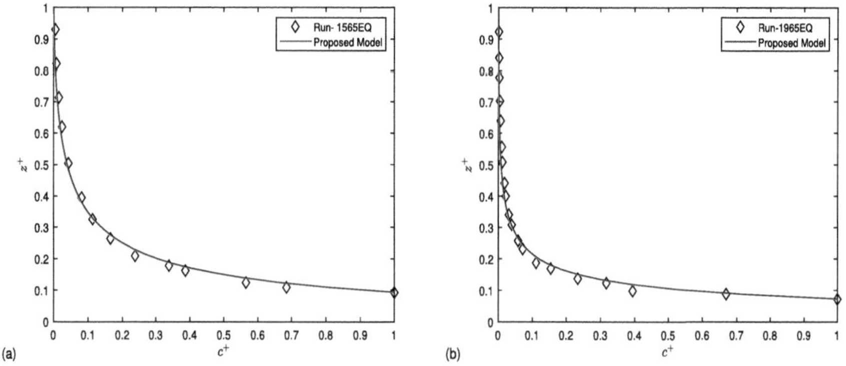

Fig. 7 Comparison of the present model with the experimental data of Lyn [12]—**a** Run-1565EQ and **b** Run-1965EQ

sediments (such as fine and coarse sands). Yet, in all the cases, the proposed model shows good agreement in general. This ensures that the proposed model is capable of predicting the distribution of suspended sediment concentration for different flows.

6 Conclusion

A theoretical model has been proposed for the steady two dimensional distribution of suspension sediment concentration in an open channel turbulent flow. The present study takes into account a number of hydrodynamic mechanisms, such as hindered settling, secondary current, fluid induced lift force and the force exerted due to the gradient of Reynolds normal stress. Consideration of these mechanisms, makes the governing equation more general as well as highly non-linear. Also, the most generalized boundary conditions, which are physically realistic, are considered at the free surface and at the bottom. Though the proposed model considers the logarithmic profile of fluid velocity and parabolic profile of eddy viscosity, it is capable of producing the result for any other profiles of both fluid velocity and eddy viscosity. The effect of the hindered settling has been shown graphically on the vertical concentration profile at far field and on two-dimensional concentration contour. The contour plot shows that the concentration distribution becomes uniform along the stream-wise direction as the distance from the channel entrance increases. Also, it has been observed that the impact of the hindered settling effect is prominent in the main flow region. Effect of secondary current has been shown by changing the value of the aspect ratio of the channel. It has been observed that secondary currents exert significant influence on the vertical concentration distribution in narrow channels. A wide range of experimental data has been considered for the verification of the proposed model and in each case, good agreement has been observed, proving the validity of the model.

References

1. Cheng, K.J.: Bottom-boundary condition for nonequilibrium transport of sediment. Journal of Geophysical Research: Oceans **89**(C5), 8209–8214 (1984)
2. Chien, N., Wan, Z.: Mechanics of sediment transport. American Society of Civil Engineers (1999)
3. Coleman, N.L.: Velocity profiles with suspended sediment. Journal of Hydraulic Research **19**(3), 211–229 (1981)
4. Einstein, H.A., Chien, N.: Effect of heavy sediment concentration near the bed on velocity and sediment distribution. University of California, Berkeley, Calif. **33**, 2 (1955)
5. Fu, X., Wang, G., Shao, X.: Vertical dispersion of fine and coarse sediments in turbulent open-channel flows. Journal of Hydraulic Engineering **131**(10), 877–888 (2005)
6. Greimann, B., Muste, M., Holly, F., Jr.: Two-phase formulation of suspended sediment transport. Journal of Hydraulic Research **37**(4), 479–500 (1999)
7. Greimann, B.P., Holly, F.M., Jr.: Two-phase flow analysis of concentration profiles. Journal of hydraulic Engineering **127**(9), 753–762 (2001)
8. Jing, H., Chen, G., Wang, W., Li, G.: Effects of concentration-dependent settling velocity on non-equilibrium transport of suspended sediment. Environmental Earth Sciences **77**(15), 1–10 (2018)
9. Kumbhakar, M., Mohan, S., Ghoshal, K., Kumar, J., Singh, V.P.: Semianalytical solution for nonequilibrium suspended sediment transport in open channels with concentration-dependent settling velocity. Journal of Hydrologic Engineering **27**(2), 04021048 (2022)
10. Kundu, S., Ghoshal, K.: Effects of secondary current and stratification on suspension concentration in an open channel flow. Environmental Fluid Mechanics **14**(6), 1357–1380 (2014)
11. Liu, X.: Analytical solutions for steady two-dimensional suspended sediment transport in channels with arbitrary advection velocity and eddy diffusivity distributions. Journal of Hydraulic Research **54**(4), 389–398 (2016)
12. Lyn, D.: A similarity approach to turbulent sediment-laden flows in open channels. Journal of Fluid Mechanics **193**, 1–26 (1988)
13. McLaughlin, J.B.: The lift on a small sphere in wall-bounded linear shear flows. Journal of Fluid Mechanics **246**, 249–265 (1993)

14. Muste, M., V.C., P.: Velocity profiles for particles and liquid in open-channel. Journal of hydraulic engineering **123**(9) (1997)
15. Nezu, I.: Turbulent structure in open channel flows. Ph. D. Dissertation, Kyoto University (1977)
16. Nezu, I., Rodi, W.: Experimental study on secondary currents in open-channel flow. 21st Congress of IAHR, Melbourne, Australia pp. 114–119 (1985)
17. Pal, D., Ghoshal, K.: Hydrodynamic interaction in suspended sediment distribution of open channel turbulent flow. Applied Mathematical Modelling **49**, 630–646 (2017)
18. Prandtl, L.: Recent results of turbulence research. Tech. rep. (1933)
19. Richarson, J., Zaki, W.: Sedimentation and fluidisation, part 1. Trans. Ins. Chem. Engrs **31**, 35–53 (1954)
20. Rouse, H.: Modern conceptions of the mechanics of fluid turbulence. Transactions of the American Society of Civil Engineers **102**(1), 463–505 (1937)
21. Wang, G., Fu, X., Huang, Y., Huang, G.: Analysis of suspended sediment transport in open-channel flows: Kinetic-model-based simulation. Journal of Hydraulic Engineering **134**(3), 328–339 (2008)
22. Wang, Z.Q., Cheng, N.S.: Time-mean structure of secondary flows in open channel with longitudinal bedforms. Advances in water resources **29**(11), 1634–1649 (2006)
23. Yang, S.Q.: Turbulent transfer mechanism in sediment-laden flow. Journal of Geophysical Research: Earth Surface **112**, F01005 (2007)
24. Yang, S.Q., Tan, S.K., Lim, S.Y.: Velocity distribution and dip-phenomenon in smooth uniform open channel flows. Journal of hydraulic engineering **130**(12), 1179–1186 (2004)

Theoretical Models of Deoxyribonucleic Acid (DEOXYRIBONUCLEIC ACID) Elasticity

Subhamoy Singha Roy and Yash Nandkuliyar

Abstract We hypothesize that when spins are aligned along the axis to form an antiferromagnetic chain, the Deoxyribonucleic Acid supercoil can act as a spin system. When spins are expressed in the Lie algebra of $\{SU(2)\}$ the spin coordination Chern-Simons topology may be used to determine the (connecting) linking number. It is possible to compute the elastic energy linked to twisting (torsion) and bending (curvature) as expressed in terms of gauge fields $\{SU(2)\}$. Bend and twist are demonstrated to be interdependent rather than two distinct entities. By such imitation, an information resource related to genetics might be spin entanglement, which is the representation of the thermodynamic entropy. The configuration of genetic information might be included in the information theory in quantum dimensions paradigm.

Keywords Linking number · Elastic energy · Torsion · Bending · Genetic information

1 Introduction

Right now commonly understood that DEOXYRIBONUCLEIC ACID functions similar to a physical, stretchy material within a viscous fluid. Two polynucleotide chains are twisted around the same axis by two alpha-helical strands of double helix, giving B-DEOXYRIBONUCLEIC ACID (Z-DEOXYRIBONUCLEIC ACID) a left-handed helical intelligence that is right-handed. We previously knew that there is supercoiled DEOXYRIBONUCLEIC ACID since it was discovered in vivo genetic chemicals DEOXYRIBONUCLEIC ACID with areas of topology that allow for super curvature [1–3]. Prokaryotes, or cells without nuclear membranes, have DEOXYRIBONUCLEIC ACID molecules that frequently embrace the interwound structures known as "plectonemic" supercoils. Chromosome it is also recognized that

S. S. Roy (✉) · Y. Nandkuliyar
Department of Physics, JIS College of Engineering (Autonomous), West Bengal University of Technology, Kalyani, Nadia, India
e-mail: subhomay.singharoy@jisacolege.ac.in

S. Bhattacharyya and H. S. Mahato (eds.), *Modeling, Analysis and Simulations of Multiscale Transport Phenomena*, Springer Proceedings in Mathematics & Statistics 491, https://doi.org/10.1007/978-981-96-3098-1_20

DEOXYRIBONUCLEIC ACID molecules are arranged into topologically distinct loops in eukaryotes, which include components with distinct inside membranes and cells with nuclei [2–5]. The qualitative mechanism of supercoiled DEOXYRIBONUCLEIC ACID has been examined by a number of writers [6]. According to Boles et al.'s tests [7], temperature shift controls the supercoil's shape. DEOXYRIBONUCLEIC ACID-DEOXYRIBONUCLEIC ACID attraction has been shown in experiments by BEDEOXYRIBONUCLEIC ACID et al. [8] to be resistant to fluctuation entropy. Since a DEOXYRIBONUCLEIC ACID molecule possibly thought of as a series of spin structures, we will be studying several multivariate mechanical features of DEOXYRIBONUCLEIC ACID molecules in this note. If a spin with a predetermined direction is added to the coils axis, we can see that each of the two coils opposing intentions of the rotating when a pair of polynucleotide manacles are twisted regarding the identical axis in a DEOXYRIBONUCLEIC ACID molecule with a certain helical sense. Every time a strand turns, since two strands go in the contradictory direction inside the axis, two neighboring coils assigned spin orientations should be facing apart. This means that a DEOXYRIBONUCLEIC ACID molecule may appear to be an antiferromagnetic chain of spins positioned along the supercoil's vertical axis. We will learn about the topological features, elastic properties, and thermodynamic parameters of a DEOXYRIBONUCLEIC ACID supercoil through study on this spin system. In fact, the Chern-Simons system topology coupled to a spin in quantum can provide topological properties like the linking number [9, 10]. When gauge currents are used to convert spins, elastic characteristics like the two terms "bending" and "torsion" refer to may be expressed regarding gauge fields [9, 10]. This assessment gives a clear indication of the mutual reliance between bending and twisting. The chains of antiferromagnetic spin entropy of entanglement is comparable to the entropy in thermodynamics of a DEOXYRIBONUCLEIC ACID superposition of coils. The entanglement entropy of a DEOXYRIBONUCLEIC ACID supercoil is simply a representation of the entire amount of information that may be transcribed. Consequently, the transcription process of genetic in sequence may be represented within the structure of quantum in sequence theory, providing a basic explanation genetic data contained within the superposition of coils. Furthermore, we will demonstrate how DEOXYRIBONUCLEIC ACID loops in the supercoil manifest in a topological entities akin to solitons, or skyrmions. We will express the elastic characteristics and calculate the supercoil's entanglement entropy.

2 Theoretical Background

A. The Elasticity of a DEOXYRIBONUCLEIC ACID Supercoil

The spin degrees of freedom analysis aids in our understanding of a DEOXYRIBONUCLEIC ACID supercoil's elastic properties. Given that spins are known to be gauge fields, gauge currents coming from an $[SU(2)]$, we project the stretchy characteristics atop the gauge potential area. To accomplish this, we relate gauge potential vector $\underset{(\mu)}{\overrightarrow{A}}$, whose mechanism communicate to the $[SU(2)]$ fields $A^{(a)}_{(\mu)}(a = 1.0, 2.0, 3.0)$, with a serving in place of the group directory and μ as the indicator of space time, to the digression vector unit $\overset{\mathbf{r}}{t} = \left[\partial_s \overset{\mathbf{r}}{r}(s)\right]$, where $\{\overset{\mathbf{r}}{r}(s)\}$ represents a freedom curvature characterized by arc distance end to end s. $\partial_s^{(2)} \overset{\mathbf{r}}{r}$ appears to match the $[SU(2)]$ gauge field power, $\overset{\mathbf{r}}{F}_{(\mu\nu)}$. Given like this, the inside energy represented by the twist provided through $\int \left\{ds.(\partial_s^2 \overset{\mathbf{r}}{r})^2\right\}$ may be expressed as $\int d^3x\left(TrF_{\mu\nu}F_{\mu\nu}\right)$. Twisting elastic energy is related to the interaction between spins and may be expressed as $m^2 \int \left(\overset{\mathbf{r}}{J}_\mu \overset{\mathbf{r}}{J}_\lambda d^3x\right)$ in the continuum limit, where $\overset{\rho}{J}_\mu$ stands for gauge in progress current [11] and m has bulk and dimension. Note that this quantity in the continuum limit related to the spin–spin interaction essentially reflects torsion [12].

According to [6], B-DEOXYRIBONUCLEIC ACID with a length of L has an elastic energy.

$$\left(E_{el}/k_B T\right) - \frac{1}{2}\int_0^L \left[ds\left\{A(\partial_s^2 \overset{\mathbf{r}}{r})^2 + C\Omega^2\right\}\right] \tag{1}$$

If the scalar domain $\{\Omega(s)\}$, where $\{A.(C)\}$ is the roundabout expandable steady, describes the twisting rate departure from ω_0.

The topological current at this time [13].

$$\overrightarrow{J_\mu} = \left(\varepsilon^{\mu\nu\lambda\sigma} \overset{\mathbf{r}}{a}_\nu \times \overset{\mathbf{r}}{f}_{\lambda\sigma}\right) = \left(\varepsilon^{\mu\nu\lambda\sigma}\right)\partial_\nu \overset{\mathbf{r}}{f}_{\lambda\sigma} \tag{2}$$

We may inscribe in the continuous boundary for the elastic energy by mapping this onto the pattern freedom of gauge prospective related to the arrangement of turn.

$$\left(E_{el}/k_B T\right) = \left(\frac{1}{2}\right)\int d^3x[ATrF_{\mu\nu}F_{\mu\nu} + \left(Cm^2\right)\overset{\mathbf{r}}{J}_\mu \overset{\mathbf{r}}{J}_\mu] \tag{3}$$

The analysis's key finding is that the curving and torsional forces related to the departure beginning ω_0 do not separate and autonomous. The two are connected, it is true. This is due to the fact that the current $\overset{\rho}{J}_\mu$ in Eq. (2), which produces torsion, is constructed using the gauge field curvature, $\left(F_{\mu\nu}\right)$. In fact, intrinsic bending may have a significant impact on the transfer of torsional stress along DEOXYRIBONUCLEIC ACID, as Nelson [14] noted in his discussion of the link between bending and torsion.

We may say that the measurement (twisting) of the base pair spatial rotation speed for a deoxyribonucleic acid that is not deformed along the central axis is ω_0. However, a deformed DEOXYRIBONUCLEIC ACID when the dual-helix twist is changed from ω_0, distortion stretchy energy is not nil. Here in instance, the chirality equilibrium breach is indicated by the torsional forces. That is seen that the contemporary topological $J_{\mu}^{(a)}$ $(a = 1.0, 2.0, 3.0)$, it has the vector current in the axial direction attached to it $J_{\mu}^{5} = \left(\overline{\psi}\gamma_{\mu}\gamma_{5}\psi\right)$ connected to a chiral spinor, is what generates torsion in the current formalism. Actually, according to Eq. (2), every element of the vector $\vec{J}_{\mu}$ is connected to the axial vector current via the relation [15].

$$\partial_{\mu}J_{\mu}^{2} = -(1/2)\partial_{\mu}J_{\mu}^{5},\ J_{\mu}^{1} = -(1/2)J_{\mu}^{2},\ J_{\mu}^{3} = (1/2)J_{\mu}^{2}, \tag{4}$$

Since the branch of this current is non-zero, chiral symmetry breaking in quantum field theory notably creates chiral. Given this, we see that the chiral anomaly in quantum field theory corresponds with the twisting elastic energy represented by torsion.

B. DEOXYRIBONUCLEIC ACID Entanglement Entropy and Supercoil Structure

Two helices with the same asymmetric, cylindrical form that are joined at the end make up an interwound plectonemic supercoil. The steady bending of the coil into a toroidal shape closes a solenoidal supercoil. The temperature fluctuation causes the radius of the supercoil to inflate up greater compared to the rigid core radius of the supercoil. Indeed, at zero temperature, this has been demonstrated has an successful entropic potential that is violent occurs opposing the supplely induced crumble [6]. It suggests that thermal entropy is a crucial factor in a DEOXYRIBONUCLEIC ACID superposition of coils. Supercoils are seen to reach the entanglement entropy when DEOXYRIBONUCLEIC ACID molecules are handled because of the interaction of spins. In actuality, it is evident that the spinning associated alongside every DEOXYRIBONUCLEIC ACID roundabout in a supercoil are prearranged down the supercoils alliance. A lattice of antiferromagnetic turn frameworks might be conceptualized as the supercoil axis., since, as we have shown, a DEOXYRIBONUCLEIC ACID molecule may appear to be an antiferromagnetic sequence of rotates situated regarding the vertical line within the supercoil. Two neighboring spins with different orientations will create a singlet in order to have the lowest possible energy level. Because chirality is strengthened by frustration, this will result in a aggravated spin arrangement. This aggravation implies that resonance valence bonding (RVB) forms spin singlets [16].

Previous research has demonstrated [17, 18] that the Berry phase [19] given by $e^{i\varphi_{B}} = e^{i(2\pi\tilde{\varphi}_{B})}$, that is obtained through a spin position when the systems spins are rotated about the quantization axis (z-axis) in a closed circuit, is related to the entanglement of configuration measure for the entanglement of two adjacent spins, which is provided by concurrence C.

Actually, we have the relationship that

$$C = |\tilde{\varphi}_{B}| = \left[|\varphi_{B}|/(2\pi)\right] \tag{5}$$

In a aggravated system, $C = 0.50$ has been found to provide the concurrence connected to two adjacent neighbor spins entanglement of configuration [17]. Consequently, the total concurrence that has built up is given in the spin sequence by

$$C = \sum_{i=1}^{L/2} |\tilde{\varphi}_B| = \{|\tilde{\varphi}_B|(0.5L)\} = \{(0.5)(0.5L)\} = (0.25L) \quad (6)$$

where $(0.5L)$ is the quantity of singlets and the figure of spins in the sequence as a whole is L. It is observed that an agreement in a mixed state provides the derivation of entanglement of formation from von Neumann entropy for a uncontaminated state entangled spin structure [20]. Therefore, the spin systems entanglement entropy is represented by the above formula (6). We now consider that DEOXYRIBONUCLEIC ACID, a hard, supercoiled polymer, is enclosed in a small tube with a radius of r. The quantity of coils in a loop, and hence the quantity of spins in the loop, determines the region of a DEOXYRIBONUCLEIC ACID ring in a supercoil.

It is feasible to connect the region of the tube's outside considering the quantity of spins in the DEOXYRIBONUCLEIC ACID ring at that outside when the supercoiled rigid polymer is contained in a narrow, radius tube. According to the holographic principle, the region of a closed outside determines the entropy. This concept states that the region of the tube's outside provides the entropy of entanglement. As a result, the form for the entanglement entropy is

$$S \approx r^2 / \left(4r_0^2\right) \quad (7)$$

r_0 is a basic area unit, while r is the tube's radius. The entanglement entropy is given by if we recognize $r \approx (2 \times A)$, where $(2A)$ is the stair distance end to end of an equivalent elastic polymer and (A) is the winding stretchy invariable and r_0 is chosen to correspond to the radial disarticulation of a particular location inside the loop, this is on the R organize [6].

$$S \approx \left|\left(A^2\right) / R^2\right| \quad (8)$$

Likewise, with regard to the change in position of a certain coil point the length of the superconductor axis, which is in accordance with order πP

$$S \approx \left\{(A)^2 / (\pi P)^2\right\} \quad (9)$$

Thus, the total entropy is determined using

$$S = \left(A^2 \div R^2\right) + \left\{A^2 / (\pi P)^2\right\} \quad (10)$$

We currently note that the thermodynamic entropy and this entanglement entropy essentially coincide. In fact, we may imagine a zero radius (straight line) in the restraining situation length of radius $(r) \rightarrow 0$ when the entire stretchy energy disappears when the temperature is 0, nevertheless, entanglement entropy cannot drop for a thin-walled tube radius. The entropy of entanglement likewise disappears in this scenario. In fact, when an elastic tube is submitted to the constraint $\left(\Delta Lk/L\right) \neq 0$ at zero temperature, it subsides into a plectonemic supercoil [21]. The twist energy decreases to zero when (ΔLk) is introduced into writhing (Wr). The plectoneme may then be made to collapse into a line $(\mathrm{Sin}(\gamma) = 1)$, which likewise results in zero bending energy. At absolute zero, the loop's area disappears. The entanglement entropy likewise disappears in this scenario. Because of this, we may determine that the thermodynamic entropy of the supercoil is equal to the entanglement entropy.

This aids in the computation of the supercoil's free energy. Observing the connection $dF = (S \times dT)$, where $\{F.(T)\}$ represents the temperature, or open energy, we may express the entropy-related free energy per division volume provided through Eq. (10) as follows

$$\left\{\Delta F/(K_B TV)\right\} \approx \left\{A^2/\left(R^2 r^3\right)\right\} + A^2/\left\{(\pi P)^2 r^3\right\} \tag{11}$$

The open energy in relation to length is shown below

$$\{(\Delta F) \div (K_B TL)\} \approx \left\{A^{2/3}/\left(R^{2/3} r\right)\right\} + A^{2/3}/\left\{(\pi P)^{2/3} r\right\} \tag{12}$$

since previously stated, r is now considered to belong to the A, and as a result, we discover

$$\begin{aligned} \left\{\Delta F/(K_B TL)\right\} &\approx \left\{A^{2/3}/\left(R^{2/3} A\right)\right\} + A^{2/3}/\left\{(\pi P)^{2/3} A\right\} \\ &= \left\{(1.0) \div \left(A^{1/3} R^{2/3}\right)\right\} + \left\{(1.0) \div \left(A^{1/3} (\pi P)^{2/3}\right)\right\} \end{aligned} \tag{13}$$

As such, finding is the same as what Marko and Sigga found [6]. One of our study's most significant findings is that, while processing quantum information, the main tool at our disposal is the entanglement of a quantum system [22–25]. The entanglement entropy controls how much information may be utilized for transcription. The amount of genetic information that can be transcribed from a supercoil is determined by the entanglement entropy, which is significant when considering DEOXYRIBONUCLEIC ACID as a linear storage of sequenced information [26].

3 Discussion

We have concluded that, under certain conditions, a DEOXYRIBONUCLEIC ACID supercoil may be worn as a structure of quantum revolve in which the spins are prearranged within an antiferromagnetic sequence along the alignment. These spins can be coupled with $\{SU(2)\}$ gauge fields situated on the connections. We have evaluated caricature (torsion) and winding (curvature) in terms of these gauge fields.

The fact that bending and twisting are related concepts, is a crucial result of this mimicry. In actuality, research supports the DEOXYRIBONUCLEIC ACID that the twisting strain propagates along when it is bent.

The consequence of evaluating DEOXYRIBONUCLEIC ACID supercoil regarding the entanglement entropy is the antiferromagnetic spin sequence, which in turn generates the entropic prospective connected the open energy per unit of length and the entropy cost of packing a rigid polymer into a tiny cylinder [27]. The terrible entropic potential, which is represented by entanglement entropy, stops a supercoil from collapsing in an elastic manner, which might occur at nil temperature. The thermodynamic entropy is well-represented by entanglement entropy. The measurement of the system's overall information content is one of the most significant functions of the entanglement entropy [28]. In fact, entanglement may be utilized to interpret information when considering DEOXYRIBONUCLEIC ACID as a linear storage of sequenced data, where entropy determines how much data is accessible for transcription. Because of this, the amount of genetic data that may be transcriptionally large can be calculated using the entanglement entropy metric. This aids in the realization of quantum information theory's genetic information transcription.

References

1. Worcel, A., Burgi, E.: J. Mol. Biol. **71**, 127 (1972)
2. Benyajati, C., Worcel, A.: Cell **9**, 393 (1976)
3. Jackson, D.A., Dickinson, P., Cook, P.R.: EMBO J. **9**, 567 (1990)
4. Wolffe, A.P.: Chromatin. Academic Press, New York (1993)
5. Freeman, L.A., Garrard, W.T.: Crit. Rev. Euk. Gene Exp. **2**, 165 (1992)
6. Marko, J.F., Siggia, F.D.: Phys. Rev. E **52**, 2912 (1990)
7. Boles, T.C., White, J.H., Cozzarelli, N.R.: J. Mol. Biol. **213**, 931 (1990)
8. BEDEOXYRIBONUCLEIC ACID, Furrer, P., Stasrak, A., Dubochet, J., Engelman, C.H., Bates, A.D.: J. Mol. Biol. **235**, 825 (1994)
9. Goswami, G., Bandyopadhyay, P.: J. Math. Phys. **34**, 749 (1995)
10. Bandyopadhyay, P.: Proc. Roy. Soc. Lond. A **466**, 2917 (2010)
11. Abanov, A.I., Wiegmann, P.B.: Nucl. Phys. B **570**, 685 (2000)
12. Bandyopadhyay, A., Chatterjee, P., Bandyopadhyay, P.: Gen. Rel. Grav. **18**, 1293 (1986)
13. Carmeli, M., Malin, S.: Ann. Phys. **103**, 208 (1977)
14. Nelson, P.: Proc. Natl. Acad. Sci. (USA) **96**, 14342 (1999)
15. Roy, A., Bandyopadhyay, P.: J. Math. Phys. **30**, 2366 (1989)
16. Anderson, P.W.: Science **235**, 1196 (1987)
17. Basu, B., Bandyopadhyay, P.: Int. J. Geo. Meth. Mod. Phys. **9**, 707 (2007)
18. Basu, B., Bandyopadhyay, P.: J. Phys. A. **41**, 055301 (2008)
19. Berry, M.V.: Proc. Roy. Soc. Lond. A **392**, 45 (1984)
20. Osborne, T.J., Nielson, M.A.: Phys. Rev. A **66**, 032110 (2002)
21. Hunt, N.G., Hearst, J.F.: J. Chem. Phys. **95**, 9329 (1991)
22. Bennett, C.H., Divincenzo, D.P.: Nature **404**, 247 (2000)
23. Nelson, M.A., Chuang, G.L.: Quantum Information and Quantum Computation. Cambridge University Press, Cambridge (2000)
24. Roy, S.S.: Theor. Phys. **2**(3), 141 (2017)
25. Roy, S.S., Bandyopadhyay, P.: Phys. Lett. A **382**, 1973 (2018)
26. Roy, S.S., Bandyopadhyay, P.: Phys. Lett. A **337**, 2884 (2013)
27. Roy, S.S., Bandyopadhyay, P.: Euro. Phys. Lett. **109**(4), 48002 (2015)
28. Basu, B., Bandyopadhyay, P., Majumdar, P.: Phys. Rev. A **86**, 022303 (2012)

GPSR Compliance

The European Union's (EU) General Product Safety Regulation (GPSR) is a set of rules that requires consumer products to be safe and our obligations to ensure this.

If you have any concerns about our products, you can contact us on ProductSafety@springernature.com

In case Publisher is established outside the EU, the EU authorized representative is:

Springer Nature Customer Service Center GmbH
Europaplatz 3
69115 Heidelberg, Germany

Batch number: 10704307

Printed by Printforce, the Netherlands